Adaptive Environmental Management

Catherine Allan · George H. Stankey
Editors

Adaptive Environmental Management

A Practitioner's Guide

Editors
Catherine Allan
Institute for Land, Water and Society
Charles Sturt University
Albury, NSW
Australia

George H. Stankey
Retired, USDA Forest Service
Pacific Northwest Research Station
Corvallis, OR
USA

Co-published by Springer Science+Business Media B.V., Dordrecht, The Netherlands and **CSIRO** PUBLISHING, Collingwood, Australia

Sold and distributed:
In the Americas, Europe and Rest of the World excluding Australia and New Zealand by Springer Science+Business Media B.V., with ISBN 978-90-481-2710-8
springer.com

In Australia and New Zealand by **CSIRO** PUBLISHING, with ISBN 978-0-643-09690-5
www.publish.csiro.au

ISBN: 978-90-481-2710-8   e-ISBN: 978-1-4020-9632-7
Springer Dordrecht Heidelberg London New York

Library of Congress Control Number: 2009928832

© Springer Science+Business Media B.V. 2009
No part of this work may be reproduced, stored in a retrieval system, or transmitted in any form or by any means, electronic, mechanical, photocopying, microfilming, recording or otherwise, without written permission from the Publisher, with the exception of any material supplied specifically for the purpose of being entered and executed on a computer system, for exclusive use by the purchaser of the work.

Cover illustration: "Ghosts of Lake Hume" - Drowned trees, Lake Hume, near Wymah Ferry, NE Victoria, Australia. Photograph by Dirk HR Spennemann (Albury)

Printed on acid-free paper

Springer is part of Springer Science+Business Media (www.springer.com)

Acknowledgements

As with any edited work, many people and organisations have contributed to the success of the endeavour. Crucial financial support has come from the Institute for Land, Water and Society, Charles Sturt University. This enabled us to hold the first workshop at Lake Hume, New South Wales in 2007 from which came the ideas, commitment and support for this volume. Participants at that session included Allan Curtis, Andrew Ross, Sue Briggs, Di Bentley, Deborah Georgievski, Rob Argent, Chris Jacobson, Will Allen, Terry Hillman, Tony Jakeman, George Wilson, Jean Chesson, Peter Ampt and Geoff Park, along with Catherine Allan and George Stankey.

Land and Water Australia provided generous financial support to cover the clerical, technical and some travel support costs for creating this book. We are very grateful for this.

In 2008, the generosity of the Institute of Environmental Studies and the FATE Program at the University of New South Wales supported our second workshop in Sydney, at which plans and draft chapters for this book were sharpened and finalised.

Allan Curtis at Charles Sturt University played a significant role in the early formulation of this book. Other obligations prevented his further participation, but we are indebted to his vision and leadership in fostering this effort.

Thanks also to Andrew Ross for manuscript review and cheerful confidence in the endeavour.

As we neared completion, the enormity and scope of the tasks before us at times seemed overwhelming, particularly with a text involving authors scattered literally around the world. Karen Retra provided yeoman support in helping finalise the manuscripts, and Harold Fry applied his computer expertise to rescue the editors who had quickly found their meagre computer skills sorely tested. We are deeply indebted to both Karen and Harold for their great contributions.

Finally, it has been a joy and an honour to work with the people who brought this volume into being; our colleagues from both workshops, as well as those separated from us and each other by vast distances. All have been contributed with good will and good humour. We share an abiding hope that readers find this text of use and value as they struggle with the challenges and the promises of inculcating the culture of exploration and experimentation that an adaptive management approach offers us.

Contents

Part I Understanding Adaptive Management

1 **Introduction**... 3
   George Stankey and Catherine Allan

2 **Components of Adaptive Management**.. 11
   Robert M. Argent

Part II Varying Contexts

3 **Lessons Learned from Adaptive Management
   Practitioners in British Columbia, Canada**... 39
   Alanya C. Smith

4 **Using Adaptive Management to Meet Multiple Goals
   for Flows Along the Mitta Mitta River
   in South-Eastern Australia**... 59
   Catherine Allan, Robyn J. Watts, Sarah Commens
   and Darren S. Ryder

5 **Adaptive Management of a Sustainable Wildlife
   Enterprise Trial in Australia's Barrier Ranges**.................................... 73
   Peter Ampt, Alex Baumber, and Katrina Gepp

6 **Learning About the Social Elements of Adaptive
   Management in the South Island Tussock Grasslands
   of New Zealand**... 95
   Will Allen and Chris Jacobson

7 **Kuka Kanyini, Australian Indigenous Adaptive Management**........... 117
   George Wilson and Margaret Woodrow

8   Crisis as a Positive Role in Implementing Adaptive
    Management After the Biscuit Fire, Pacific Northwest, U.S.A. 143
    Bernard T. Bormann and George H. Stankey

Part III   Tools for Adaptive Management

9   Modelling and Adaptive Environmental Management 173
    Tony Jakeman, Serena Chen, Lachlan Newham,
    and Carmel A. Pollino

10 Lessons Learned from a Computer-Assisted
    Participatory Planning and Management Process
    in the Peak District National Park, England 189
    Klaus Hubacek and Mark Reed

11 Signposts for Australian Agriculture 203
    Jean Chesson, Karen Cody, and Gertraud Norton

12 Environmental Management Systems as Adaptive
    Natural Resource Management: Case Studies
    from Agriculture 209
    George Wilson, Melanie Edwards, and Genevieve Carruthers

13 The Adaptive Management System for the Tasmanian
    Wilderness World Heritage Area – Linking Management
    Planning with Effectiveness Evaluation 227
    Glenys Jones

Part IV   The Importance of People

14 Adaptive Management of Environmental Flows – 10 Years On 261
    Tony Ladson

15 Collaborative Learning as Part of Adaptive Management
    of Forests Affected by Deer 275
    Chris Jacobson, Will Allen, Clare Veltman, Dave Ramsey,
    David M. Forsyth, Simon Nicole, Rob Allen,
    Charles Todd, and Richard Barker

16 Effective Leadership for Adaptive Management 295
    Lisen Schultz and Ioan Fazey

17 Institutionalising Adaptive Management: Creating
    a Culture of Learning in New South Wales Parks
    and Wildlife Service 305
    Peter Stathis and Chris Jacobson

18 Adaptive People for Adaptive Management 323
Ioan Fazey and Lisen Schultz

Part V Conclusion

19 Synthesis of Lessons 341
Catherine Allan and George Stankey

Index 347

Contributors

Catherine Allan is Senior Lecturer, Environmental Planning and Sociology in the School of Environmental Sciences, Charles Sturt University, Albury, Australia. A member of the Institute for Land, Water and Society she is fascinated by the idea, and sometime practice, of adaptive management.

Will Allen is an action researcher with more than 15 years experience in natural resource management. He currently works for Landcare Research, a New Zealand Crown Research Institute. He has particular interests in the development of collaborative adaptive natural resource management approaches, along with the multi-stakeholder management information systems (MIS) and participatory monitoring and evaluation (PM&E) programs required to support these. His has regional experience with participatory approaches in Australasia, East Africa and the Pacific.

Rob Allen is a plant ecologist and the science leader for ecosystem processes at Landcare Research Ltd., New Zealand. He leads research programs in the areas of ecosystem resilience and indigenous forestry.

Peter Ampt manages the Future of Australia's Threatened Ecosystems (FATE) Program at the University of New South Wales. He has a background in agricultural science, rural sociology, participatory research and education and is undertaking a Ph.D. on the use of Landscape Function Analysis (LFA) for adaptive management. His focus is on integrating socio-economic and conservation priorities to generate incentives for conservation on private land. His work involves exploring the potential for the commercial use of native species to develop a mix of alternative conservation-friendly land uses.

Robert Argent is the Australian Water Resources Information System, Project Director for the Bureau of Meteorology, Australia. Robert has over 20 years international experience in water and natural resources research, teaching, design, consulting and project management. He has focussed on interpretation and delivery of research outcomes, data and information into forms that meet the needs of end users, particularly in environmental model development and integration. During 1990–2007, at the University of Melbourne he investigated a variety of adaptive management applications on water quality and land management.

Richard Barker is a Professor of Statistics, at the University of Otago and is Head of the Department of Mathematics and Statistics, New Zealand.

Alex Baumber has a first class honours degree in environmental science from the University of Wollongong. He was a targeted graduate for the Australian Government in the Department of Environment and Heritage where he worked on the harvesting of native species of plants and animals. He then joined the FATE team at UNSW where he was project manager of the Barrier Ranges Sustainable Wildlife Enterprise Trial. He is currently undertaking a Ph.D. in Environmental Policy and Management at UNSW.

Bernard Bormann is Principal Forest Ecologist and Team Leader, Ecosystem Processes Program, with the USDA Forest Service, Pacific Northwest Research Station, Corvallis, and Professor (courtesy), Department of Forest Ecosystems and Society, College of Forestry, Oregon State University. His research interests include management and natural processes controlling long-term ecosystem productivity studies in Oregon, Washington, and Alaska and speeding learning with active adaptive management.

Serena Chen is a Ph.D. scholar at the Fenner School of Environment and Society at the Australian National University. Her current research involves developing Bayesian network models for predicting the ecological condition of river systems in the Murray-Darling Basin.

Jean Chesson leads the Decision Sciences Team in the Bureau of Rural Sciences, Australian Government Department of Agriculture, Fisheries and Forestry. Her particular interest is in evaluating performance with respect to complex objectives such as sustainable development.

Karen Cody is a senior policy officer with the Department of Agriculture Fisheries and Forestry. She has over a decade of experience in designing and implementing monitoring and evaluation strategies at the corporate, program and project scale for the environment and agricultural sector. She was recently seconded to the National Land & Water Resources Audit.

Sarah Commens is a freshwater ecologist (B. App. Sci. (Hons)) trained in River Murray System operations. She is the Technical Coordinator for a comprehensive system-wide Review of River Murray System Operations being undertaken by the Murray-Darling Basin Authority. Sarah has almost a decade of experience in the water industry which includes working on the high profile 'Living Murray' initiative. Sarah combines her policy and river operations knowledge, and liaises with scientists, river managers and policy makers, to facilitate the adoption of practical and more balanced approaches to river management that meet both human and environmental needs.

Melanie Jane Edwards is a Research Officer, Australian Wildlife Services. She provides research support for a range of projects. These include such diverse topics as quantifying and comparing the amount of greenhouse gases produced by kangaroos and cattle, collating population estimates of kangaroos to achieve a national estimate, scoping the potential for the reintroduction of brushtail possums in Central Australia, and outlining the legislation associated with the use of wildlife resources for Indigenous enterprise development.

Ioan Fazey is a Lecturer in Sustainable Rural Development at the Institute of Biological, Environmental and Rural Sciences, Aberystwyth University in Wales. His research includes investigations of environmental values, adaptation and vulnerability to environmental change, and learning in complex social and ecological systems. He uses his teaching as a test-bed for improving interventions aimed at promoting more effective learning in rural communities. These ideas are mostly applied in his work with the Kahua people in the Solomon Islands.

David M. Forsyth is a scientist involved in terrestrial wildlife management, particularly the ecology and management of mammalian herbivores. He is based at Arthur Rylah Institute of Environmental Research, Department of Environment and Sustainability, Victoria.

Katrina Gepp was born and raised in Broken Hill on sheep and cattle stations, and spent ten years as a local pastoralist and kangaroo harvester. She has worked in regional business development, business tourism, youth and community management, and rural reporting for ABC radio Broken Hill. For the past 3 years Katrina was the local research officer for the Barrier Ranges Sustainable Wildlife Enterprises Trial in parallel with a position with the Western Catchment Management Authority as a Community Support Officer. She is completing an Ecological Agriculture degree from Charles Sturt University.

Klaus Hubacek is a Reader/Associate Professor at the School of the Environment, University of Leeds in the UK. His expertise is in conceptualizing and measuring interactions between nature and society, scenario analysis and futures research, and ecologic-economic modeling. Klaus has conducted studies for national agencies in Austria, China, Japan, and the UK, and international institutions such as the European Statistical Office (EUROSTAT), the OECD, and UNESCO. Recently, he was awarded a grant from the UK research councils to lead an interdisciplinary project group working with stakeholders on developing visions for a sustainable future and sustainable land management practices.

Chris Jacobson's Ph.D. focused on different discourses of adaptive management and ways in which to build reflection on its practice in the conservation management sector. At the same time, she worked on contract for Landcare Research NZ on the adaptive management of forests affected by deer and tussock grassland management. She is currently at The University of Queensland, Australia, working on research partnered by the New South Wales Department of Environment and Climate Change, Parks Victoria and Parks Australia exploring the back-loop of adaptive management: how to support the integration of new knowledge into decision making.

Tony Jakeman is Professor, Fenner School of Environment and Society, and Director of the Integrated Catchment Assessment and Management Centre, The Australian National University. He has been an Environmental Scientist and Modeller for 32 years. His interests include integrated assessment methods and decision support for water and associated land resource problems, including modelling and management of water supply and quality problems in relation to climate, land use and policy changes and their effects on biophysical and socioeconomic outcomes. He has been Editor-in-Chief, Environmental Modelling and Software (Elsevier) since 1996 and was President of the International Environmental Modelling and Software Society.

Glenys Jones is the Planner for performance evaluation and reporting with the Tasmanian Parks and Wildlife Service. Glenys holds a first class honours degree in science (University of NSW) and is qualified in landscape design. Her career spans the fields of ecological and taxonomic research, scientific editing and landscape design. Glenys joined the Parks and Wildlife Service in 1989. She was part of a team of management planners who prepared the statutory management plans for the Tasmanian Wilderness World Heritage Area. She coordinated the first comprehensive evaluation of management effectiveness for the Tasmanian Wilderness World Heritage Area and prepared the 'State of the Tasmanian Wilderness World Heritage Area Report'.

Tony Ladson has more than 20 years experience in hydrology and water resource management and has worked on projects throughout Australia, the US and Taiwan. He has worked extensively on environmental flows, and has contributed to strategy plans and policies to improve river health. In 2000 he was awarded a Victorian Fellowship to undertake international research into adaptive management of environmental flows and in 2005 received the GN Alexander Medal from Engineers Australia for work on evaluation of stream rehabilitation projects.

Lachlan Newham is a Research Fellow at the Fenner School of Environment and Society at the Australian National University. His research interests are in techniques for assessing water quality, particularly the development of models and decision support systems to aid management decision making. He has interests also in how to incorporate participatory processes into the development of environmental models. Lachlan is the Treasurer of the Modelling and Simulation Society of Australia and New Zealand Inc. and the convener of the Water Quality and Environmental Flow Modelling and Assessment course at the Australian National University.

Simon Nicol is the Principal Fisheries Scientist (Tuna biology and ecology), Secretariat of the Pacific Community, New Caledonia. Before moving to New Caledonia, he worked on adaptive management of freshwater fisheries in Australia.

Gertraud Norton is a scientist at the Bureau of Rural Sciences, Australian Government Department of Agriculture, Fisheries and Forestry. She recently completed profiling six agricultural industries with respect to their contributions to sustainable development. Her current focus is on monitoring and evaluation in the fields of marine and other pests and fisheries.

Carmel Pollino has a Ph.D. in Environmental Toxicology from RMIT University, and a Masters in Environmental Law from Macquarie University. Between 2000 and 2002, Carmel worked at City University of Hong Kong, investigating the effects of sediment-derived and food borne pollutants on native marine fish species. In 2002, she joined Monash University, specialising in the development of risk management methodologies, particularly focusing on stakeholder engagement and Bayesian models. In 2006, she moved to Australian National University to focus on integrated assessment, ecological modelling, climate change and risk management.

David Ramsey is a scientist specialising in statistical analyses and ecological modelling. He is based at the Arthur Rylah Institute for Environmental Research, Department of Sustainability and Environment, Victoria.

Mark Reed is a senior lecturer at the University of Aberdeen. His expertise is in participatory conservation focuses on land degradation, sustainability indicators and participatory processes. He gained his Ph.D. in 2005, working on land degradation assessment with communities in the Kalahari, Botswana. He co-manages the Sustainable Uplands project in the UK and leads part of an EU-funded global land degradation remediation project.

Darren Ryder is a Senior Lecturer in Ecosystem Rehabilitation and Aquatic Ecology at the University of New England, Armidale NSW, Australia. Darren has over 10 years experience in research on aquatic ecosystems, with a focus on developing ecological indicators and environmental flow regimes for the effective management and restoration of freshwater ecosystems. His research examines the links between flow regime and stream processes such as nutrient and organic matter cycling, algal and microbial metabolism, and food web structure, with an aim to better understand the link between ecosystem structure and function in regulated river systems.

Lisen Schultz is a doctoral student at Stockholm Resilience Centre and Department of Systems Ecology, Stockholm University, Stockholm. She is interested in what enables some people, at certain points in time, and in certain places, to catalyze transformations in management systems and societies towards sustainability and resilience thinking. She is currently analyzing the management of Biosphere Reserves across the globe, and she plans to defend her doctoral thesis during 2009.

Alanya C. Smith A.Ag. Ministry of Forests and Range, Forest Practices Branch Research Officer. Her current professional focus is on monitoring and evaluation of forest and range practices.

George H. Stankey earned his B.S. and M.S. degrees at Oregon State University and a Ph.D. from Michigan State University. He worked as a Research Social Scientist with the US Forest Service in Montana and Oregon, USA, taught at university level in the US and Australia, and worked with the New South Wales National Parks and Wildlife Service. His research interests include recreation and other amenity uses of forests, adaptive management, and institutional structures and processes. Now retired, he remains active in resource management issues.

Peter Stathis is Manager of the Management Effectiveness Unit, Parks & Wildlife Group, Department of Environment & Climate Change, Hurstville NSW, Australia. His role includes running a triennial broad scale adaptive management program across the entire formal reserve system in NSW.

Charles Todd is a scientist specialising in Ecological Modelling, particularly in developing models for the management of wildlife. He is based at Arthur Rylah Institute of Environmental Research, Department of Environment and Sustainability, Victoria.

Clare Veltman is an animal pest ecologist working for the Research and Development Group of the New Zealand Department of Conservation. She is also the leader and project manager for the project: Adaptive management to restore forests affected by deer.

George Wilson, M.VSc. Ph.D. operates a consultancy firm which specialises in wildlife management, veterinary services, support for Aboriginal communities, ecotourism and emerging rural industries. He has contracts to manage research programs for the Rural Industries Research and Development Corporation. He has previously worked with Australian Government agriculture and environment departments and a leading economic and policy company. He also operates a small farm on NSW south coast. His company is Australian Wildlife Services.

Margaret Woodrow has a Bachelor of Arts with majors in Economics and Geography, and Diploma of Education. She provides administrative and research support for the wide range of AWS activities and projects. She has 15 years public service experience and 11 years as a subject coordinator in a large school.

Robyn Watts is Associate Professor and Principal Researcher with the Institute for Land, Water and Society at Charles Sturt University, Australia. She has over 15 years research and tertiary teaching experience in the fields of aquatic ecology, restoration ecology and river management. Her research focuses on biodiversity and connectivity in aquatic ecosystems and ecological responses to flow regimes in regulated rivers. Robyn is a member of the Environmental Water Scientific Advisory Committee for the Australian Government Department of Environment, Water, Heritage and the Arts.

Part I
Understanding Adaptive Management

Chapter 1
Introduction

George Stankey and Catherine Allan

Abstract The increasing complexity and uncertainty surrounding the management of natural resource systems, combined with the complex interactions that occur between those systems and people, over multiple jurisdictional and temporal scales, have revealed the limits to traditional, reductionist scientific inquiry. In response to this, there has been increasing interest in the concept of adaptive management – the purposeful and deliberate design of policies in such a way as to enhance learning as well as to inform subsequent action. Yet despite the great promise such an approach holds, experiences across multiple resource systems and social–political settings suggest that major barriers confront efforts to implement adaptive management effectively. Nonetheless, major progress is occurring. In an effort to explicate the developments taking place between the intuitive simplicity of the adaptive management concept and the elegant theoretical dispositions that have been offered in the literature, this chapter introduces a set of operational applications across a range of biophysical and institutional settings that reveal the concept's potential. Although not a handbook or set of "how to do" rules, the chapters offer important insight and principles upon which adaptive enterprises might be productively employed.

Introduction

People who manage water, soil, air, vegetation and animals face many challenges. High levels of complexity and uncertainty, combined with secondary and tertiary scale impacts that cross multiple disciplinary, geographic, and political boundaries, make our ability to produce effective policies and programs problematic. Traditional forms and methods of scientific inquiry and management, which seek answers and assume stasis and simplicity, are pressed to provide managers and policymakers

---

G. Stankey
Private consultant, Seal Rock, Oregon, USA (Retired research social scientist),
Pacific Northwest Research Station, USDA Forest Service, Corvallis, Oregon, USA

C. Allan
Institute for Land, Water and Society, Charles Sturt University, Albury, Australia

with the quantity and types of knowledge they need. Conventional scientific inquiry is often challenged to produce new understandings at a pace that meets the needs of managers and policymakers. Moreover, the typical reductionist model of scientific inquiry often means that the scope – geographic, temporal and disciplinary – of inquiry is narrow, thereby potentially limiting the utility of results.

In light of this complexity and uncertainty, a growing interest in the notion of adaptive management has emerged. Adaptive management is especially appealing to those who recognise that "the answer" is rarely simple or wholly attainable. Holling (1995, p. 8) has argued that the burgeoning interest in adaptive management has been driven by three interlocking elements:

> The very success in managing a target variable for sustained production of food or fiber apparently leads inevitably to an ultimate pathology of **less resilient and more vulnerable ecosystems, more rigid and unresponsive management agencies, and more dependent societies.** This seems to define the conditions for gridlock and irretrievable resource collapse [emphasis added].

The notion of learning from management experiences has been with us for a long time. Some would trace the idea to Lindblom's (1959) discussion of "disjointed incrementalism" or, as more commonly described "muddling through." In a 1973 text (*On Learning to Plan – and Planning to Learn*), Michael linked the ideas of action and learning explicitly. Later planning texts (e.g., Friedmann, 1987; Lee, 1993) further explored the dialectic between action (policy implementation) and learning, and how that learning could shape and direct subsequent action.

Explicit interest in adaptive management can be traced to the mid-1970s, when it was recognised that the very process of framing policies and implementing them could be the source of increased knowledge and understanding that could be used to inform subsequent action (Holling, 1978; Walters, 1986). Initially portrayed in a technical and statistically rigorous manner, Kai Lee's 1993 text, *Compass and Gyroscope*, expanded the concept to be inclusive of a wider, socio-political context, one that explicitly acknowledged the value-based nature of natural resource decision-making. In subsequent years, efforts to apply adaptive management across a range of natural resource settings and in various political settings began to appear; fisheries management in Canada (Hilborn, 1992) and the United States (Butler et al., 2001), water resource policy and management in South Africa (MacKay et al., 2003) and Australia (Ladson & Argent, 2002), and riparian and coastal ecosystems (Walters, 1997). Large regional scale applications began to appear; the Columbia River region of the U.S. and Canada (Lee, 1995); Everglades National Park (Light et al., 1995); Grand Canyon National Park, Australian multispecies fisheries (Sainsbury et al., 1997) and the Colorado River (National Research Council, 1999) among them. The importance of a participatory and collaborative framework within which adaptive management was undertaken gained added intellectual attention (Buck et al., 2001). A series of synthetic analyses, striving to capture the experience and lessons gained through this range of experiments and policies, began to appear from the 1980s: Environmental and Social Systems Analysts, Ltd. (1982), Barriers & Bridges (Gunderson et al., 1995), and Stankey et al. (2005). Collectively, these efforts began to help assemble, describe, and evaluate the widening effort to make adaptive management an effective strategy.

In its simplest sense, adaptive management advocates argued "policies are experiments; learn from them!" (Lee, 1993). Adaptive management is characterized by both a compelling and intuitive simplicity (we learn by doing) as well as a growing sophisticated and elegant theoretical discourse (for example Environmental and Social Systems Analysts, Ltd. (1982); Dovers et al., 1997; Gunderson et al., 1995; Lee, 1999; Light, 2002). Yet, at the same time, there is a disquieting sense that adaptive management has become little more than a rhetorical notion, constructed more by assertion than by demonstration. Lee (1999, p. 1) concludes "adaptive management has been more influential, so far, as an idea than as a practical means of gaining insight into the behaviour of ecosystems utilized and inhabited by humans."

At the beginning of the twenty-first century we face a situation where the requirement to use "adaptive management" is routinely inserted into strategies and plans with little appreciation of what might be needed to fulfill this requirement, and/or little will to provide it. In the US, an evaluation of efforts to implement an adaptive approach in the management of a 10 million hectares forest region in the Pacific Northwest concluded that a host of barriers – institutional inertia, lack of organisational capacity, an absence of leadership and inadequate resources – constrained efforts to implement adaptive management (Stankey & Clark, 2006). Similar stories have been reported in a wide range of settings and sectors (for example, Allan & Curtis, 2005; Briggs, 2003). At the root of these potentially dispiriting reports lies the idea that there remains a failure to acknowledge that adaptive management represents a fundamental and systemically different approach. Adaptive management explicitly acknowledges that we often lack sufficient knowledge to act with a full understanding of consequences and implications. It accepts that our knowledge of appropriate interventions is limited. And it elevates the role of monitoring and evaluation beyond the cosmetic and superficial attention often given these activities to a level at which they become the mechanisms through which significant changes in policy and practice in light of outcomes can occur.

The promise of adaptive management – of using the management process as a way of gaining increased understanding of complex processes – remains worthy of attention and support. Ideally, adaptive management offers both a scientifically sound course that does not make action dependent on extensive, traditional scientific inquiry and a strategy of implementation designed to enhance systematic evaluation of actions (Lee & Lawrence, 1986). One way to gaining insight into how to capture this potential is to examine the realm of operational experience between "intuitive appeal" and "theoretical elegance" in a thoughtful, critical, and comprehensive fashion.

With this ideal in mind, in April 2007, a workshop was convened at Lake Hume in southern New South Wales, Australia, organised by staff at Charles Sturt University and involving 16 people from Australia, New Zealand and the United States. A variety of resource sectors were represented by the participants, as were a variety of institutional homes. Drawing on their own practice this group agreed on the need for a book that revealed the range of experiences, offered insights about challenges and opportunities, and suggested strategies for successful implementation of adaptive management in real-world settings. Practitioners and theorists alike at the workshop also agreed that, as effective participatory practice is essential for dealing with complex issues, so the book should be developed as an integrated whole, by an actively engaged team. The intended audience was identified as policy makers and managers seeking to undertake or enhance adaptive approaches to environmental or natural resource management. Participants used the workshop to develop the overall framework of this book to meet the needs of that audience, determined through personal experience and research such as that from Allan & Curtis (2003). Draft contributions were prepared and shared, and the final form of the book consolidated, at a second workshop held in Sydney in August 2008. Chapter 2, which provides the theoretical basis of adaptive management, was determined by the participants of the Sydney workshop to be the reference point for the subsequent chapters, so this was independently peer reviewed. The remaining chapters, presented as experience rather than theory, were reviewed in draft form by the workshop participants and by the editors prior to publication.

Participatory approaches, built around sharing and iteration, take time, and risk loss of focus. The authors of this book kept sight of the framework and intent of the work by keeping the readers and their needs in mind. This book was designed and written and polished for all managers – of protected areas, farms, forests, waterways, catchments, oceans – who need to build adaptive capacity into their operations. How to start? How to keep going? How to know if you've done any good? are some of the questions addressed for this audience. This book is also for policy makers and strategists who seek to include adaptive management in future plans. What support will you need to provide to see that it really happens? What can you realistically expect adaptive management to look like? What changes will need to be made to process and expectations? "Traditional' environmental scientists are also catered for, with examples of large scale, real world enquiries to compare with reductionist experiments. And, this book is for the future managers and policy makers who are learning to understand and work with complex and changing socio-ecosystems.

We do not provide prescriptions and guidelines, but rather present the distilled lessons learned from a range of real adaptive management projects. This approach acknowledges that context is a critical feature of managing complex systems, and encourages readers to apply and modify the lessons to their own situations and needs.

How This Book Is Presented

All the contributions in this book are written against the backdrop of the concepts and principles outlined in Chapter 2, so we suggest readers become familiar with the ideas and terms introduced in that chapter. Each of the following chapters then present some detailed description and reflection on the real world adaptive management with which the authors are, or have been, involved. These chapters conclude with the lessons the authors feel are important to share. Within many chapters there are also information boxes that provide brief discussions of some of the key ideas referred to throughout the book.

The book concludes with a critical, synthetic chapter that captures and distils the reported experiences, highlighting both necessary and sufficient conditions for successful implementation of adaptive management, the importance of organisational capacity, the social-political nature of the challenges facing adaptive management, and the critical role of context.

References

Allan, C., and Curtis, A. 2003. Learning to implement adaptive management. *Natural Resource Management*, 6(1): 23–28.

Allan, C., and Curtis, A. 2005. Nipped in the bud: Why regional scale adaptive management is not blooming. *Environmental Management*, 36(3): 414–425.

Briggs, S. 2003. Command and control in natural resource management: revisiting Holling and Meffe. *Ecological Management and Restoration*, 4(3): 161–162.

Buck, L.E., Geisler, C.C., Schelhas, J., and Wollenberg, E. (Eds.). 2001. Biological diversity: balancing interests through adaptive collaborative management. New York: CRC Press. 465 p.

Butler, M.J., Steele, L.L., and Robertson, R.A. 2001. Adaptive resource management in the New England groundfish fishery: implications for public participation and impact assessment. *Society and Natural Resources*, 14(9): 791–801.

Dovers, S.R., Mobbs, C.D., and Lunt, I. 1997. An alluring prospect? Ecology, and the requirements of adaptive management. In Klomp, N. (Ed.). Frontiers in Ecology: Building the links (pp. 39–52). Albury, NSW: Elsevier.

Environmental and Social Systems Analysts, Ltd. [ESSA]. 1982. Review and evaluation of adaptive environmental assessment and management. Ottawa, ON: Environment Canada. 116 p.

Friedmann, J. 1987. Planning in the public domain: from knowledge to action. Princeton, NJ: Princeton University Press. 501 p.

Gunderson, L.H., Holling, C.S., and Light, S.S. (Eds.). 1995. Barriers & bridges to the renewal of ecosystems and institutions. New York: Columbia University Press.

Hilborn, R. 1992. Institutional learning and spawning channels for sockeye salmon *(Oncorhynchus nerka)*. *Canadian Journal of Fisheries and Aquatic Sciences*, 49: 1126–1136.

Holling, C.S. 1978. Adaptive environmental management and assessment. Chichester: Wiley.

Holling, C.S. 1995. What barriers? What bridges? In Gunderson, L.H., Holling, C.S., Light, S.S. (Eds.). Barriers & bridges to the renewal of ecosystems and institutions (pp. 3–34). New York: Columbia University Press.

Ladson, A.R., and Argent, R.M. 2002. Adaptive management of environmental flows: lessons for the Murray-Darling Basin from three large North American rivers. *Australian Journal of Water Resources*, 5(1): 89–101.

Lee, K.N. 1993. Compass and gyroscope: integrating science and politics for the environment. Washington, DC: Island Press. 243 p.

Lee, K.N. 1995. Deliberately seeking sustainability in the Columbia River basin. In: Gunderson, L.H., Holling, C.S., Light, S.S. (Eds.). Barriers & bridges to the renewal of ecosystems and institutions (pp. 214–238). New York: Columbia University Press.

Lee, K.N. 1999. Appraising adaptive management. *Conservation Ecology*, 3(2): 3. http://www.consecol.org/vol3/iss2/art3.

Lee, K.N., Lawrence J. 1986. Adaptive management: learning from the Columbia River basin fish and wildlife program. *Environmental Law*, 16: 431–460.

Light, S. 2002. Adaptive management: a valuable but neglected strategy. *Environment*, 44(5): 42.

Light, S., Gunderson, L.H., C.S. Holling. 1995. The Everglades: evolution of management in a turbulent ecosystem. In Gunderson, L.H., Holling, C.S., Light, S.S. (Eds.). Barriers & bridges to the renewal of ecosystems and institutions (pp. 103–168). New York: Columbia University Press.

Lindblom, C. 1959. The science of muddling through. *Public Administration Review*, 19(2): 79–99.

MacKay, H.M, Rogers, K.H., and Roux, D.J. 2003. Implementing the South African water policy: holding the vision while exploring an uncharted mountain. *Water SA*, 29(4): 353–358.

Michael, D.N. 1973. On learning to plan—and planning to learn. San Francisco, CA: Jossey-Bass. 341 p.

National Research Council. 1999. Downstream: adaptive management of Glen Canyon Dam and the Colorado River ecosystem. Washington, DC: National Academy Press. 230 p.

Sainsbury, K. J., Campbell, R.A., Lindholm, R., and Whitelaw, A.W. 1997. Experimental management of an Australian multispecies fishery: examining the possibility of trawl induced habitat modification. In Pikitch, E.L., Huppert, D.D., and Sissenwine, M.P. (Eds). Global trends: fisheries management (pp. 107–112). American Fisheries Society Symposium, Bethesda, Maryland.

Stankey, G.H., and Clark, R.N. 2006. Adaptive management: facing up to the challenges. In G.H. Stankey, B.T. Bormann and R.N. Clark (Eds.). Learning to manage a complex ecosystem: adaptive management and the Northwest Forest Plan (pp. 137–180). Portland, OR: U.S. Department of Agriculture, Forest Service, Pacific Northwest Research Station.

Stankey, G.H., Clark, R.N., and Bormann, B.T. 2005. Adaptive management of natural resources: theory, concepts, and management institutions. General Technical Report PNW-GTR-654. Portland, OR: U.S. Department of Agriculture, Forest Service, Pacific Northwest Research Station.

Walters, C.J. 1986. Adaptive management of renewable resources. New York: Macmillan. 374 p.

Walters, C.J. 1997. Challenges in adaptive management of riparian and coastal ecosystems. *Conservation Ecology*, 1(2): 1. http://www.consecol.org/vol1/iss2/art1.

Words Matter

Catherine Allan

We communicate through our words and images, but usually behave as if the act of conversing is irrelevant to the topic, content and aim of the dialogue. However, as language both represents and constructs reality (Penman et al., 2001) our use and understanding of words and terms can have major influences on practice. Adaptive management is invoked partly in response to acknowledged epistemic uncertainty. Carey and Burgman (2008) suggest that understandings of uncertainty which focus on the variability of a system, and incertitude about the system ignore the important role of linguistic uncertainty; i.e., the uncertainty of definition that comes from ambiguity, vagueness, underspecificity or loss of context. 'Adaptive management' is used in many disciplines, and is applied to many apparently different practices, so the chances for deliberate or unintentional linguistic uncertainty is high. Self aware use of, and enquiry into adaptive management should therefore seek to articulate meaning as clearly and as contextually dependently as possible. Clear articulation of the form of the adaptive management being attempted evaluated or discussed can help to avoid false expectations and disappointments. Precision helps to head off the implied judgements that lurk behind many apparently benign word (Lakoff & Johnson, 1980). Precision in defining the form of adaptive management will also assist development of appropriate evaluation regimes.

References

Carey, J. M. and Burgman, M.A. 2008. Linguistic uncertainty in qualitative risk analysis and how to minimize it. *Annals of the New York Academy of Sciences*, 1128: 13–17.

Lakoff, G. and Johnson, M. 1980. Metaphors we live by. Chicago, IL: The University of Chicago Press.

Penman, R., Fill, A., and Muhlhausler, P. 2001. The ecolinguistics reader: Language, ecology and environment. London: Continuum.

Chapter 2
Components of Adaptive Management

Robert M. Argent

Abstract Adaptive Management of ecosystems for production and preservation is a cyclical process with four components: learning, describing, predicting and doing. Learning involves monitoring and evaluation, describing uses models to summarise and represent systems, prediction and gaming are used to test policies and proposed actions, and the doing is done through management experiments. Successful adaptive management needs clear objectives, data and knowledge, the right participants, science skill, willing partners, and money and time. Additionally, in doing adaptive management it is necessary to drive and steer the process, keep momentum, embrace uncertainty, and beware of the danger of half measures. Adaptive management supports decisions and resource allocation, and provides a framework for action directed to changing ecosystem state while learning through and from such change. It focuses conversation and reduces arguments and finger pointing; and also reduces excuses for inaction, provides system understanding, identifies data and knowledge gaps, and sets up a time and space framework for explanation of key processes. Adaptive management won't make decisions, won't do the work or the thinking, and has scientific, social, political and economic aspects that may cause failure, but which, when understood and embraced, provide the framework for successful ecosystem improvement.

Introduction

The previous chapter provided an overview of the role of adaptive management in managing natural resources for production and preservation. Although good management of natural systems has been adaptive over history, it has only been in the last 30 years that adaptive management of natural systems has become a recognised process, with methods defined, explored, tested and refined (ESSA, 1982; Holling, 1978; IIASA, 1979; Walters, 1986). This chapter describes the components of adaptive management, and provides a framework and strategies for

R.M. Argent
AWRIS Project Director, Bureau of Meteorology, Australia

applying these to complex natural resources management problems. It provides the building blocks for successful adaptive management, describes the conditions required to make an adaptive management process fly, gives pointers to keeping the process under control, and identifies some of the key points of failure – and possibly even how to avoid them. In doing so it acts as a primer for the detailed case studies which follow.

Over the 30 years of formal publication on adaptive management various authors have described it in ways that vary widely from the broadest philosophy of ecosystem management, to a sub-component of a management framework, to the narrowest interpretation of specific management actions (e.g. Brussard et al., 1998; Grumbine, 1997; Keough & Blahna, 2006; MacDonald & Coe, 2007; Matsuda, 2003; Moir & Block, 2001; Prato, 2007; Richter et al., 2003). In this text we consider adaptive management with respect to complex and complicated environmental management problems, often with:

- Multiple uses and multiple objectives
- A mix of scales of interest and boundaries of responsibility
- Divergent needs and desires of stakeholder groups
- Tight economic imperatives around ecosystem exploitation
- Reduced ecosystem health and ecosystem services
- Significant technical information on parts of the system, with information gaps on other parts and
- Competing or open mandates, with different policy options and system targets

Adaptive management comes in a variety of flavours (see Box), often described as evolutionary, passive and active (Walters, 1986; Walters & Holling, 1990). To clarify the definition of adaptive management used in this chapter, the following hypothetical scenario is offered.

A social ecological 'Problem' exists. The Problem has no clear causative factors, and no clear management path to 'solving the problem'. Studies of the Problem and associated factors have occurred over the years, and there is a rich field of data available. A new initiative or impetus arrives, prompting a responsible authority to tackle the Problem in an inclusive and integrated manner. The decision is taken to use a structured management approach that is intended to be adaptive. Monitoring data, knowledge of ecosystem dynamics and informal observations are combined through a consultative model building process. Testing and exploration of the model shows that it describes reasonably observed historical responses, and it is agreed by all parties that it makes fairly good use of the available data. Three critical factors, one relating to scale and two relating to timing of natural events, are identified. A 7-year cycle of management experiments is proposed to test these critical factors, in order to determine a management course that reduces the nature of the problem to 'sustainable' levels. Four years of combined push from industry, agencies and environmental groups results in acceptance, funding and initiation of the 7-year experimental plan. This proceeds, and the success after 5 years (due to fortuitous conditions and some early wins), when combined with improved understanding of aspects of the problem, provide both a healthier ecosystem and an agreed platform and process for managing and continual learning into the future.

This accords with 'active' adaptive management concepts, and the following sections explore the components of active adaptive management practice, and inform both the choice of approach (Gregory et al., 2006b) and the possible levels of application of the components to a problem situation (*sensu* Checkland, 1981).

Cycling Through Adaptive Management

Adaptive management is a cyclical process, relying on the results of prior actions to inform future actions. At its most basic level it is described as 'learning by doing' (Walters & Holling, 1990), so Fig. 2.1 gives the simplest description of the process.

For environmental managers, scientists and others with a stake in 'solving' an ecosystem problem, this simple construction raises questions:

- Learn what?
- Do what?
- What do we aim to learn from doing?
- How do we use what we have learned in deciding what to do?

The primary position of learning in Fig. 2.1 is purposeful – in all ecosystem problems the managers, scientists and other participants in, and observers of, the system have already learned a lot about the system – certainly enough to know that it is declining, and needs improvement. These simple points cut to the heart of adaptive management – the specific intention to learn from the responses to management actions (e.g. McDaniels & Gregory, 2004; Torell, 2000). Although this sounds easy, Allan and Curtis (2005) note that it is often difficult for managers and other ecosystem stakeholders to go forward while looking back. Adaptive management offers a way to do this, and also to look forward in an effective and structured way while going forward with ecosystem management. In regards to effective ecosystem management, Ludwig et al. (1993) reflect many of the principles of adaptive management, stating:

> ....We must consider a variety of plausible hypotheses about the world; consider a variety of possible strategies; favor actions that are robust to uncertainties; hedge; favor actions that are informative; probe and experiment; monitor results; update assessments and modify policy accordingly; and favor actions that are reversible

In learning by doing, the questions above give rise to further considerations and implications, informing more detailed understanding of the adaptive management components and the processes by which adaptive management can be undertaken.

Fig 2.1 Adaptive management in a nutshell – learning by doing

Learn What?

Each attempt to manage a system adaptively involves learning. Learning not only includes the system responses, at various temporal and spatial scales, to the management levers that we manipulate, but also learning to fill gaps or improve understanding of key parts of the system where our knowledge is lacking – noting that such key parts always exist!

Do What?

Deciding which actions to take is generally easier than deciding exactly what and how to learn from these actions. Management is an ongoing process involving planning and implementation of actions based upon varying levels of available information, so managers are adept at decision making under uncertainty. The difference with adaptive management, compared with less 'intentional' forms of management, lies in the explicit planning for the 'doing' to be information-driven in addition to being result-driven. It can be quite challenging for managers to expand from choosing actions with the greatest likely effect, or possibly the least harm, to inclusion of actions which provide the greatest learning about the system (Lessard, 1998; McDaniels & Gregory, 2004; Ojha & Bhattarai, 2003).

What Do We Aim to Learn from Doing?

When the aim is to learn, the parallel requirement is to specify what is needed in terms of learning (Armitage et al., 2008), and how the outputs will be captured and added to the knowledge structure around the problem (Allen et al., 2001). Such structuring of knowledge requires a method for holding or retaining system information, and is generally done through models and system reporting.

How Do We Use What We Have Learned in Deciding What to Do?

This final question brings us back to the start of the cycle. At this step the things learned from 'doing' are evaluated, then fed back to the knowledge base of the system and support the understanding brought to bear in the next round.

This expands the adaptive management cycle to four components (Fig. 2.2). Although these components are given as discrete entities, in reality, they occur simultaneously as part of adaptive management, to a lesser or greater degree, with multiple feedback loops. It is, rather, that the *focus* shifts around the cycle as the adaptive management process progresses, and often the trick is to know where to start.

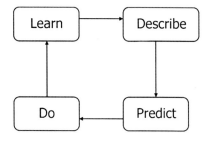

Fig. 2.2 Expanding the adaptive management cycle to include system description, as a place holder for things learned, and prediction, as part of the process of deciding what to do and learn

For example, when the focus is on describing the system, it is still necessary to have a eye and ear to the political and financial processes that underpin, for example, the 'doing' of adaptive management. The core components of adaptive management thus include monitoring or learning from the system, modelling or system description, proposed actions and prediction of changes, and the 'doing' of adaptive management in the form of implementation as experiments (Walters & Holling, 1990). This process has been described, and re-described, in as few as two steps (learning and doing) and as many as five or seven or even more, in varying levels of detail and with different granularity of the components (e.g. Grafton & Kompas, 2005; Haney & Power, 1996; Herrick et al., 2006; Thom, 2000; Thom et al., 2005; Wilson & Lantz, 2000).

In this chapter a four-component model of adaptive management is used, and each of the components is described in sufficient detail to support slicing and dicing the steps into as many as suit a given application, analysis or description.

LEARN: Evaluating, Monitoring, Observing, Data Capture, Learning

Evaluation of ecosystem responses requires data on key indicators of system state, either directly measured or determined from a variety of approaches in situations where responses of interest and monitored variables are not well aligned (Kneeshaw et al., 2000; Pik et al., 2002; Smyth et al., 2007; Stauffer, 2008). Core data for adaptive management applications or processes often come from traditional or institutional monitoring where there is sufficient long term data over multiple sites or parameters to indicate trends. Research-based and operational monitoring (e.g. Converse et al., 2006), which tends to shorter time-frames and a narrower focus, is also invaluable for providing key information on ecosystem processes, as is targeted monitoring within an adaptive management evaluation framework (Bisbal, 2001; Caruso, 2006; Plummer & Armitage, 2007). In addition to these sources, 'snapshot' monitoring provides a balance of resolution and extent by giving multiple samples from multiple sites over a short time. If the selected time is a quiescent part of the system cycle, then snapshot samples add great depth to baseline data. Final sources for data to support learning are non-systematic and anecdotal observations. These are sometimes useful, and have been shown to provide significant insights into the problem situation (Ballard & Huntsinger, 2006).

In addition to learning from data, the modelling and policy exploration described in the following also provide avenues for learning.

DESCRIBE: Describing, Summarising, Modelling

Models for explanation and prediction are part and parcel of all adaptive management processes (e.g. Walters et al., 2000), and although this is one area where there have been significant developments over recent decades, it is also an area where technical dramas often come to the fore. Developing adaptive management models is an art in and of itself, and methods vary across systems modelling (e.g. Rowntree, 1998; Walters, 1974) such as stocks and flows and predator–prey relationships, agent-based modelling (Pahl-Wostl, 2002), Bayesian approaches (Dorazio & Johnson, 2003; McCarthy & Possingham, 2007; Nyberg et al., 2006), and Markov decision processes (Sharma & Norton, 2005).

The core models for an adaptive management application vary in complexity from conceptual or relationship models (e.g. Gentile et al., 2001) on the back of an (electronic) envelope, through long term static or dynamic stocks and flows analysis, to multi-component 3D dynamics of system processes at sub-hour time steps and running over multiple seasons or years. Models not only provide the vehicle to describe the system, but also impose structure on the description of the science or phenomena of interest. They also help to constrain the behaviour of the participants in the process by providing a neutral home for debate on process representation. Description of the system requires a focus on *actions* and *indicators* (Walters, 1986) In many adaptive management processes a turbulent history of managing 'the Problem' results in accusative finger pointing, entrenched views and strong disagreements. People involved at this level sometimes come to an adaptive management process seeking to focus on a particular issue (such as timber harvesting or river flow control). These foci are generally unhelpful. By turning to the indicators of system state and possible actions, the process moves ahead by looking at *how* an action will affect the indicators, rather than the emotions of the action. Indicators such as employment, harvested area, species richness or habitat sparseness (e.g. Boddicker et al., 2002; Kremen et al., 1998; Kremsater et al., 2003; Pik et al., 2002) can be related to actions such as changing road building policy to reduce forest gaps, or releasing flood waters from a dam (e.g. Walters et al., 2000). This approach identifies the dynamics that need to be included in the system description, and the measures that will be used to compare outcomes of alternative actions.

It is tempting to over-parameterise an adaptive management model to include all things that may be of relevance. As suggested by Grayson and Blöschl (2000), an appropriate level of complexity is found in the balance between detail and testability. Limiting the system description complexity in a model via reduced numbers of parameters and relationships reduces the 'equifinality' problem (Beven & Binley, 1992) that occurs when a desired indicator change is achieved through multiple alternative combinations of actions, with no clear causative links. However, too simple a model may reduce the explanatory value of variables that were aggregated or excluded.

System simulation through an adaptive management model is broader than simply describing the system. Shannon (1975), in one of the earlier texts on the art and science of simulation, notes that simulation is intended to describe the behaviour of the system, support hypotheses to account for such behaviour, and to predict future behaviour. In terms of ecosystems, Walters (1997) notes the weakness of the latter point, as historical behaviour is often a poor predictor of the future, especially when compensatory ecosystem responses occur. Adaptive management models support not only these simulation roles, but also aid thought and communication, underpin training and instruction, and provide the test-bed for experimentation (Shannon, 1975), thereby extending learning aspects of adaptive management across all components. See Chapter 9, Jakeman et al., this volume for more on models.

PREDICT: Predicting, Scenarios, Game Playing

Beyond system description, and all the data analysis and knowledge integration that go into creating a sound and valid model, key roles for an adaptive management model include exploration of system understanding and behaviour, and provision of a 'playpen' for thinking about and testing policy options. These 'gaming' activities are often the most informative of the descriptive processes around a system, and are the home for a lot of knowledge, policy and experiment generation amongst all participants.

Gaming involves generating ideas about different possible policies and implementation actions, then testing these with the model. The process of testing proposed actions leads participants to consider the reality of trying to model and predict ecosystem responses (e.g. Melbourne et al., 2004), such as (i) the sensitivity of the system responses to the management levers, (ii) the types of responses being seen, the ecosystem relationships expressed within a model, and the stability and predictability of these responses (McLain & Lee, 1996; Pimm, 1984), (iii) the things that are not represented in the model, that may become influential as the system deviates from historical behaviour in non-linear and compensatory ways (Walters, 1997), and (iv) the ways that the key processes in the model may change as the system changes.

Modelling and gaming is a mental as well as a numerical process – experts, observers and policy makers still test the outcomes from models against the results from mental models. The primary power of the *process* by which adaptive management models are created and run is the development of mental models and system understanding, so that knowledge of all participants is advanced, decisions are informed by consideration of uncertainties and assessment of risks (Pittock & Jones, 2000; Prato, 2005), and those with divergent views converge on some systems aspects at some scales. Such structuring of knowledge is an essential aspect of any learning process.

DOING: Doing, Enacting, Experimenting

Once various scenarios, policies, actions and implementations have been tested in both the numerical 'mind' of the model and the natural minds of the participants, and possible management experiments have been identified and designed, focus shifts to implementation. This shift commonly sees the balance of adaptive management activity move from scientific to political, as the machinery of bureaucracy and management processes kick into action. This is particularly true if expensive, possibly risky, large scale management experiments are proposed, and also if changes are required at high levels, such as in regulations and legislation. As noted earlier, all components of the cycle are active to a greater or lesser degree throughout application of adaptive management, so adoption and enactment of appropriate participative processes (Dovers & Mobbs, 1997; Edwards-Jones, 1997; Schindler & Aldred Cheek, 1999) provides a significant benefit here, and the strategies and procedures already put in place, such as communication and lobbying, will contribute to a successful 'doing' phase.

The Doing of adaptive management requires careful experimental design (Gerber et al., 2007; Gregory et al., 2006a) in addition to careful management planning. Two ways that adaptive management experiments often vary from traditional scientific experiments are (i) the limited ability to have controls and replicates, and (ii) the risk of harm in the case of significant unintended consequences. Often the 'system' being managed is the *only* system available, so assessment of experiments is sometimes designed around changes in the trajectory of indicators from those predicted under the business-as-usual or incremental management approaches.

In 'Doing', approaches such as the multiple lines and levels of evidence (MLLE) offer a way to provide scientific weight to experimental findings. In this approach, it is the weight of evidence rather than any particular statistical measure that carries the case. Either way, monitoring the results of management actions provides the primary data on system indicators (Kremen et al., 1998), and contributes the key feedbacks that close the adaptive cycle by contributing to Learning.

Component Interfaces – Where the Fun Really Begins

An important aspect of adaptive management, and often of science and society, is that challenges and rewards occur at the margins or interfaces between the components that we feel comfortable putting in 'boxes'. Often the boxes in conceptual diagrams such as Fig. 2.2 are the knowledge, domains, sciences, or beliefs that we are most willing to label, and where we feel a sense of commonality and community. In connecting or integrating between these boxes we cross into the unknown, where we cross paths between knowledge domains or beliefs (Szaro et al., 1998). In the case of Fig. 2.2, the interactions are:

- Between learning and describing we have the processes of theorising, analysis and synthesis that bring data together with ideas to form new knowledge and understanding.
- Between describing and predicting we have assumptions about linearity, representativeness, policy interpretation, and scenario formulation. In the context of a system-describing model, this also represents the processes of interpreting our desired or proposed actions into model parameters that act as the levers on model behaviour.
- Between predicting and doing occur the political, moral, motivational and financial processes of deciding on a course of action, as well as the operational aspects of working out who, how and when things will be done.
- Finally, the link from doing to learning encompasses the evaluation component, with monitoring, testing and feedback processes that arise as the system responds, or fails to respond, to the selected interventions.

As simple as the above explanations are, a key point for practitioners is that for any system diagram presented in an adaptive management application, attention needs to be given to both the labelled 'boxes' in a diagram and the connections between.

These components, and their connections, form the core parts of adaptive management. In much of the reflection and analysis around adaptive management (e.g. Ladson & Argent, 2002; Stankey et al., 2003; Walters, 2007) these components provide the framework for action, and it is the 'how' of adaptive management, rather than the 'what', that garners attention.

Framing Adaptive Management – Necessary Conditions for 'Success'

The successes and failures of adaptive management (e.g. Lee, 1999; Torell, 2000; Walters, 1997; Walters, 2007) show that although it is easy to understand the 'what' of the components, it is much harder to do these successfully. Defining success is rarely easy with dynamic systems (Kentula, 2000) and many adaptive management practitioners focus on success in ecosystem improvement rather than successful experiments or successful restoration.

A key starting point for success, however defined, is to have clear objectives for the adaptive management activity. As with many aspects of adaptive management, identification of objectives works on multiple levels, and may be viewed differently by different participants (e.g. Porter & Underwood, 1999). In significant expansion to the hypothetical case presented earlier, ESSA (1982) identify seven high level 'process' objectives that have been set at different times for adaptive management activities:

- Identification of issues and unknowns
- Identification of impacts
- Communication
- Information synthesis
- Research planning

- Policy analysis
- Project management

One, or more, or all, of these objectives are possible in any application of adaptive management, and it is therefore necessary to clearly understand and communicate which of these are being sought (Lynch et al., 2008; Wilson & Lantz, 2000). If parties to an adaptive management process do not clearly understand the high level objectives, or the possibility of expanding, contracting or changing objectives as the process moves along, misalignment and misunderstanding is nearly guaranteed. Also, clear objectives are a core part of the risk management approaches that operates throughout adaptive management.

Clear Objectives

Specific objectives are required for a given adaptive management activity, and for each of the components of adaptive management, such as 'restoring' ecosystem components (Coen & Luckenbach, 2000), predicting the effects of factor X on problem Y (Converse et al., 2006), or designing an experiment to test hypothesis Q (Theberge et al., 2006). Spelling these out clearly at the start of an adaptive management process helps answer the critical question of when to end the process. Although a cycle has no end, there can be periods during, say, the learning component, when day-to-day adaptive management activity is low. Clear objectives also help identify exit strategies, and help to focus participants on the 'why' of adaptive management while doing the 'how' (Lee, 1999). In one of the author's adaptive management activities investigating landscape nutrient loads, four clear objectives (e.g. "To determine, on a broad scale, the relative nutrient inputs and outputs of the various activities taking place in the catchment") were identified as part of choosing to use adaptive management. These were subsequently presented as the first points in communications regarding the activity, so there was never any doubt about why the process was occurring.

Data and Knowledge

Data and knowledge are essential in the descriptive component of adaptive management. Data include the raw measurements of indicators of interest (e.g. Kneeshaw et al., 2000; Kremen et al., 1998), while knowledge requires understanding the important and the negligible processes that need to be included in the prediction phase. If either or both are lacking, the adaptive management activity is either delayed while data are gathered or knowledge is generated, or is frustrated by an inability to describe the system. Although a lack of data is a self-evident truth in all modelling processes, a drive for more precise detail may not improve either description or prediction (Grayson & Blöschl, 2000), so practitioners must be realistic about how much data are sufficient to describe key system relationships.

Participants

Another key element in successful adaptive management is having the right people with the right attitude (Cote et al., 2001). One of the reasons that adaptive management is sometimes referred to as a philosophy is the attitude of those involved. Generally, these people are technically competent, capable communicators, who, whilst holding potentially strong beliefs about aspects of the system, are open to both the thoughts of others and the understanding that systems are generally larger than the sum of the parts. They are able to recognise disagreements and to work to understand and overcome these, as well as to communicate the disagreements and resolutions to others. Early adopters often find a home within an adaptive management activity. As a generalisation, the social, ecological and physical science aspects of adaptive management attract seekers of greater knowledge and understanding, rather than greater volumes of data. These are people frustrated by the limitations of reductionism, who believe that informing management is a key role for science and that applying science can rarely be done in isolation from other sciences, peoples' opinions or the practicalities of institutions and politics (e.g. Olsson et al., 2004; Pahl-Wostl, 2002). Similarly, managers and other stakeholders drawn to adaptive management often are those who recognise the limitations of previously tried management approaches, who see that adaptive management might offer a way to build bridges and get around walls, and who are comfortable working in the uncertain world of adaptive management application (see also Chapter 18, Fazey and Schultz, this volume).

Science Practitioners

Scientific skill is essential for all of the components of adaptive management. This includes not only domain experts capable of simplifying, translating and communicating their domain knowledge, but also those experienced in integration, or interpreting science from one field to make it fit with that of another. The ability to express scientific concepts in understandable ways is also necessary, as the learning, describing, predicting and doing components draw heavily upon communication – up, down and sideways. These skills extend across the social (Dovers & Mobbs, 1997; Roe, 1996), ecological (Crossley, 1996; Gentile et al., 2001; Walters, 1997; Williams, 2001) and physical (MacDonald & Coe, 2007; Viney et al., 2007) science domains, as adaptive management has needs for all these in different measure for the different components. Equally, risk, uncertainty and complexity science skills have essential roles to play (McDaniels & Gregory, 2004; Pahl-Wostl, 2007; Prato, 2005; Thom et al., 2005; Williams, 2001).

Partners

Willing partners provide the raw power of successful adaptive management. Much has been written of the barriers to, and structures for, adaptive management (Gunderson et al., 1995; Lee, 1993; Pinkerton, 1999; Stankey et al., 2003; Stubbs & Lemon, 2001; Szaro et al., 1998; Walters, 1997, 2007), and the importance not only of participation (Berkes, 2004; Dovers & Mobbs, 1997), but *willing* participation of stakeholders across the board, from those most able to be effective to those most affected – as these are commonly not one and the same (Chavez, 2002; Cote et al., 2001; Skogen, 2003). It is difficult to identify the factors that contribute to willingness – the most impacted parties (apart from the unvoiced and unheard ecosystem) can be unwilling due to previous failures, while the potentially most effective, in terms of political or financial influence, may be unwilling due to perceived risks of failure or unacceptable levels of uncertainty. Willingness is often influenced strongly by the nature of institutions and the barriers they raise (Butler & Koontz, 2005), and it is desirable to seek or construct institutions that are open to adaptive management (Habron, 2003; Stankey et al., 2003; Steyer & Llewellyn, 2000), and that are able to learn and respond (Allan & Curtis, 2005; Dovers, 2001; Hughes et al., 2007; Olsson et al., 2004; Stubbs et al., 2000) as they go along.

Money and Time

Successful adaptive management is generally neither a cheap nor a short exercise (Walters, 1997), although this depends upon the high level objectives. Adaptive management activities that aim to identify impacts and understand unknowns are sometimes completed in short, intense bursts that cost relatively less than those aimed at planning research and analysing policy (e.g. Hennessey, 1994; Wilson et al., 1996). There is a natural relationship between willingness to participate and cost – the more willing partners are, the less direct and transactional costs will be incurred. This is not to say that these costs are not real, and so should be included in accounting for the activity, but rather that the multipliers involved in externalising costs can be significantly reduced. The maxim of taking the number your first thought of, and doubling or tripling it, often holds true.

There are a host of other factors that have been identified as being influential in the success of adaptive management activities. These include:

- The presence of a 'champion' for the activity
- Effective coordination bodies and processes
- The previous history of management or dispute between parties
- The political climate, and structure of the participating institutions
- Timing of the activity in relation to natural and institutional cycles
- Extreme natural events such as tropical storms, droughts or floods
- Other external factors drawing attention from, or to, the problem situation and
- Economic health of the region, state or nation

The best adaptive management process, well planned and supported, with good science, data and knowledge, and many willing partners, can founder on the timing of an election or an economic recession.

Bounding Adaptive Management – Making It Work

The factors mentioned in the previous section, if accounted for, do not guarantee the success of adaptive management. There are other, possibly more subtle, aspects of the *way* that adaptive management is undertaken that can swing the pendulum across from failure to success, or vice versa.

Momentum is possibly the most useful, but elusive, factor around successful adaptive management. In situations where there are many participants, with a hundred calls on their time, 'push' email in their hands, and another budget cycle coming up, the sense (and actuality) of momentum is important, particularly in ensuring that 'learning' occurs along with the 'doing' (Allan & Curtis, 2005). As with any body of work directed to a specific end, a task left undone tends to grow, and the difficulties, hurdles and uncertainties of adaptive management grow to unsurmountable proportions if they are left alone. If progress is made and communicated, hurdles are overcome, and momentum maintained, then the 'unsurmountable' problem becomes another small challenge to conquer before the close of business.

Momentum needs both a driver and a steerer. Although a critical mass is required for adaptive management to have momentum, this can go astray if it is neither well driven nor well steered. Holling (1978) identifies the roles of program manager (champion, key player) to drive, and core group (steering committee) to steer in a typical adaptive management process. The champion's role can be played variously by contractors, communicators, scientists, local stakeholders, managers or bureaucrats, with the latter 'institutional champion' often significantly influencing success (Gilmour et al., 1999). It matters less where they are from, than how much passion and freedom they have, and whether they are driven by their concern for the adaptive management process and the outcomes, than for any personal gain or loss. Steering committees need a combination of wise heads and purse string holders, to set direction and focus, and to provide a buffer to outside negative influences and detractors.

Within the application of adaptive management there are also strategies that increase the chances that decisions will be made and enacted, and that useful things will be learned. Adaptive management authors over the years have repeatedly urged (e.g. Lister, 1998; Ludwig et al., 1993; Walters & Holling, 1990) that uncertainty be embraced rather than avoided or ignored, as this provides the greatest opportunities to learn. This not only includes hedging of actions by choosing paths not of least risk, but also considering ranges of options that enhance the chance of early learning and improved turnaround and response to unpredicted negative effects. 'Expecting the unexpected' (IIASA, 1979; Lister, 1998) is easy to say, but planning and responding to the unexpected is an extremely useful adaptive management skill (Lessard, 1998). Other useful and practical bounding methods include a strong focus on inclusion and participation (Pahl-Wostl, 2002), and clear and shared understanding of the things most and least likely to be well predicted.

In approaching the doing component, one of the cautionary notes (Walters, 1997; Walters & Holling, 1990) is that of the danger of half-measures. The size of the interdiction or action should be appropriately scaled to ensure that influences at higher or lower scales do not outweigh the measure being sought. Small or half measures, which may be more politically or socially palatable (Irvine & Kaplan, 2001) or of perceived lower risk than a big experiment, come with the risk of not providing the ecosystem the correct context and setting for the experiment to return a clear result. One thing more frustrating than a half-measure experiment is a quarter-measure result - one which gives no clear indication of the key relationships leading to an observed outcome. Similarly, experimental design also requires consideration of the predictive capability of linear models when applied to ecosystems with non-linear or resilient behaviours (Folke, 2006; Folke et al., 2002; McLain & Lee, 1996; Walters, 1997) (See Box this chapter)

What Adaptive Management May Do for You

Undertaking adaptive management in the ways described above, fulfilling key roles, maintaining healthy social processes and producing beneficial responses in ecosystems brings benefits beyond those of improved ecosystem health. These benefits occur in the most interesting and unlikely of places, and contribute to the 'hard to describe' aspects that draw many passionate people to adopt and promote adaptive management.

Two direct benefits for managers lie in the area of supporting decisions on resource allocation, and in selecting management options. By building understanding of the social and ecological systems and the reasoning behind decisions, as well as building support amongst stakeholders, these common decision process barriers are overcome.

Adaptive management also provides a framework for action. One of the often frustrating things in ecosystem management is the sea of uncertainty (e.g. social, institutional, regulatory, economic, ecological, and biophysical), with no solid foundation to act upon, and no structure for action. Adaptive management provides this, and provides it in a way that is readily describable in terms of mandatory management planning and budgeting procedures (Herrick & Sarewitz, 2000). The framework also allows people to act with the clear and stated intention of moving from a current situation to a new and better position, and learning along the way.

As noted earlier the 'action', 'indicator' and 'interaction' focus diffuses or redirects long held disagreements, and moves conversation from blame and argument to desired social and ecosystem outcomes. An extended benefit of this is the development of a more centred, cohesive and consistent system description around which debate can occur. Conversely, this central system description often provides a target for more focussed querying, debate, and attack – not necessarily a bad thing. An agreed and centred structure also reduces excuses for inaction – there are fewer opportunities for people to use lack of consensus or need for more study as a way of avoiding action.

In systems with a history of management activities, strategies, structures, institutions, plans, and initiatives, one challenge is that of retaining 'corporate' knowledge – the knowledge and understanding not only of the system, but also of the beliefs, prior thoughts and activities around the system. The more diffuse and poorly captured is corporate knowledge, the harder and longer new participants and processes have to work to re-learn. Adaptive management provides both the process and the structure for improving corporate knowledge, and even the most abysmal adaptive management failure provides knowledge that can be accessed in future. Accession includes not only getting the information, but also understanding it, due to the formalism of information that occurred when the system was described. In successfully developing a model that encapsulates system and process understanding, the model and its documentation provide a valuable knowledge foundation. When looking back at the model, and decisions based upon its predictions, the things learned from management experiments automatically have a home.

This raises the question of the things learned that don't fit with the agreed picture. Provided people live the open minded philosophy and have a supportive and open framework for debate, created by the adaptive management process, the provision of a structured home for knowledge also makes it easier to identify what doesn't fit, and to identify more readily misconceptions within the framework – changing the thesis rather than the data.

One of the high level objectives noted by ESSA (1982) was the identification of unknowns, which commonly end up being translated into knowledge gaps and data gaps. Knowledge gaps include absence of data to explain a process, conflicting data about a process, and conflicting hypotheses about how the data explain the process. Data gaps include no data, limited data, or little specific data on key parts of the system, such as on extreme or extended parts of an event distribution. Examples include species behaviour under rare conditions such as extended drought, extremely abundant resources, depleted or increasingly scarce habitat, or response to new and significant perturbations. By helping to identify these, adaptive management provides the framework for both undertaking reversible experiments within the bounds of the adaptive management process, and also for further data exploration or gathering within or without. Exposure and exploration of data is also useful in identifying what data are not necessarily useful. Weighing another tree diameter measurement or another low flow water quality sample against other monitoring investments is supported by adaptive management processes.

A final added benefit from adaptive management is that of providing a space and time framework for key processes. Cross-scale relationships form some of the potential pitfalls of adaptive management (Hobbs, 2003; Walters, 1997), but an understanding of the scale of processes is naturally needed to allow clear identification of what is 'cross-scale'. By forcing an exploration and explicit description of ecosystem processes these scales become more clear, and debates about scales are better supported.

What Adaptive Management May Not Do for You

The above are all well and good, but there are also many problems and pitfalls in adaptive management (McLain & Lee, 1996). A broad examination of the published history of adaptive management shows a sweeping path from discovery, extension, exploration, failure, re-visitation and development, to the point where current and recent publications reflect people at all stages of the cycle:

- Learning – of or from adaptive management (e.g. Allan & Curtis, 2005; Bunnell & Dunsworth, 2004; Contador, 2005; Porter & Underwood, 1999; Thayer & Kentula, 2005; Theberge et al., 2006; Thom, 2000)
- Describing – adaptive management processes, pitfalls and promises (Boesch, 2006; Moir & Block, 2001; Rogers, 2006; Salafsky et al., 2002)
- Predicting – how adaptive management would be useful for a certain problem or how adaptive management could be assisted by various tools and techniques (Bunch, 2003; Dorazio & Johnson, 2003; O'Rourke, 2006; Richter et al., 2006; Shea et al., 2002) and
- Doing – doing the doing! (MacDonald & Rice, 2004; Marttunen & Vehanen, 2004)

The literature contributions (e.g. Stankey et al., 2005) discuss many of the faults, failings and misconceptions of adaptive management at many levels, and provide guidance on what adaptive management can and cannot do for you. A primary one, which draws from decision theory and support literature and reflects the role of adaptive management in informing decisions, is that it is not a process of decision making. In many cases it can make the decision process harder because it embraces complexity and presents and evaluates alternative options on the assumption that decisions will be made and enacted, rather than, as occasionally occurs, avoided.

Adaptive management is also not a process that can be selected as an easy way to get the work or thinking around a problem done – it will not provide an easy way to shift a problem to another place or time. Engaging in adaptive management requires time and effort, often in excess of current activities due to the needs to transact, translate and understand beyond the bounds of our compartmentalised daily boxes of operation. Adaptive management will also not do the learning – it is engagement with the process that provides the learning.

Faults and failings of adaptive management include those that arise through the process, and also those of the process. For example, if learning about the system does not occur or is unclear, if a consistent and agreed description of the relevant aspects of the ecosystem is unattainable, if predictions are inconclusive or indescribable, and if management experiments are seen as too risky or expensive, then it is possible that the process has failed or that the process has succeeded, and that these are valid outcomes of the process. The question is how to distinguish between the two, as the former are more readily addressed than the latter.

To decide this, practitioners need to understand the processes and possible pitfalls, including:

- Divergent scientific debate
- Closed or limited engagement
- Competing or incompatible social and physical views
- Constrained institutional structures or processes and
- The myriad of motivations, drivers, distractions and blockages that can affect those involved

In the Description component, the war of models seen in the Columbia River fish management debate (McLain & Lee, 1996) destroyed the Description process and broke the cycle, raising awareness of the risks in this area.

Predictions have been found, for example in wolf population dynamics (Theberge et al., 2006), where ecological stochasticity and management timeframes limit the opportunity to learn. Understanding the falsifiability of hypotheses was critical here.

The Doing component of adaptive management involves exactly that – manipulating weir gates, harvesting in certain ways or places, spraying pests at particular times, constructing or destroying habitat or breeding grounds – so there are many ways this component fails. There is where the theory–practice nexus is most keenly felt, as management practitioners put these in place. Hearing and heeding practitioners when they say 'it can't be done that way' is important, as is understanding the limitations of on-ground activity.

Divergent Learning, where trials could lead to different conclusions on the effectiveness of treatments, such as with the response of small mammals to forest fuel reduction fires (Converse et al., 2006) highlight the need for clear structures around the Learning phase.

Thus, all components, and interactions and interfaces between components, have the capacity to lead or promote the failure of adaptive management.

Learning from the Doing of Adaptive Management

In many ways 'Learning' occurs throughout the adaptive management process, both as it cycles, and in the cycles within cycles. This chapter has raised the things learned from applications of adaptive management, provided some description of the adaptive management components, predicted the likely ways of succeeding or failing with adaptive management, but has not delved into the Doing of adaptive management. The following chapters take us more deeply into the actual doing and learning of adaptive management, providing examples of the variations that can occur in adoption, and adaptation, of the components described.

References

Allan, C., & Curtis, A. (2005). Nipped in the bud: Why regional scale adaptive management is not blooming. *Environmental Management, 36*(3), 414–425.

Allen, W., Bosch, O., Kilvington, M., Oliver, J., & Gilbert, M. (2001). Benefits of collaborative learning for environmental management: Applying the integrated systems for knowledge management approach to support animal pest control. *Environmental Management, 27*(2), 215–223.

Armitage, D., Marschke, M., & Plummer, R. (2008). Adaptive co-management and the paradox of learning. *Global Environmental Change-Human and Policy Dimensions, 18*(1), 86–98.

Ballard, H. L., & Huntsinger, L. (2006). Salal harvester local ecological knowledge, harvest practices and understory management on the Olympic Peninsula, Washington, DC. *Human Ecology, 34*(4), 529–547.

Berkes, F. (2004). Rethinking community-based conservation. *Conservation Biology, 18*(3), 621–630.

Beven, K., & Binley, A. (1992). The future of distributed models - model calibration and uncertainty prediction. *Hydrological Processes, 6*(3), 279–298.

Bisbal, G. A. (2001). Conceptual design of monitoring and evaluation plans for fish and wildlife in the Columbia River ecosystem. *Environmental Management, 28*(4), 433–453.

Boddicker, M., Rodriguez, J. J., & Amanzo, J. (2002). Indices for assessment and monitoring of large mammals within an adaptive management framework. *Environmental Monitoring and Assessment, 76*(1), 105–123.

Boesch, D. F. (2006). Scientific requirements for ecosystem-based management in the restoration of Chesapeake Bay and Coastal Louisiana. *Ecological Engineering, 26*(1), 6–26.

Brussard, P. F., Reed, J. M., & Tracy, C. R. (1998). Ecosystem management: What is it really? *Landscape and Urban Planning, 40*(1–3), 9–20.

Bunch, M. J. (2003). Soft systems methodology and the ecosystem approach: A system study of the Cooum River and environs in Chennai, India. *Environmental Management, 31*(2), 182–197.

Bunnell, F. L., & Dunsworth, B. G. (2004). Making adaptive management for biodiversity work - the example of Weyerhaeuser in coastal British Columbia. *Forestry Chronicle, 80*(1), 37–43.

Butler, K. F., & Koontz, T. M. (2005). Theory into practice: Implementing ecosystem management objectives in the USDA Forest Service. *Environmental Management, 35*(2), 138–150.

Caruso, B. S. (2006). Effectiveness of braided, gravel-bed river restoration in the Upper Waitaki Basin, New Zealand. *River Research and Applications, 22*(8), 905–922.

Chavez, D. J. (2002). Adaptive management in outdoor recreation: Serving Hispanics in southern California. *Western Journal of Applied Forestry, 17*(3), 129–133.

Checkland, P. B. (1981). *Systems Thinking, Systems Practice*. Chichester: Wiley.

Coen, L. D., & Luckenbach, M. W. (2000). Developing success criteria and goals for evaluating oyster reef restoration: Ecological function or resource exploitation? *Ecological Engineering, 15*(3–4), 323–343.

Contador, J. F. L. (2005). Adaptive management, monitoring, and the ecological sustainability of a thermal-polluted water ecosystem: A case in SW Spain. *Environmental Monitoring and Assessment, 104*(1–3), 19–35.

Converse, S. J., White, G. C., Farris, K. L., & Zack, S. (2006). Small mammals and forest fuel reduction: National-scale responses to fire and fire surrogates. *Ecological Applications, 16*(5), 1717–1729.

Cote, M. A., Kneeshaw, D., Bouthillier, L., & Messier, C. (2001). Increasing partnerships between scientists and forest managers: Lessons from an ongoing interdisciplinary project in Quebec. *Forestry Chronicle, 77*(1), 85–89.

Crossley, J. W. (1996). Managing ecosystems for integrity: Theoretical considerations for resource and environmental managers. *Society & Natural Resources, 9*(5), 465–481.

Dorazio, R. M., & Johnson, F. A. (2003). Bayesian inference and decision theory - A framework for decision making in natural resource management. *Ecological Applications, 13*(2), 556–563.

Dovers, S. (2001). *Institutions for Sustainability* (No. 7), TELA: Environment, Economy and Society, Melbourne: Australian Conservation Foundation, with Environment Institute of Australia and Land and Water Australia [http://www.acfonline.org.au/uploads/res/res_tp007.pdf]

Dovers, S. R., & Mobbs, C. D. (1997). An alluring prospect? Ecology, and the requirements of adaptive management. In N. Klomp & I. Lunt (Eds.), *Frontiers in Ecology. Building the Links*. Oxford: Elsevier.

Edwards-Jones, E. S. (1997). The river valleys project: A participatory approach to integrated catchment planning and management in Scotland. *Journal of Environmental Planning and Management, 40*(1), 125–141.

ESSA (1982). *Adaptive Environmental Assessment and Management*: ESSA Environmental and Social Systems Analysts Ltd.

Folke, C. (2006). Resilience: The emergence of a perspective for social-ecological systems analyses. *Global Environmental Change-Human and Policy Dimensions, 16*(3), 253–267.

Folke, C., Carpenter, S., Elmqvist, T., Gunderson, L., Holling, C. S., & Walker, B. (2002). Resilience and sustainable development: Building adaptive capacity in a world of transformations. *Ambio, 31*(5), 437–440.

Gentile, J. H., Harwell, M. A., Cropper, W., Harwell, C. C., DeAngelis, D., Davis, S., et al. (2001). Ecological conceptual models: A framework and case study on ecosystem management for South Florida sustainability. *Science of the Total Environment, 274*(1–3), 231–253.

Gerber, L. R., Wielgus, J., & Sala, E. (2007). A decision framework for the adaptive management of an exploited species with implications for marine reserves. *Conservation Biology, 21*(6), 1594–1602.

Gilmour, A., Walkerden, G., & Scandol, J. (1999). Adaptive management of the water cycle on the urban fringe: Three Australian case studies. *Conservation Ecology, 3*(1), [online] Art. 11.

Grafton, R. Q., & Kompas, T. (2005). Uncertainty and the active adaptive management of marine reserves. *Marine Policy, 29*(5), 471–479.

Grayson, R. B., & Blöschl, G. (Eds.). (2000). *Spatial Patterns in Catchment Hydrology: Observations and Modelling* (1 ed.). Cambridge: Cambridge University Press.

Gregory, R., Failing, L., & Higgins, P. (2006a). Adaptive management and environmental decision making: A case study application to water use planning. *Ecological Economics, 58*(2), 434–447.

Gregory, R., Ohlson, D., & Arvai, J. (2006b). Deconstructing adaptive management: Criteria for applications to environmental management. *Ecological Applications, 16*(6), 2411–2425.

Grumbine, R. E. (1997). Reflections on 'what is ecosystem management?'. *Conservation Biology, 11*(1), 41–47.

Gunderson, L. H., Holling, C. S., & Light, S. S. (Eds.). (1995). *Barriers and Bridges to the Renewal of Ecosystems and Institutions*. New York: Columbia University Press.

Habron, G. (2003). Role of adaptive management for watershed councils. *Environmental Management, 31*(1), 29–41.

Haney, A., & Power, R. L. (1996). Adaptive management for sound ecosystem management. *Environmental Management, 20*(6), 879–886.

Hennessey, T. M. (1994). Governance and adaptive management for estuarine ecosystems - The case of Chesapeake Bay. *Coastal Management, 22*(2), 119–145.

Herrick, C., & Sarewitz, D. (2000). Ex post evaluation: A more effective role for scientific assessments in environmental policy. *Science, Technology, & Human Values, 25*(3), 309–331.

Herrick, J. E., Bestelmeyer, B. T., Archer, S., Tugel, A. J., & Brown, J. R. (2006). An integrated framework for science-based arid land management. *Journal of Arid Environments, 65*(2), 319–335.

Hobbs, N. T. (2003). Challenges and opportunities in integrating ecological knowledge across scales. *Forest Ecology and Management, 181*(1–2), 223–238.

Holling, C. S. (1978). *Adaptive Environmental Assessment and Management*. Chichester: Wiley.

Hughes, T. P., Gunderson, L. H., Folke, C., Baird, A. H., Bellwood, D., Berkes, F., et al. (2007). Adaptive management of the great barrier reef and the Grand Canyon world heritage areas. *Ambio, 36*(7), 586–592.

IIASA. (1979). *Expect the Unexpected. An Adaptive Approach to Environmental Management* (Executive Report No. 1). Laxenburg: International Institute for Applied Systems Analysis.

Irvine, K. N., & Kaplan, S. (2001). Coping with change: The small experiment as a strategic approach to environmental sustainability. *Environmental Management, 28*(6), 713–725.

Kentula, M. E. (2000). Perspectives on setting success criteria for wetland restoration. *Ecological Engineering, 15*(3–4), 199–209.

Keough, H. L., & Blahna, D. J. (2006). Achieving integrative, collaborative ecosystem management. *Conservation Biology, 20*(5), 1373–1382.

Kneeshaw, D. D., Leduc, A., Drapeau, P., Gauthier, S., Pare, D., Carignan, R., et al. (2000). Development of integrated ecological standards of sustainable forest management at an operational scale. *Forestry Chronicle, 76*(3), 481–493.

Kremen, C., Raymond, I., & Lance, K. (1998). An interdisciplinary tool for monitoring conservation impacts in Madagascar. *Conservation Biology, 12*(3), 549–563.

Kremsater, L., Bunnell, F., Huggard, D., & Dunsworth, G. (2003). Indicators to assess biological diversity: Weyerhaeuser's coastal British Columbia forest project. *Forestry Chronicle, 79*(3), 590–601.

Ladson, A. R., & Argent, R. M. (2002). Adaptive management of environmental flows: Lessons for the Murray-Darling Basin from three large North American rivers. *Australian Journal of Water Resources, 5*(1), 89–101.

Lee, K. N. (1993). *Compass and Gyroscope: Integrating Science and Politics for the Environment*. Washington, DC: Island Press.

Lee, K. N. (1999). Appraising adaptive management. *Conservation Ecology, 3*(2), [online] Art. 3.

Lessard, G. (1998). An adaptive approach to planning and decision-making. *Landscape and Urban Planning, 40*(1–3), 81–87.

Lister, N. M. E. (1998). A systems approach to biodiversity conservation planning. *Environmental Monitoring and Assessment, 49*(2–3), 123–155.

Ludwig, D., Hilborn, R., & Walters, C. (1993). Uncertainty, resources exploitation, and conservation: Lessons from history. *Science, 260*(2 April), 17, 36.

Lynch, H. J., Hodge, S., Albert, C., & Dunham, M. (2008). The greater yellowstone ecosystem: Challenges for regional ecosystem management. *Environmental Management, 41*(6), 820–833.

MacDonald, G. B., & Rice, J. A. (2004). An active adaptive management case study in Ontario boreal mixedwood stands. *Forestry Chronicle, 80*(3), 391–400.

MacDonald, L. H., & Coe, D. (2007). Influence of headwater streams on downstream reaches in forested areas. *Forest Science, 53*(2), 148–168.

Marttunen, M., & Vehanen, T. (2004). Toward adaptive management: The impacts of different management strategies on fish stocks and fisheries in a large regulated lake. *Environmental Management, 33*(6), 840–854.

Matsuda, H. (2003). Challenges posed by the precautionary principle and accountability in ecological risk assessment. *Environmetrics, 14*(2), 245–254.

McCarthy, M. A., & Possingham, H. P. (2007). Active adaptive management for conservation. *Conservation Biology, 21*(4), 956–963.

McDaniels, T. L., & Gregory, R. (2004). Learning as an objective within a structured risk management decision process. *Environmental Science & Technology, 38*(7), 1921–1926.

McLain, R. J., & Lee, R. G. (1996). Adaptive management: Promises and pitfalls. *Environmental Management, 20*(4), 437–448.

Melbourne, B. A., Davies, K. F., Margules, C. R., Lindenmayer, D. B., Saunders, D. A., Wissel, C., et al. (2004). Species survival in fragmented landscapes: Where to from here? *Biodiversity and Conservation, 13*(1), 275–284.

Moir, W. H., & Block, W. M. (2001). Adaptive management on public lands in the United States: Commitment or rhetoric? *Environmental Management, 28*(2), 141–148.

Nyberg, J. B., Marcot, B. G., & Sulyma, R. (2006). Using Bayesian belief networks in adaptive management. *Canadian Journal of Forest Research-Revue Canadienne De Recherche Forestiere, 36*(12), 3104–3116.

O'Rourke, E. (2006). Biodiversity and land use change on the Causse Mejan, France. *Biodiversity and Conservation, 15*(8), 2611–2626.

Ojha, H., & Bhattarai, B. (2003). Learning to manage a complex resource: A case of NTFP assessment in Nepal. *International Forestry Review, 5*(2), 118–127.

Olsson, P., Folke, C., & Berkes, F. (2004). Adaptive comanagement for building resilience in social-ecological systems. *Environmental Management, 34*(1), 75–90.

Pahl-Wostl, C. (2002). Towards sustainability in the water sector - The importance of human actors and processes of social learning. *Aquatic Sciences, 64*(4), 394–411.
Pahl-Wostl, C. (2007). The implications of complexity for integrated resources management. *Environmental Modelling & Software, 22*(5), 561–569.
Pik, A. J., Dangerfield, J. M., Bramble, R. A., Angus, C., & Nipperess, D. A. (2002). The use of invertebrates to detect small-scale habitat heterogeneity and its application to restoration practices. *Environmental Monitoring and Assessment, 75*(2), 179–199.
Pimm, S. L. (1984). The complexity and stability of ecosystems. *Nature, 307,* 321–326.
Pinkerton, E. (1999). Factors in overcoming barriers to implementing co-management in British Columbia salmon fisheries. *Conservation Ecology, 3*(2), [online] Art. 2.
Pittock, A. B., & Jones, R. N. (2000). Adaptation to what and why? *Environmental Monitoring and Assessment, 61,* 9–35.
Plummer, R., & Armitage, D. (2007). A resilience-based framework for evaluating adaptive co-management: Linking ecology, economics and society in a complex world. *Ecological Economics, 61*(1), 62–74.
Porter, W. F., & Underwood, H. B. (1999). Of elephants and blind men: Deer management in the US national parks. *Ecological Applications, 9*(1), 3–9.
Prato, T. (2005). Accounting for uncertainty in making species protection decisions. *Conservation Biology, 19*(3), 806–814.
Prato, T. (2007). Evaluating land use plans under uncertainty. *Land Use Policy, 24*(1), 165–174.
Richter, B. D., Mathews, R., & Wigington, R. (2003). Ecologically sustainable water management: Managing river flows for ecological integrity. *Ecological Applications, 13*(1), 206–224.
Richter, B. D., Warner, A. T., Meyer, J. L., & Lutz, K. (2006). A collaborative and adaptive process for developing environmental flow recommendations. *River Research and Applications, 22*(3), 297–318.
Roe, E. (1996). Why ecosystem management can't work without social science: An example from the California northern spotted owl controversy. *Environmental Management, 20*(5), 667–674.
Rogers, K. H. (2006). The real river management challenge: Integrating scientists, stakeholders and service agencies. *River Research and Applications, 22*(2), 269–280.
Rowntree, R. (1998). Modeling fire and nutrient flux. *Journal of Forestry, 96*(4), 6–11.
Salafsky, N., Margoluis, R., Redford, K. H., & Robinson, J. G. (2002). Improving the practice of conservation: A conceptual framework and research agenda for conservation science. *Conservation Biology, 16*(6), 1469–1479.
Schindler, B., & Aldred Cheek, K. (1999). Integrating citizens in adaptive management: A propositional analysis. *Conservation Ecology, 3*(1), [online] Art. 9.
Shannon, R. E. (1975). *Systems Simulation: The Art and Science.* Englewood Cliffs, NJ: Prentice-Hall.
Sharma, M., & Norton, B. G. (2005). A policy decision tool for integrated environmental assessment. *Environmental Science & Policy, 8*(4), 356–366.
Shea, K., Possingham, H. P., Murdoch, W. W., & Roush, R. (2002). Active adaptive management in insect pest and weed control: Intervention with a plan for learning. *Ecological Applications, 12*(3), 927–936.
Skogen, K. (2003). Adapting adaptive management to a cultural understanding of land use conflicts. *Society & Natural Resources, 16*(5), 435–450.
Smyth, R. L., Watzin, M. C., & Manning, R. E. (2007). Defining acceptable levels for ecological indicators: An approach for considering social values. *Environmental Management, 39*(3), 301–315.
Stankey, G. H., Bormann, B. T., Ryan, C., Shindler, B., Sturtevant, V., Clark, R. N., et al. (2003). Adaptive management and the Northwest Forest Plan – Rhetoric and reality. *Journal of Forestry, 101*(1), 40–46.
Stankey, G. H., Clark, R. N., & Bormann, B. T. (2005). *Adaptive Management of Natural Resources: Theory, Concepts, and Management Institutions.* Portland, OR: U.S. Department of Agriculture, Forest Service, Pacific Northwest Research Station.
Stauffer, H. B. (2008). Application of Bayesian statistical inference and decision theory to a fundamental problem in natural resource science: The adaptive management of an endangered species. *Natural Resource Modeling, 21*(2), 264–284.

Steyer, G. D., & Llewellyn, D. W. (2000). Coastal Wetlands Planning, Protection, and Restoration Act: A programmatic application of adaptive management. *Ecological Engineering, 15*(3–4), 385–395.

Stubbs, M., & Lemon, M. (2001). Learning to network and networking to learn: Facilitating the process of adaptive management in a local response to the UK's National Air Quality Strategy. *Environmental Management, 27*(3), 321–334.

Stubbs, M., Lemon, M., & Longhurst, P. (2000). Intelligent urban management: Learning to manage and managing to learn together for a change. *Urban Studies, 37*(10), 1801–1811.

Szaro, R. C., Berc, J., Cameron, S., Cordle, S., Crosby, M., Martin, L., et al. (1998). The ecosystem approach: Science and information management issues, gaps and needs. *Landscape and Urban Planning, 40*(1–3), 89–101.

Thayer, G. W., & Kentula, M. E. (2005). Coastal restoration: Where have we been, where are we now, and where should we be going? *Journal of Coastal Research*, Special Issue 40, 1–5.

Theberge, J. B., Theberge, M. T., Vucetich, J. A., & Paquet, P. C. (2006). Pitfalls of applying adaptive management to a wolf population in Algonquin Provincial Park, Ontario. *Environmental Management, 37*(4), 451–460.

Thom, R. M. (2000). Adaptive management of coastal ecosystem restoration projects. *Ecological Engineering, 15*(3–4), 365–372.

Thom, R. M., Williams, G., Borde, A., Southard, J., Sargeant, S., Woodruff, D., et al. (2005). Adaptively addressing uncertainty in estuarine and near coastal restoration projects. *Journal of Coastal Research*, Special Issue 40, 94–108.

Torell, E. (2000). Adaptation and learning in coastal management: The experience of five East African initiatives. *Coastal Management, 28*(4), 353–363.

Viney, N. R., Bates, B. C., Charles, S. P., Webster, I. T., & Bormans, M. (2007). Modelling adaptive management strategies for coping with the impacts of climate variability and change on riverine algal blooms. *Global Change Biology, 13*(11), 2453–2465.

Walters, C. (1974). An interdisciplinary approach to development of watershed simulation models. *Technological Forecasting and Social Change, 6*, 299–323.

Walters, C. (1997). Challenges in adaptive management of riparian and coastal ecosystems. *Conservation Ecology (online), 1*(2), 1 ff.

Walters, C., Korman, J., Stevens, L. E., & Gold, B. (2000). Ecosystem modeling for evaluation of adaptive management policies in the Grand Canyon. *Conservation Ecology (online), 4*(2), [online] Art 1.

Walters, C. J. (1986). *Adaptive management of renewable resources*. New York: Macmillan.

Walters, C. J. (2007). Is adaptive management helping to solve fisheries problems? *Ambio, 36*(4), 304–307.

Walters, C. J., & Holling, C. S. (1990). Large-scale management experiments and learning by doing. *Ecology, 71*(6), 2060–2068.

Williams, B. K. (2001). Uncertainty, learning, and the optimal management of wildlife. *Environmental and Ecological Statistics, 8*(3), 269–288.

Wilson, C., Argent, R., Grayson, R., Bunn, S., Davies, P., Hairsine, P., et al. (1996). A framework for integrating the impacts of riparian revegetation on stream values in large catchments. *Eos, Transactions., 77*(22), W38.

Wilson, M. V., & Lantz, L. E. (2000). Issues and framework for building successful science management teams for natural areas management. *Natural Areas Journal, 20*(4), 381–385.

Passive/Active Adaptive Management

Catherine Allan and Chris Jacobson

Walters and Holling (1990) identified three types of adaptive management. They used the term 'evolutionary' adaptive management to describe changes developed through trial and error, or the act of learning from management actions without purposeful direction. The case studies in this book are not based on evolutionary adaptive management, but rather on the other types identified by Walters and Holling; 'passive' and 'active' adaptive management. These purposeful learning approaches can be conceptualised as occupying positions on a continuum from almost "pure" implementation (doing) to "pure" research (learning).

Passive adaptive management sits towards the management end of the continuum. Managers/implementors learn and improve by using past experience and learning to develop a current best policy/practice. After some time, implementation of the practice is reviewed, possibly resulting in changes to policy and the acceptance of a new 'best' practice. Passive adaptive management thus uses a cyclical plan, act, monitor and assess cycle process in management to gradually improve practice. Passive adaptive management is appropriate in simple or tame management situations, especially when single use or exploitation of a resource is the goal. The focus of passive adaptive management is the management outcome, rather than the learning *per se*, so passive approaches cannot discriminate between different options for achieving management goals.

Active adaptive management is closer to the learning end of the continuum. It is the conscious and purposeful use of policy and its implementation as experiments designed to enable people to learn about systems as they manage them (Johnson, 1999; Lee, 1993; Walters & Green, 1997). For active adaptive management, past learning is used to develop and test a number of alternative policy and management responses. These alternative responses are tried, monitored, reviewed and compared, and subsequent management and policy are altered in response to what is learned. This description of active adaptive management is broad enough for many different nuances and interpretations. An early model of active adaptive management involved development of multiple mathematical models of an ecological system from existing data, enabling predictions to be made about the outcome of a range of different management options, from which one would be implemented, monitored and assessed (Walters, 1986). Schreiber et al. (2004) noted that this differs from many current conceptions of active adaptive management in that it requires considerable skills for modelling and experimental design, and hence is likely to be an expert-driven management process. Other forms of active adaptive management are variants on scientific research, in that a number of management options are trialled against each other, or natural variance in management is used for quasi-treatments. Generally active adaptive management is about testing hypotheses on a real world scales. What separates this from pure scientific research is that it occurs in the field, is often undertaken with or by managers, and is designed to inform future management actions.

References

Johnson, B. L. (1999). Introduction to the special feature: Adaptive management – scientifically sound, socially challenged? *Conservation Ecology, 3*(1), art. 10.

Lee, K. N. (1993). *Compass and Gyroscope: Integrating Science and Politics for the Environment.* Washington, DC: Island Press.

Schreiber, E. S. G., Bearlin, A. R., Nicol, S. J., & Todd, C. R. (2004). Adaptive management: A synthesis of current understanding and effective application. *Ecological Management and Restoration, 5*(3), 177–182.

Walters, C. J. (1986). *Adaptive management of renewable resources.* New York: Macmillan.

Walters, C. J. & Green, R. (1997). Valuation of experimental management options for ecological systems. *Journal of Wildlife Management, 61*(4), 987–1006.

Walters, C. J. & Holling, C. S. (1990). Large-scale management experiments and learning by doing. *Ecology, 71*(6), 2060–2068.

Adaptive Management, Resilience, Hierarchy Theory and Thresholds

Robert Argent

Understanding and explanation of some of the intricacies of ecosystem behaviour, and our ability or inability to predict these, have developed considerably in recent decades. One of the cornerstones of traditional technical-rational management models has been the capacity to predict the responses of systems to various interventions. However, under conditions of complexity and uncertainty, the capacity to predict becomes significantly compromised. A particularly attractive of an adaptive management model is that it focuses on enabling rigorous ex post facto evaluation, primarily because of its emphasis on rigorous monitoring and multi-party evaluation. That is, it promotes the formulation of reasonable and testable hypotheses and uses the process of implementation as a treatment whose various outcomes are then subject to assessment and evaluation. From this process, a much sounder basis for establishing how system variables act and interact in the face of alternative interventions occurs. Emerging theories of ecosystem dynamics offer improved ways to explain some observed behaviours and to move back the limits to prediction. The resilience perspective and the notion of multiple stable states in ecosystems (e.g. Folke, 2006; Pimm, 1984) offers the opportunity in adaptive management to understand both the resistance to change, the ability of the system to respond to perturbation, and the nature or magnitude of perturbation that might be needed to move a system from a less attractive to a more attractive state. Hierarchy theory (e.g. O'Neill et al., 1986) with its exploration of scaling and relationships between entities across scales and levels, as well as the positions of observer and observed, offers insights into ecosystem response to perturbations that impact our ability to describe and predict systems. Finally, threshold-based management (Roe & van Eeten, 2001) encourages managers to work within a framework that encompasses adaptive management and offers a pathway to step between management methodologies as institutional, knowledge and other situational aspects develop.

References

Folke, C. (2006). Resilience: The emergence of a perspective for social-ecological systems analyses. *Global Environmental Change – Human and Policy Dimensions, 16*(3), 253–267.

O'Neill, R. V., DeAngelis, D., Waide, J., & Allen, T. F. H. (1986). *A hierarchical Concept of Ecosystems*. Princeton, NJ: Princeton University Press.

Pimm, S. L. (1984). The complexity and stability of ecosystems. *Nature, 307*, 321–326.

Roe, E., & van Eeten, M. (2001). Threshold-based resource management: A framework for comprehensive ecosystem management. Environmental Management, *27*(2), 195–214.

Part II
Varying Contexts

Chapter 3
Lessons Learned from Adaptive Management Practitioners in British Columbia, Canada

Alanya C. Smith

Abstract Four adaptive forest management case studies from British Columbia, Canada, show an interesting diversity in the approach and provide an excellent source of "lessons learned." Included are: the Coast Forest Strategy, the Forest and Range Evaluation Program, the Pine-Lichen Woodlands and Northern Caribou Adaptive Management Project, and the Ospika Mountain Goat Trial. Practitioners revealed the demands faced in their adaptive management projects and shared their insights and advice about implementing these projects. Common themes included leadership, partnerships, "closing the loop" to management, and organizational commitment and resources.

Introduction

In Canada, the province of British Columbia's Ministry of Forests and Range is exploring how adaptive management can be applied to help continuously improve forestry practices and policies on the Crown forest and range lands. Adaptive management is an approach that has promise for application to various issues and scales, from testing alternative silvicultural practices in forest stands, to ecosystem-based management for entire watersheds or landscape units. The active co-operation of resource managers, forest professionals, scientists, the forest industry, First Nations, and other partners is crucial to the success of this approach, as is the support of public groups and individuals. Although adaptive management is a fairly intuitive concept, it can be a complex approach to put into practice effectively. When applied in forestry situations, implementing adaptive management is particularly challenging because its effects may require decades to materialize and often appear at many spatial scales, with many confounding factors to recognize and isolate.

A review of four adaptive forest management case studies from British Columbia shows an interesting diversity in the approach and provides an excellent source of

---

A.C. Smith
Research officer, Forest Practices Branch, British Columbia Ministry of Forests and Range

"lessons learned." Included are: the Coast Forest Strategy, the Forest and Range Evaluation Program, the Pine–Lichen Woodlands and Northern Caribou Adaptive Management Project, and the Ospika Mountain Goat Trial.

Through written surveys and telephone interviews, practitioners revealed the demands faced in their adaptive management projects and shared their insights and advice about implementing these projects. Common themes included leadership, partnerships, "closing the loop" to management, and organizational commitment and resources. Each of these themes is discussed after the case studies.

Case Study 1: Coast Forest Strategy

In coastal British Columbia, forestry planning and management activities are challenging because of the high cost of operations, and effective environmental awareness campaigns which have targeted harvesting in the natural temperate rainforest. In 1998, growing marketplace displeasure with the clear-cutting of old-growth forest led the forest sector giant MacMillan Bloedel (subsequently owned by Weyerhaeuser, then Cascadia Forest Products, and now led by Western Forest Products) to examine how it could maintain a safe, respected, and profitable business and also sustain biological diversity (native species richness and associated values) within its coastal land base.

MacMillan Bloedel announced it would no longer clear cut coastal forests, but would implement a "Coast Forest Strategy" (the "Strategy") of zoning and variable retention, a silvicultural approach that retains trees as structural elements of a harvested stand for at least the next harvest rotation in an effort to maintain species and forest processes. This technique retains varying numbers of trees either in patches or uniformly throughout a stand. Under the Strategy, company tenures were divided into three zones, each with a different management emphasis. Variable retention was phased-in over 5 years with different standards for each zone.

To monitor the implementation and effectiveness of the Strategy, the company's adaptive management program uses both passive and active approaches (see the information box in Chapter 2, Robert M. Argent, this volume). A set of five experimental comparisons (100 ha per site, replicated three times) is the focus of monitoring. Each site compares two to three retention alternatives to clear cut and uncut areas (e.g., percent group/dispersed retention, group size, riparian retention, group removal) (see Figs. 3.1 and 3.2). To date, 9 of 15 sites have been established. Effectiveness monitoring includes indicators of biodiversity (ecosystem representation, habitat structure, and organisms) and silviculture (growth and yield, and windthrow). The Strategy began on 1.1 million hectares of public and private forest lands in coastal British Columbia. It currently covers 1.4 million hectares on Vancouver Island, Haida Gwaii (an island off the Province's north coast), and the province's mainland coast. Annual program costs are over Can$0.5 million.

The company conducts the Strategy in conjunction with a core group of consultants and academic researchers, assisted by company scientists and forestry staff.

3 Lessons Learned from Adaptive Management Practitioners in British Columbia

Fig. 3.1 Forestry planning and management in coastal British Columbia

Several groups were established to provide input:

- Science Panel: Local and international experts from academia and government reviewed the Strategy and provided strategic advice. Environmental non-governmental organizations (ENGOs) participated in the panel and nominated scientists.
- Adaptive Management Working Group: Company, government, and contract biologists meet to guide monitoring and facilitate information exchange and extension activities.
- Forest Strategy Working Group: Company staff develops guidelines and policy.
- Community Advisory Groups: Established as part of Sustainable Forest Management Certification to provide input from public and other stakeholders.

Three indicators of success are used to focus the Strategy's goals and monitoring:

- Ecologically distinct ecosystem types are represented in the non-harvestable land base to maintain lesser-known species and ecological functions.
- The amount, distribution, and heterogeneity of stand and forest structures important to sustain biological richness are maintained over time.
- Productive populations of forest-dwelling species are well distributed.

An adaptive management framework provides guidance and criteria to evaluate new information related to the indicators. Contractors conduct most of the monitoring; universities, government agencies, or other forest companies collaborated on several projects. Many projects used the experimental sites for active adaptive

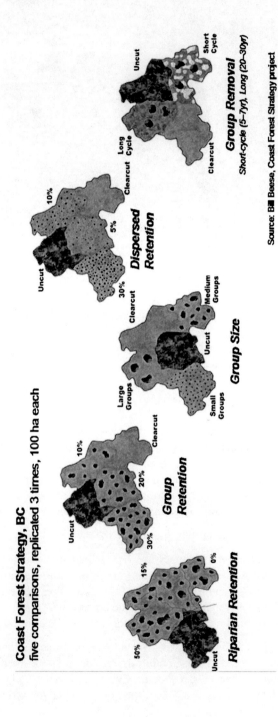

Fig. 3.2 Experimental comparisons in the Coast Forest Strategy

management comparisons. Other projects (e.g., forest structure, windthrow, bird surveys) rely primarily on passive monitoring of operational sites using a project-specific design. The company's adaptive management team, working groups, and Science Panel evaluate new information. A database stores project data files and metadata. Lead project scientists provide data to the technical coordinator at the completion of each field season.

The company made an effort to build understanding about the project among the public, ENGOs, and their peers through a website, brochures, project summaries, presentations at workshops and meetings, and published articles. Communications activities resulted in improved understanding of the company's forest management approach, and greater acceptance of its harvesting practices in general.

The success of the Strategy and its adaptive management approach is due in part to continued support by senior management. Results and issues are ongoing, and many questions remain unresolved; however, initial findings led to management-approved changes in practices.

Thus far, results indicate:

- Variable retention shows potential to maintain species from the original forest that would otherwise not be present in a clear cut (0–5 years post-harvest).
- Group retention maintains habitat better than dispersed retention (within the 0–30% range tested).
- Species respond differently to harvesting; pilot studies identified candidate "indicator" species for monitoring (those that prefer or require older forest conditions).
- Windthrow remains a significant issue in many parts of coastal British Columbia, leading to revised standards and more reliance on larger patches for stand-level retention.

The initial design of the Coast Forest Strategy increased the amount of old-growth forests set aside in both landscape- and stand-level reserves, and shifted the company's harvesting from predominantly clear-cutting to 100% variable retention. Monitoring results and a substantial shift in the company's land base, led the company to revise the zoning scheme and retention standards. Use of the retention system varies by zone from 30% to 90% or more, with an increase in the average group size retained. These changes are intended to reduce wind damage and provide improved habitat for "lifeboating" species that require old-forest attributes.

In July 2007, Western Forest Products' managers approved the revised "Western Forest Strategy." Although new standards will be phased-in by 2010, the adaptive management program continues to inform the Strategy. To develop a 5-year plan, the company is reviewing progress to date.

The practitioner, Bill Beese, Forest Ecologist for Western Forest Products and project lead, identified the following challenges.

- Implementing a major shift in practice – Accomplished with corporate commitment, leadership, and training
- Funding for monitoring and research – Accomplished with government funding programs and company support

- Maintaining the adaptive management program through changes to company ownership – Accomplished with leadership from key people in the organization with a strong commitment to the approach

The practitioner offered the following reflections and advice.

- Don't undertake adaptive management unless managers voice a strong commitment to the approach and provide the necessary funding.
- Operational staff must support the approach (this will not happen immediately, so training and leadership are essential).
- Partner with others; seek outside advice and opportunities to collaborate.
- Don't try to monitor everything.
- Feedback to management often takes place informally, outside of the formal channels devised for this purpose.
- Long-term data management requires significant resources.

Extensive background on this case study can be found in Bunnell and Dunsworth (2004), Beese et al. (2005), Bunnell (2005) and at http://www.forestbiodiversityinbc.ca/forest_strategy/default.htm. It is also the topic of a forthcoming book by Bunnell et al., from University of British Columbia Press.

Case Study 2: The Forest and Range Evaluation Program

The Forest and Range Evaluation Program (FREP) was created in 2003 to investigate the effectiveness of the *Forest and Range Practices Act (FRPA)*, its regulations and resulting practices in achieving the British Columbia provincial government's environmental stewardship objectives. The *FRPA* outlines government objectives for 11 resource values: biodiversity, cultural heritage, fish/riparian, forage and associated plant communities, recreation, resource features, soils, timber, visual quality, water, and wildlife. To assess whether these objectives are met through current management practices, FREP consulted with policy and resource experts to define specific management questions for each value. For each value, a FREP team provides technical expertise to:

- Identify priority questions
- Develop evaluation and monitoring indicators and protocols
- Conduct analysis
- Provide interpretation of monitoring and evaluation data and
- Develop recommendations for forest practices and policy improvements

FREP is a partnership of the provincial Ministry of Forests and Range, the Ministry of Environment, the Ministry of Agriculture and Lands, and the Ministry of Tourism, Culture and the Arts. Government forestry, range and biology professionals manage and implement the program for about Can$4 million annually.

The program exemplifies *passive adaptive management*: best forest practices are evaluated according to specific evaluation questions or issues and the results are then used to improve practices over time. It encompasses both routine resource stewardship monitoring and intensive evaluation. The latter perhaps aligns most closely with adaptive management as these evaluations are triggered by "red flags" emerging from resource stewardship monitoring results. Intensive evaluations provide input for science-based recommendations designed to improve forest policy and practices (e.g., legislation, guidelines, and best management practices).

The program is still in the early years of monitoring, but some initial results have been interpreted from the data. For example, FREP investigated the amount of retention on large cutblocks in areas affected by the mountain pine beetle *(Dendroctonus ponderosae)* in the province's Interior and compared this to pre-harvest baseline data. This revealed how well forest tenure holders were meeting the Chief Forester's guidance for retention on large salvage cutblocks (Snetsinger, 2005). These comparisons showed that the sampled retention had:

- Similar numbers of tree species present – a good trend for biodiversity
- Greater density of large trees ($\geq$50 cm diameter breast height) – a good trend for biodiversity
- A higher density of large snags ($\geq$30 cm diameter breast height and $\geq$10 m high) – potentially a good trend for biodiversity, although this needs further study

Coarse woody debris (CWD) indicators found in the harvested area were compared to the same indicators for CWD found in the retention patches. These comparisons showed that the sampled harvested areas had:

- Similar volumes of CWD compared to CWD in patch retention – a good trend for biodiversity
- Lower density of long ($\geq$10 m) CWD pieces compared to CWD in patch retention – a concern for biodiversity (British Columbia Ministry of Forests and Range, 2007)

FREP results are publicly communicated through peer-reviewed documents and the program's website (Government of British Columbia, 2007). Although no formal legislative changes have resulted thus far, practitioners report a change in some practices due to information communicated by FREP. As the program continues, the full adaptive management cycle will result in adjustments to policy, practices, and legislation. The strong framework now in place will likely ensure the program is a successful, long-term initiative.

Peter Bradford, FREP Provincial Lead, B.C. Ministry of Forests and Range offered the following advice.

- Start communications with a broad cross-section of individuals both internal and external to the organization early and communicate frequently using diverse approaches.
- Build a broad community of partnerships.
- Start slowly and take the time to develop the project with the right people.
- Capture and use lessons learned to improve over time.
- Tie results and recommendations implicitly to management.
- Find a motivated leader to champion the project.

Case Study 3: Pine–Lichen Woodlands and Northern Caribou Adaptive Management Project

Terrestrial lichens occurring in pine forests on dry, nutrient-poor sites of north-central British Columbia constitute an important source of winter forage for northern caribou (*Rangifer tarandus*), a species at risk legislated in the Canadian Species at Risk Act (SARA). These forests also supply fibre for local mills, providing a major source of income for many forestry-dependent communities. Forestry activities are generally considered to have detrimental effects on lichen development, although a retrospective study conducted in the region discovered that this was not always the case (Sulyma, 2001). This finding led to discussions about the ways in which forestry activities and silvicultural methods could enhance or maintain terrestrial lichens. Specific questions related to how various actions affect lichen development, included:

- Disturbance or displacement of the organic mat
- Debris accumulation and
- Forest stand development (e.g., forest stands influence interception of solar radiation, air flow, and other factors believed to contribute to the ecological succession of terrestrial forage lichens)

In 2001, a major forest tenure holder and the B.C. Ministry of the Environment formed a partnership to investigate these questions. An adaptive management project was designed to consider management options based on three replicates of nine treatments predicted to create different growing conditions for lichen. The treatments varied in timber harvesting method, harvesting season, site preparation, and regeneration method. Effectiveness indicators, which included vegetation cover, as well as the percentage of exposed mineral soil, coarse woody debris, and litter, were all monitored before and after harvesting.

The forests in these winter ranges are easily developed for timber values because the region is topographically flat with low-elevation sites that require relatively easily constructed infrastructure. These sites also tend to be fairly warm, shed snow early in spring, and are drier compared to adjacent sites. Therefore, the sites are attractive to timber licensees attempting to re-establish fibre supply after the depletion period of the previous winter. By comparison, biologists were interested in rejuvenating sites with potential to grow terrestrial lichens, but that had progressed to a seral stage in which bryophytes (non-vascular terrestrial plants) tended to compete more successfully as the understorey plant community.

Scientific evidence of the relationships among terrestrial lichens, site factors, and silviculture was anecdotal and retrospective. Use of a Bayesian Belief Network modelling (see Chapter 9, Jakeman et al. this volume) approach captured the expected ecological relationships and management interests. The model was then used to construct hypotheses and to organize treatments that would presumably deliver the anticipated products: a fibre source for the industry and rejuvenated lichen sites for caribou.

Initially, only a few basic standards were in place to help develop a monitoring design. Because terrestrial lichens grow slowly the project team had time to develop specific protocols (Sulyma, 2008). The project cost of Can$40,000–$50,000 per year did not include routine operations.

The project team made efforts to communicate the purpose of the project and its results. Extension occurred primarily through technical reports, newsletters, and occasional presentations to regional committees responsible for recovery of the area's caribou populations.

Results to date confirm predictions that winter harvesting reduces damage to terrestrial lichens as the snowpack protects the organic mat on the forest floor. A summer whole-tree harvest system also maintained conditions suitable to perpetuate lichen communities; however, more organic mat displacement occurred during summer compared to the winter harvesting regime, although the negative impacts of this may be short term. In forests at later stages of succession, some displacement may actually produce favourable microsites for lichen recruitment. Therefore, winter harvesting may not provide a significant benefit compared to the costs of restricting the seasonal scheduling of forestry activities on these sites.

Because terrestrial lichens are slow to respond to treatment, this project will continue to provide new information for managers over the next decade. Preliminary results from the adaptive management trials are already raising questions about an apparent variance in ecological succession of terrestrial forage lichens. This has resulted in further adaptation and implementation of new management hypotheses and the formulation of guidelines for forestry operations conducted within winter ranges established for caribou (McNay et al., 2008). In addition, the project will now include fire as an alternative silvicultural approach for managing lichens. The fire-based prescription requires terrestrial lichen sites to be distinguished at a finer resolution than has been previously attempted; ongoing research is helping to achieve this distinction.

The project team has benefited from:

- A relatively close mentoring relationship with government personnel who were champions of adaptive management and who assisted greatly in the design of the project
- A Master of Science candidate interested in adaptive management as part of a post-graduate learning experience and
- Ongoing research on caribou that was relatively well funded through commitment from forest licensees

Managers had to accept that products and deliverables would accrue primarily to one interest group (forest licensees) in the short term and that the understanding of caribou habitat would take place more slowly over the longer term. No special method was in place to overcome this challenge – the participants involved generally accepted this position, recognizing that long-term results would be better than no information. Another challenge was finding sites large enough to permit relatively regular forestry activities and environmentally consistent enough to facilitate systematic and statistical comparison. Although three sites were located, these were distributed in different parts of the province, and in at least one case, the site and condition of the lichen community was less than desirable.

Scott McNay and Randy Sulyma, of Wildlife Infometrics were involved in the development and implementation of the project. They offered the following advice.

- Good communication with operational staff is necessary to see that harvest operations are actually changed. Forestry operations are expensive and the implications of changes to operational practices (as a result of adaptive management) may not be sufficiently considered when adaptive management is conducted.
- Emphasis must be placed on closing the loop through feedback and adjustments based on what has been learned. This last step is important; otherwise, the exercise is just a large-scale experiment.

Case Study 4: Ospika Mountain Goat Project

In northern British Columbia, mountain goats (*Oreamnos americanus*), a species of special concern, live at high elevations but descend through the forest seasonally to low-elevation mineral licks typically exposed along creek or river systems (Ministry of Environment, 2007). These forests can be subject to heavy harvesting. It is currently assumed that harvesting has a negative effect on the migratory movements of mountain goats as they travel from alpine areas through cutblocks to valley-bottom salt licks. If goat trails are found within the proposed harvest block, regulations require the block boundary to be moved. Although not a legislated species at risk in Canada, mountain goats are a species of concern in BC due to concerns about hunting and habitat loss, requiring special management consideration (Fig. 3.3).

This project aims to produce an inventory that will help develop effective policy for management of mountain goats and their habitats. The project consists of four phases: modelling, resource inventory, policy development, and active adaptive management. The adaptive management portion will test assumptions, articulated at workshops held with forest mangers and wildlife biologists, about goat behaviour.

In 1999–2000, mineral licks were mapped throughout the northern portion of the Williston watershed in north-central British Columbia. Several mineral licks in the Ospika River drainage were slated for forest harvesting, presenting an opportunity to test assumptions. Project planning took place in 2001. The original design included harvesting on either side of a known goat trail to a mineral lick, with the retention of a 150-m forested buffer strip on either side of the trail. At a different mineral lick, harvesting was planned for the entire area up to and including one side of the trail.

The Ospika Goat Project involved several partnering agencies and is overseen by the Mackenzie Mountain Goat Management Team, a collaboration between Peace/Williston Fish and Wildlife Compensation Program (an initiative of BC Ministry of Environment, BC Hydro, and Department of Fisheries and Oceans), B.C. Ministry

Fig. 3.3 Mountain goats

of Forests, Abitibi Consolidated, Canadian Forest Products, university scientists, and a consultant (Wildlife Infometrics Inc.).

The Team focussed on developing and implementing an informed and effective management strategy for mountain goats in the local forest development unit. The Team:

- Maintains effective communication among stakeholders
- Creates a forum for knowledge transfer between participants through meetings, workshops, and presentations
- Ensures that all stakeholder interests are considered in developing the project and applying the results and
- Ensures that overall project objectives are met

Team members are expected to attend regularly scheduled meetings to update members on project activities, facilitate the development of mountain goat habitat supply models, and develop and implement regional mountain goat management policies.

Staff and contractors implemented several monitoring approaches at various time intervals. Remote telemetry stations located along the trail to the lick monitored the radio-collared goats to determine the number of lick visits per goat, duration of visits, and seasonal and daily timing of visits. Monitoring of non-collared goats (the "population") took place by remote cameras positioned along the trail to determine the seasonal timing of lick use, diurnal timing of visits, group size, age/sex composition of goats, and use of trails by other species (predators, other ungulates). Aerial monitoring of collared goats also occurred monthly to bi-weekly to confirm summer range use, distances moved to licks, and mortality status. Regular site visits occurred between April and November for 6 years to download and maintain remote cameras and data-loggers. "Interpretation" protocols were developed after the first year of data collection to evaluate both camera and telemetry data. The project is in the last phase of the adaptive management cycle; fieldwork is complete and the data are currently undergoing analysis.

This project illustrates the importance of longer-term monitoring – if monitoring had ceased after only 2 years, very different conclusions would have resulted. For instance, it was previously believed that goats would abandon use of a trail that was entirely logged over, or would continue to use it only if a significant adjacent buffer was left intact. Preliminary results from the first 2 years post-logging (but with a buffer retained along the trail) did confirm that goats continued to visit the lick using the trail through the buffer strip. However, by the third year post-logging, a shift in goat use away from the buffer strip was observed, with about 50% of the movements to the mineral lick occurring through the clear-cut areas. This may be because goats rely more on vision than smell to detect the presence of predators; the clear-cuts therefore provided greater visibility for goats to detect predators while moving to and from the lick.

Longer-term negative effects (20 or 30 years later) on the goat population are likely once regeneration occurs, which will result in a denser, immature stand that will impede the ability of goats to detect predators. Predator numbers have increased in the clear-cuts, which raises further questions about risks to goats after regeneration. This is expected, as early seral stage habitat created during harvesting benefits moose, deer, and elk, which also come with predators (wolves primarily). Bears benefit both from the early seral stage vegetation and berries, and the increase in ungulates.

After data analysis is complete, the results will be communicated to the forest industry and policy makers in the form of reports and recommendations. Changes to guidelines, policies, and best management practices will be discussed with practitioners.

The project cost $Can250,000 to start-up, and $Can175,000 per year for monitoring, management team costs, and one full-time employee. The implementation of the Ospika Mountain Goat Adaptive Management Trial faced three important challenges.

Lack of personnel: Only two people were designated to implement the project, both of whom had other ongoing duties and projects. Although contractors were brought in to assist, data interpretation and analyses fell behind, as did communication activities with partners and others.

Remoteness of the site: Equipment failures and bear attacks on equipment resulted in lost data. Staff and contractors couldn't get to the site frequently enough to deal with breakdowns (sometimes only once per month).

Changes in scheduling of harvesting, selection of cutblocks, and priorities of the forest company affected the planned study design: Forest company plans changed a couple of years into the project, and harvesting of the second treatment site did not occur. The practitioners then adjusted the study design to implement a second "scenario" (clear-cut up to and including the goat trail) at the first, single treatment site, removing a forested buffer previously retained along the goat trail (Fig. 3.4).

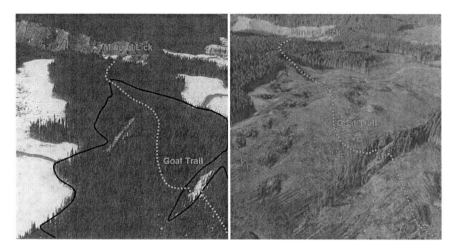

Fig. 3.4 Goat trial aerial. Study site with buffer and with buffer removed (Photos by Mari Wood)

Other design shortcomings also stemmed from constraints introduced by the need to co-operate with the forest company's block layout and harvest plan. For example, the first harvesting occurred 1 year into the project, although a delay would have enabled the acquisition of better pre-harvest baseline data.

In addition, the reality of large animal experiments is that sample sizes are small: project organizers had planned to radio-collar 90 goats in the vicinity of the licks, but were only able to successfully capture 22 due to lower densities than anticipated and difficult terrain and weather conditions. Also, the goats could move over long distances, and could therefore move between licks, confounding results.

Mari Wood, Senior Wildlife Biologist with the Peace/Williston Fish and Wildlife Compensation Program who was involved in the project offered the following advice.

– When one partner depends on another to conduct the management action according to a plan, establish a memorandum of understanding at the outset.
– Plan harvest timing and block location together to improve the project design.

Discussion

Despite some challenges, these case studies illustrate the success of the adaptive management approach in dealing with forestry problems at various scales and differing complexity. Success is often defined differently by the various people involved in or interested in the project. Success is likely measured by managers as the ability to answer the question you set out to answer in the first place. In the absence of that result, the learning that did occur and the relationships built through the process

Table 3.1 Lessons learned from four adaptive management practitioners in British Columbia, Canada

Project	Lessons learned
Coast forest strategy	Adaptive management requires strong commitment from managers to both the approach and the resources ($) to do it.
	Operational staff must have sufficient "buy-in" to the approach (this will not happen immediately, so training and leadership are essential).
	Partner with others; seek outside advice and collaboration.
	Don't try to monitor everything.
	Feedback to management occurs in a lot of ways, many of which are rather informal and often happen outside of any 'formal' process that may be devised for this purpose.

are also factors in success that may benefit an organization long into the future. By investigating the lessons learned from past and current projects, managers and practitioners can inform their own initiatives.

Practitioners confirmed that certain common factors ensured the success of adaptive management projects undertaken in British Columbia (see Table 3.1). For instance, underlying many of the themes raised by practitioners was the necessity of effective communication between all parties involved in the project. The Pine–Lichen Woodlands Project illustrated that good communication with operational staff was essential to revise harvest practices according to the project design.

Leadership is essential to initiate an adaptive management project and to sustain it over time. As a project progresses through the adaptive management cycle, leadership may change. For example, a project initiated at the executive level may eventually be led by a program-level manager or an operations supervisor. In fact, leadership at all levels is important for project success. The Forest and Range Evaluation Program achieved success when a dynamic leader, whose energy, communication skills, and influence with people at all levels of the organization, took on the task and achieved the support needed. One important factor was that the program was built in large part "from the bottom up," with support of operational field staff. At the field level, a leader can ensure that operational staff fully understand the value of their role in implementing the adaptive management design according to the plans. The Ospika Mountain Goat Project was derailed at one point when plans were not followed. A supervisor was not brought into this project early enough to ensure that the design was implemented properly. This shows that "when adaptive management projects are initiated from the top down, it is important to also create the conditions that will enable success by securing support at lower levels of the organization" (Marmorek et al., 2006).

Participating organizations must provide the funding and human resources necessary to achieve results over the full term of the project. Marmorek et al. (2006) found that having adequate funding to properly design an adaptive management initiative, to implement the needed management actions, and to monitor and evaluate the outcome is important to its success, does not in itself guarantee success. Many organizations, especially government agencies, experience frequent changes in priorities. This can make long-term adaptive management projects challenging to sustain. The Ospika Mountain Goat Project struggled because minimal human resources were committed to it. The Coast Forest Strategy project survived reduced funding several times as the forest company changed corporate hands. The lead practitioner was able to keep the project going by "selling" its benefit to each new administration. If budget is a consistent problem, this may reflect a lack of executive support or changing priorities. Securing long-term funding may also be a challenge if benefits are not expected for several years. This is a common complaint in forestry projects, as forest processes often occur over extended time frames and require a commitment to long-term monitoring.

Practitioners from the two large-scale adaptive management programs (the Coast Forest Strategy and the Forest and Range Evaluation Program) cited the building of effective partnerships as a key success factor. Conversely, success of the smaller Ospika Mountain Goat Trial was challenged by the lack of understanding among the partners involved. Partnerships help ensure the long-term success of projects by offering a diversity of funding sources and securing broader support. A cornerstone of a good partnership is effective communication; however, in adaptive forest management, this can be a significant challenge. The people implementing forestry practices are often not the ones who design the adaptive management project. A lack of common understanding may create situations in which implementation does not proceed exactly as planned, often resulting in design adjustments. For example, a couple of years into the Ospika Mountain Goat Project the partnering forest company's plans changed, so the second treatment site was not harvested. Retrospectively, a memorandum of understanding would have ensured that project roles and responsibilities were better understood. Accountability is an important part of a successful partnership. Partners should be aware of their respective responsibilities and obligations.

Throughout the adaptive management process, it is important to maintain the garnered support and to "close the loop" so that project outcomes are incorporated into policy and future management. This is the ultimate goal of adaptive management; often, however, closing the loop does not occur at a pre-determined end point. Therefore, it is worthwhile during the project's planning phase to articulate how the results will be used and ultimately integrated into the decision-making process. New information needs to be effectively communicated as soon as it becomes available. If new information is not accepted and incorporated into policy, it may be that the right question was not addressed, that the context had changed enough to make the question no longer relevant, or that the will is no longer there (especially if the management recommendations came with a high cost). The emphasis in adaptive management projects should be on producing results relevant to management decision making; therefore, asking the right questions is essential.

Organizational cultures, the strength of relationships with partners and stakeholders, and the scale and scope of adaptive management projects will be unique in every case. Therefore, the specific situation will determine the factors necessary for project success. The suitability of the institutional and social context should be considered when initiating a project, even if the problem itself is well suited to an adaptive management approach. The context surrounding a project may cause it to develop in different ways. For adaptive management to be adopted in an organization, people at all levels must understand the basic concepts and approach, which will require training and extension. Choosing adaptive management means choosing to do things differently, and also choosing to accept a level of uncertainty. This may require "training" the organization's culture to be more receptive of risk-taking and appreciative of the learning that can occur.

Acknowledgements The author is grateful to W.J. (Bill) Beese, Forest Ecologist, Western Forest Products; Peter Bradford, FREP Provincial Lead, B.C. Ministry of Forests and Range; Scott McNay, Wildlife Infometrics Inc.; Randy Sulyma, Wildlife Infometrics Inc.; Mari Wood, Senior Wildlife Biologist, Peace/Williston Fish and Wildlife Compensation Program – the practitioners who openly shared their experiences and insights participating in adaptive forest management projects for this chapter. Many thanks as well to Carol Murray, Senior System Ecologist, ESSA Technologies Ltd. for her help assembling the case study information.

References

Beese, W. J., Dunsworth, B. G., & Smith, N. J. (2005). Variable retention adaptive management experiments: Testing new approaches for managing British Columbia's coastal forests. In C.E. Peterson and D.A. Maguire (Eds.) *Balancing ecosystem values: Innovative experiments for sustainable forestry – proceedings of a conference.* (Gen. Tech. Rep. PNW-GTR-635). Portland, OR: U.S. Department of Agriculture, Forest Service, Pacific Northwest Research Station. Retrieved October 31, 2008 from http://www.fs.fed.us/pnw/publications/gtr635/GTR635a.pdf

Bunnell, F. L. (2005). Adaptive management for biodiversity in managed forests: It can be done. In C.E. Peterson and D.A. Maguire (Eds.) *Balancing ecosystem values: Innovative experiments for sustainable forestry – proceedings of a conference.* (Gen. Tech. Rep. PNW-GTR-635). Portland, OR: U.S. Department of Agriculture, Forest Service, Pacific Northwest Research Station.

Bunnell, F. L. & Dunsworth, B. G. (2004). Making adaptive management for biodiversity work: The example of Weyerhaeuser in coastal British Columbia. *Forestry Chronicle*, 80, 37–43.

British Columbia Ministry of Forests and Range (2007). *Stand-level biodiversity monitoring in 44 large cutblocks in the Central Interior of British Columbia, 2007.* Victoria, BC: Forest Practices Branch. Retrieved October 31, 2008 from http://www.for.gov.bc.ca/hfp/frep/site_files/reports/FREP_Report_10.pdf

Government of British Columbia. (2007). *Forest and Range Evaluation Program*, Ministry of Forests and Range. Retrieved December 2008 from http://www.for.gov.bc.ca/hfp/frep

Marmorek, D. R., Robinson, D. C. E., Murray, C., & Greig, L. (2006). *Enabling adaptive forest management – final report.* (Prepared for the National Commission on Science for Sustainable Forestry). Vancouver, BC: ESSA Technologies.

McNay, S., Heard, D., Sulyma, R., & Ellis, R. (2008). A recovery action plan for northern caribou herds in north-central British Columbia. (FORREX Series Report 22).

Ministry of Environment (2007). *British Columbia Conservation Data Centre.* Retrieved December 15, 2008 from http://www.env.gov.bc.ca/cdc/

Snetsinger, J. (2005). *Guidance on landscape- and stand-level structural retention in large-scale mountain pine beetle salvage operations.* Retrieved October 31, 2008 from http://www.for.gov.bc.ca/hfd/library/documents/bib95960.pdf

Sulyma, R. (2001). Towards an understanding of the management of pine-lichen woodlands in the Omineca region of British Columbia. (M.Sc. thesis). Prince George, BC: University of Northern British Columbia.

Sulyma, S. (2008). Terrestrial lichen evaluation tool: Determining how much terrestrial lichen cover exists to help caribou survive the winter. *LINK*, 10(1), 10–11. Retrieved October 31, 2008 from http://www.forrex.org/publications/link/ISS51/vol10_no1_art7.pdf

Adaptive Management and the Precautionary Principle

Chris Jacobson

Adaptive management and the precautionary principle are seemingly at odds: adaptive management calls for risk taking in order to learn, whilst the precautionary principle is evoked to avoid risk. Whether or not adaptive management is actually at odds with the precautionary principles depends on whether precaution is interpreted as avoiding/preventing *any* risk or avoiding/preventing *serious or irreversible* risk.

Three types of uncertainty give rise to risk in adaptive management: (1) uncertainty about status of entities such as species populations (statistical uncertainty), (2) uncertainty about the relationships between entities (structural uncertainty), and (3) uncertainty about unprecedented and random events (stochastic uncertainty) (Charles, 1998). The last of these is particularly significant given that it is extremely difficult, if not impossible, to predict or quantify (Holling, 1978).

The precautionary principle can hinder the application of active adaptive management, which seeks to accelerate learning for better outcomes over the long-term (Bormann et al., 2007). Over the short-medium term, a management program that produces smaller-sized but more likely returns is preferable to a program that produces larger but less likely returns (Hauser & Possignham, 2008). The risk associated with experimentation under active adaptive management can be reduced by allocating management options to different land units to spread the risk of uncertain outcomes. However, managers might choose to avoid risk associated with their allocated management option and withdraw from experimentation (for examples, see Jacobson, 2007).

While managing in a precautionary way can minimise risk over the short term, there is potential for it to serve as an excuse for not taking the (perhaps risky) steps needed to learn. In cases where there is sensitivity about experimentation, Polacheck (2002) suggests that it is important to collectively agree on (1) a response framework for results (e.g., predetermined decision rules) (2) that scientific advice will change as a result of experimentation and (3) that experimental design is appropriate. Gustavson (2003) suggests that adaptive management should not be applied in cases where there is high irreversibility of impacts and either uncertainty about the type of impacts is high or the size of impact and its likelihood are high. In cases where there is high certainty about the type of impact, impacts are largely reversible and the size of impact and its likelihood are low, then adaptive management should always be applied. In other combinations, the case for adaptive management depends on how precautionary a manager is; i.e., whether they choose to avoid/prevent *any* risk or avoid/prevent *serious or irreversible* risk. In any case, managers should realise that uncertainty cannot be resolved entirely. There is no such thing as a "no action" alternative, and the decision to not act can itself carry significant risks to values judged to be important.

References

Bormann, B.T., Haynes, R.W., and Martin, J.R. 2007. Adaptive management of forest ecosystems: Did some rubber hit the road? *BioScience*, 57(2): 186–191.

Charles, A.T. 1998. Living with uncertainty in fisheries: Analytical methods, management priorities and the Canadian groundfishery experience. *Fisheries Research*, 37: 37–50.

Gustavson, K.R. 2003. Applying the precautionary principle in environmental assessment: The case of reviews in British Columbia. *Journal of Environmental Planning and Management*, 43(3): 365–379.

Hauser, C.E. and Possignham, H.P. 2008. Experimental or precautionary? Adaptive management over a range of time horizons. *Journal of Applied Ecology*, 45: 72–81.

Holling, C.S. 1978. *Adaptive Environmental Assessment and Management*. Chichester: Wiley, 377 p.

Jacobson, C.L. 2007. *Towards Improving the Practice of Adaptive Management in the New Zealand Conservation Sector*. Unpublished Ph.D. thesis, Lincoln University, Christchurch, New Zealand.

Polacheck, T. 2002. Experimental catches and the precautionary approach: The Southern Bluefin Tuna dispute. *Marine Policy*, 26: 283–294..

Chapter 4
Using Adaptive Management to Meet Multiple Goals for Flows Along the Mitta Mitta River in South-Eastern Australia

Catherine Allan, Robyn J. Watts, Sarah Commens, and Darren S. Ryder

Abstract In this chapter we reflect on a relatively small but influential example of adaptive management which seeks to enhance the environmental benefits of the flow regime in the highly regulated Mitta Mitta River in Australia's Murray-Darling Basin. In 1999 an operational review recommended the reintroduction of greater in-stream flow variability in the Mitta Mitta River in an attempt to improve river health. The river managers have worked towards this through managed variable releases from Dartmouth Dam. These variable releases have been trialled four times from 2001–2008, with the explicit intention of learning more about the ecological impacts of variable flows while still achieving operational goals for the River Murray System overall. The ecological impact of the variable releases was studied via a series of consultancies by a University freshwater ecology team. They concluded that variable flow improved ecological condition compared with the condition after periods of relatively constant flow for greater than 1 month, although the benefits of it are relatively short-lived. Principles were developed over time through discussions between river managers and the research team. These principles are being progressively refined and incorporated into the current operational plan for the river, and learning continues. We suggest that three key ingredients enabled and supported adaptive management in this particular case; aspects of the operational context, the people involved and the trusting relationships that developed.

C. Allan and R.J. Watts
Institute for Land, Water & Society, Charles Sturt University, Albury, Australia

S. Commens
Murray-Darling Basin Authority, Canberra, Australia

D.S. Ryder
School of Environmental and Rural Science, University of New England Armidale, Australia

Background

Water management has come to be recognised as one of the Earth's 'wicked' issues. What was an apparently tame project of storing and redistributing water has spawned numerous ecological, social and economic challenges that require increasing levels of interdisciplinary collaboration and integration of different types of knowledge (Freeman, 2000). Australia's Murray-Darling system exemplifies the complexity of water management as numerous governments and citizens work to balance the wealth and well-being gained from the waters of the Basin (Department of the Environment Water Resources, 2004) with the serious degradation that has put the Murray-Darling Basin into World Wildlife Foundation's top ten international rivers at risk list (Wong et al., 2007). Choosing appropriate management actions is further complicated by uncertainties related to climate change (Khan, 2008).

The management of water resources in Australia has been undergoing reform since 1992, when the heads of all Australian governments adopted the National Strategy for Ecologically Sustainable Development, which is a commitment to more effective and integrated water management policies and practices (Pigram, 2006). In recognition of the complexity and uncertainty of water management the National Water Initiative, launched in 2004, aims to "*provide for adaptive management of surface and groundwater systems in order to meet productive, environmental and other public benefit outcomes*" (National Water Commission, 2005).

In this chapter we reflect on a relatively small but influential example of adaptive management occurring within the broader context of Australian water reform. The management aim in this case is to enhance the environmental benefits of the flow regime for the highly regulated Mitta Mitta River. Regulation has impacted on this river to a greater extent than most others in the Murray-Darling Basin (Jacobs et al., 1994). Opportunities for variable release exist during transfers of water from Dartmouth Reservoir to Hume Reservoir, and also during periods of 'minimum release' when inflows to the dam are being stored. We provide a brief description of the context of the variable release trials since 2001, before exploring what we have learned about undertaking adaptive management in this particular case.

Case Study

The Murray-Darling Basin, a catchment of over 1 million square kilometres in the Southeast of Australia, is an important source of wealth and wellbeing for Australia. The huge area covers numerous social and physical landscapes, and jurisdictions, which prompted the creation of the River Murray Commission (RMC) in 1917, and its successors the Murray-Darling Basin Commission (MDBC) in 1988 and the Murray-Darling Basin Authority (MDBA) in 2008. This unique organisation is a partnership of the Australian, New South Wales, Victorian, South Australian, Queensland and Australian Capital Territory governments. The purpose of this partnership, enabled by the Murray-Darling Basin Agreement 1992, is to "*promote*

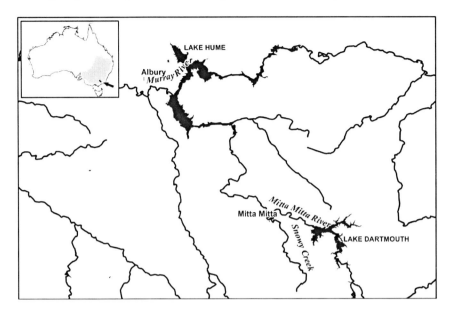

Fig. 4.1 Lakes Dartmouth and Hume, located in the south east of the Murray-Darling Basin (shown in grey in inset). Map courtesy of MDBC

and coordinate effective planning and management for the equitable, efficient and sustainable use of the water, land and other environmental resources of the Murray-Darling Basin" (Murray-Darling Basin Commission, 2006). The Mitta Mitta River is a tributary of the River Murray and is an important source of water within the Murray-Darling Basin (Fig. 4.1).

Hume Dam, on the Murray River was constructed between 1919 and 1936, and enlarged between 1950 and 1961 to re-regulate additional water from the Snowy Mountains Scheme. Dartmouth Dam was constructed between 1973 and 1979 on the Mitta Mitta River, a major tributary entering Hume Reservoir. Dartmouth Reservoir has a larger capacity (3908GL) than Hume Reservoir (around 3000GL) and is primarily used as "drought reserve" to supplement storage in Hume, the primary regulating storage for the River Murray system. Dartmouth Reservoir can take several years to fill because of its large storage capacity relative to its catchment size. Hume typically fills and empties more frequently, sometimes annually (Hume and Dartmouth Dams Operations Review Reference Panel, 1999). Although the primary purpose of Hume and Dartmouth Reservoirs is to store water for irrigation, and stock, domestic and town use, dam operations also mitigate flooding in the valleys below these reservoirs. Both dams are operated as part of the River Murray System by the River Murray Division of the MDBA.

Soon after Dartmouth's completion, downstream Mitta Mitta farmers reported declining pastures and reduced milk production, attributed to reduced floodplain watering (Allan et al., 2006). The public discussion over the operation of the dams continued for some time. In early 1997 the MDBC undertook a review of the operation of Hume and Dartmouth Dams, establishing an independent stakeholder Reference Panel to assist with this task. The Reference Panel consulted widely with impacted communities and the Review gained wide community acceptance (Hume and Dartmouth Dams Operations Review Reference Panel, 1999).

One of the many issues considered in the Review was relatively steady flows being maintained for long periods of time in the Mitta Mitta River immediately downstream of Dartmouth Dam, to which some of the ecological deterioration of that section of the river was attributed. This echoed similar concerns from regulated river systems around the world, including the Colorado River in the USA, where it was suggested that some variation be reintroduced through managed flow patterns (for details of that well-represented case see, for example, Jacobs & Wescoat, 2002; Light, 2002). When the Hume and Dartmouth Dams operation Review was completed in 1999 it recommended addressing the impacts of Dartmouth operation on river health by reintroducing greater in-stream flow variability in the Mitta Mitta River, viz: *"Strategies to increase the variability of in-stream flows below Dartmouth should be developed, and should not await solution of the water temperature problem."* The Scientific Reference Panel on Environmental Flows also commented that *"introduction of variability would have some value even if the water temperature issue was not addressed immediately. It will reduce the current level of bed and bank erosion and should create more bank habitat for bank vegetation to re-establish"* (Hume and Dartmouth Dams Operations Review Reference Panel, 1999).

In response to this recommendation, MDBC have worked towards increasing the variability of flows in the Mitta Mitta River through managed variable releases from the Dartmouth Dam. These releases have been trialed four times in the 8 year period 2001–2008 with the explicit intention of learning by doing; i.e. adaptive management.

The first trial of variable releases from Dartmouth Dam was during late spring/ early summer 2001/2002. This trial consisted of three successive large volume (approaching bankfull) 'pulses' over approximately a month, following an extended period of water transfers with low variability. The MDBC commissioned ecological monitoring and evaluation of the event via an open tender. The tender documents suggested a suite of environmental indicators based on previous reviews (e.g., Fairweather & Napier, 1998) that could be examined to provide an indication of ecosystem response to the variable releases. This tender was won by researchers from Charles Sturt University (CSU) and included field and laboratory experiments and monitoring at four sites on the Mitta Mitta River and at a reference site in the nearby unregulated Snowy Creek. The monitoring program was devised to test multiple hypotheses for suites of indicators and the findings were documented in a 150 page report (Sutherland et al. 2002). A further trial took place in the 2004/2005 summer, which consisted of a single large pulse following an extended water transfer period. The CSU research team was again contracted to monitor and evaluate the trial from an ecological perspective (Watts et al., 2005). The CSU team monitored and evaluated a variable low flow trial during a period of minimum release in autumn 2006 (Watts et al., 2006), and a single larger flow pulse in late spring 2007 following an extended period of low constant flow (Watts et al., 2008b).

The first study monitored the response of a comprehensive set of environmental indicators which included water quality, water column microbial activity, biofilm composition and metabolism and macroinvertebrates. For efficiency and effectiveness, successive monitoring was progressively refined to use only water quality, and biofilm biomass and composition as ecological indicators of the success of variable flows in improving the ecological condition of the river.

Key conclusions of the four flow trials were:

- Variable flow is ecologically more beneficial than relatively constant flow.
- The benefits of variable flows are relatively short-lived (one or two weeks), if relatively constant flow resumes.
- Some environmental dis-benefits start to become apparent if flows are relatively constant for more than one month.

The outcome of the adaptive management process is clearly evident when we compare hydrographs for a water transfer period preceding the variable flow trials (e.g. in 1987/1988) with the proposed water transfer plan for 2008/2009 which incorporates variable releases Fig. 4.2. Traditionally, operational practice was to delay water transfers from Dartmouth Reservoir for as long as possible to minimise the risk of unnecessarily transferring water to Hume Reservoir. Consequently, when transfers were required the river managers were compelled to manage releases near bankfull flows with limited variability, often for extended periods of time (Fig. 4.2). The river operators incorporated the learnings and principles developed from the four flow trials into the 2008/2009 flow plan to 'mimic' some elements of the natural flow regime (Fig. 4.2). In the case shown in the Fig. 4.2 the 'design' of the pulses fully complies with existing operating rules, for example the maximum rate of rise and fall in water level, as well as meeting the fundamental requirement to transfer a given volume of water to Hume Reservoir that season.

Post the variable flow trials (2001–2008) the University research team continue to work collaboratively and iteratively with MDBA to develop operational principles and recommendations. This consolidates the substantial financial and intellectual investments in this work. A feature of the series of trials was the openness, honesty and transparency in communication among the researchers and operators. For example, before and during the trials the researchers and MDBC discussed, informally, the emerging results and possible implications. The close contact between the research team and the dam operators also enabled the researchers to be informed of changes to proposed discharge patterns, allowing better preparation for research. Following each trial the CSU researchers presented their results at seminars for MDBC, and formal meeting were held to discuss findings and future directions and potential activities. MDBC engaged CSU to prepare a written "Synthesis" to consolidate key findings and operational recommendations arising from the trials, an exercise that added significant value to the river managers' prior investment in this work because it facilitated the adoption and extension of outcomes. Central to the sense of shared commitment that developed among the researchers and the river managers was that the work was mutually beneficial to both parties.

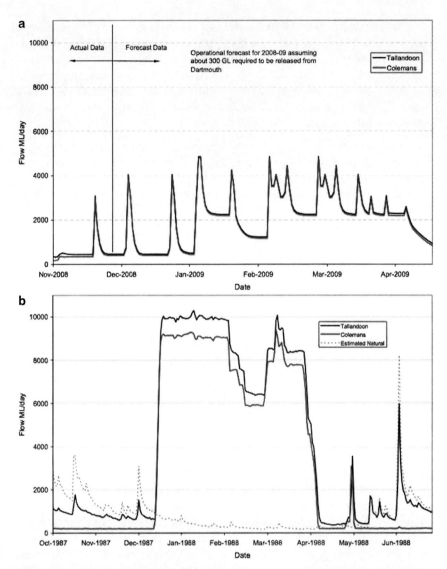

Fig. 4.2 Flow in the Mitta Mitta River below Dartmouth Dam. The top figure shows near regulated channel capacity flow (around 10,000 ML/d) for an extended period. The bottom figure shows that when average flow rates are lower than this there is greater operational flexibility to vary flows. Courtesy of MDBA

Honesty and transparency were also features of the communication with the local community. For instance, Mitta Mitta landholders were regularly informed during trials through "Flow Advices" sent by fax or email from MDBC so they could prepare as necessary by, for example, moving pumps. Informative articles, written jointly by MDBC and CSU, were published in a local newsletter, the "Bush and Bulldust", to provide context for the trials. This open communication facilitated the maintenance and further building of trust during this time.

This case study provides an example of the classic form of active adaptive management described in Chapter 2, with a cycle of learning from a series of monitored and evaluated variable releases and their outcomes. These cycles of *"Learn what? Do what?* and *What have we learned from doing?"* have led to sufficient understanding of the situation in the Mitta Mitta River for managers to now be asking *How do we decide what to do, from what we have learned*? However, this case varies from the description in Chapter 2 in this volume in its understanding of the type of problem being addressed. The variable release trials had a simple focus, in a bounded environment, and in this respect the trials are very like traditional scientific enquiry. Allan (2008) notes that reduction and simplification of complex problems is part of traditional scientific inquiry rather than adaptive management. However, this case study is clearly adaptive because the lessons from the Mitta Mitta variable flow case study are not *confined* to answering the simple, tightly focused question, but rather are being incorporated into the broader system operation and water reform framework, including the system wide review of River Murray Systems operations which commenced in 2007/2008. This Mitta Mitta case study is also informing a recently commissioned NWI report (Watts et al., 2008a) which reviews extant understandings and knowledge of pulsed flows in Australia. This case study, then, provides an example of how an operational review can initiate research to inform changes to local and system wide management, and national water management policy development.

Learning to Operate Differently

Reflecting on this case study, we suggest that learning and informed changes to management practice can occur even without a long-term, neatly articulated, all encompassing 'adaptive management' project. In this instance, the river operators, supported by a larger organisation, used University expertise in focused bursts to provide scientific information to guide their adaptive management. The ecological research projects themselves are indistinguishable from countless other studies – what makes them part of adaptive management is the framework within which the studies are viewed and used.

We suggest three key ingredients fostered the adaptive management in this particular case; aspects of the operational context, the people involved and the trusting relationships that developed.

Operational Context

The regional context and the nature of the issue each enabled and encouraged adaptive management in this case. The regional context (itself part of the larger water reform context in Australia and globally) was strongly influenced by the nature of the review of operations undertaken in the late 1990s. The consultation process for this review was genuinely inclusive, so the desire to learn about the impacts of variable flows had some local legitimacy and relevance. The inclusive nature of the review and the acceptance of its outcomes also created a social climate in which local people were at least not antagonistic, and were often supportive, of activities undertaken by MDBC, including these trails of variable flows.

The issue itself – centred on the environmental impacts of dam to dam water management – was tightly bounded in both its intellectual and physical scope. This is because the learning was narrowly focused on impacts of variable flows on instream parameters, and because the trials were exploring flexibility within the current operating rules and changing variability not volume. A far more elaborate process of negotiation and approvals would be required for testing hypotheses outside of current operating rules. The issue in this case was also one in which action in response to learning could be taken fairly quickly, as those who commissioned and received the scientific reports were the people with the authority and capacity to act on them.

People

The role(s) of individuals and their institutional arrangements have also clearly played an enabling role in this case study. A key point is that people within MDBC were committed to learning – both about the impacts of their activities, and about how to do things better. Their desire to learn was supported and championed by key people within MDBC. This enthusiasm for learning was matched by that of the University research team, who were more committed to the long-term learning than might be implied by noting that a series of consultancies was undertaken. Discussion among water managers and members of the research team is ongoing, with mutual benefits and learning continuing to accrue to both parties. All of this was facilitated greatly by the continuity of involvement of key personnel in both the University research and MDBC teams over the eight years. Reflecting on the importance of the people involved suggests a key role for structures and processes to enhance and protect organisational memory, and the importance of nurturing and encouraging adaptive people within organisations (see Chapter 18, Fazey and Schultz, this volume, for discussion on ways to support adaptive people).

Trust

The trust between individuals and organisations that developed in this case is related to the individual people involved, but it seems to be such an important enabling factor that we have highlighted it in its own section (refer also to Box in this chapter for a general discussion of trust). The initial open tender process facilitated the commissioning of a competent research team. Trust was then developed over time as each party delivered anticipated outcomes and, most importantly, developed shared questions and approaches. Trust almost invariably needs time to develop between people, and within and between organisations, so people remaining in their professional positions, and their organisations remaining stable, were clearly factors that enabled 'internal' trust to develop. However, in this case the wider public must also have trust in the process, via trust in the key organisations. Local trust in the MDBC was facilitated by the history of the inclusive consultation processes associated with the Hume and Dartmouth Dams Operations Review, and by the regular communication of river operations as described above. However, trust in the process could be threatened by perceptions of the nature of the internal relationships that developed. Cynical 'readings' of the case study could conclude that the researchers were feathering their own nests by always concluding their reports with recommendations for future work. The maintenance of transparent records (relating for instance, to why the subsequent tenders were awarded to CSU) is thus important, as is explaining the nature of adaptive management and continuous learning to people who may be impacted.

Trust is also developed through shared language, and this is taking longer to play out in this case.

It is becoming clear that the language of the ecological reports does not necessarily provide everything that is needed by operations managers to usefully inform their everyday decisions. A feature of this case study is the willingness for linguistic ambiguity to be raised and discussed among the parties. The statement of need for the "Synthesis report" is an expression of genuine desire on the part of the river managers to improve their operations and to consolidate previous investment. That the work is scientifically rigorous and undertaken by respected practitioners provides a sound basis to proceed as required to effect permanent changes to river operation rules.

Potential Risks with Incremental Approaches to Adaptive Management

The enabling factors discussed above suggest some potential risks with approaching adaptive management in small stages. The paradigms and adaptive capacities of the people involved in the project will impact on how inquiry is undertaken and how the results of that enquiry are understood and incorporated, and unsuitable people may inhibit adaptive management at many points in the cycle. An even greater risk of an incremental approach is that funding is not guaranteed, and must be secured at every stage. A supportive operational context is clearly necessary for the approach to adaptive management described in this case study; in an institutional context that is hostile to long-term learning, or is undergoing change, individual research projects may be isolated, and be confined to one off inquiries. Without a larger learning framework information from such inquiries is likely to remain local and restricted.

Conclusion

Effective adaptive management of flows from Dartmouth to Hume Reservoirs has occurred through a series of small research consultancies that reflect a broader desire by water managers to provide environmental benefits from river operations, which in turn fits into the longer-term decision for water reform in Australia which seeks multiple benefits from every drop of water. The success of this project (in terms of improving understanding the system, informing operational activities, and informing the wider water reform process) results from factors which combined to promote a desire to learn, to listen and to change behaviour. Some of these factors may be specific to this case and the people involved, and may seem fortuitous, but many should be reproducible in other projects where goodwill and capacity for trust reign.

Acknowledgments The work presented was enabled by a number of people whose enthusiasm and skills have supported the trials since 2001 to the present. Trevor Jacobs, Bruce Campbell, Neville Garland and Damian Green (MDBC) and Peter Liepkalns (Goulburn Murray water, Dartmouth Dam) are key personnel in operational planning and implementation, and Mac Paton, a local landholder has facilitated local information exchange.

References

Allan, C. (2008). Can adaptive management help us embrace the Murray-Darling Basin's wicked problems? In C. Pahl-Wostl, P. Kabat & J. Moltgen (Eds.), *Adaptive and Integrated Water Management: Coping with Complexity and Uncertainty* (pp. 61–73). Berlin/Heidelberg: Springer.

Allan, C., Curtis, A., & Mazur, N. (2006). Understanding the social impacts of floods. In A. Poiani (Ed.), *Floods in an Arid Continent* (pp. 159–174). San Diego, CA: Elsevier.

Fairweather, P.G. & Napier, G.M. (1998). Environmental indicators for national state of the environment reporting – inland waters. Department of Environment, Canberra.

Freeman, D.M. (2000). Wicked water problems: Sociology and local water organizations in addressing water resources policy. *Journal of the American Water Resources Association, 36*(3), 483–491.

Hume and Dartmouth Dams Operations Review Reference Panel. (1999). *Hume and Dartmouth Dams Operations Review Final Report*. Canberra: MDBC.

Jacobs, T.A., Koehn, J.D., Doeg, T.J., and Lawrence, B.W. (1994). *Environmental Experience Gained from Operation of Reservoirs in the Murray-Darling Basin, Australia*. Durban: Commission Internationale Des Grands Barrages.

Jacobs, J.W. & Wescoat Jr, J.L. (2002). Managing RIVER resources. (Cover story). *Environment, 44*(2), 8.

Khan, S. (2008). Managing climate risks in Australia: Options for water policy and irrigation management. *Australian Journal of Experimental Agriculture, 48*(3), 265–273.

Light, S. (2002). Adaptive management: A valuable but neglected strategy. *Environment, 44*(5), 42.

Murray-Darling Basin Commission. (2006). Murray-Darling Basin Agreement. Retrieved December, 2008, from http://www.mdbc.gov.au/about/the_mdbc_agreement.

National Water Commission. (2005). Intergovernmental agreement on a national water initiative. Retrieved from http://www.nwc.gov.au/nwi/index.cfm#overview.

Pigram, J.J. (2006). *Australia's Water Resources*. Collingwood, Australia: CSIRO.

Sutherland, L., Ryder, D.S., & Watts, R.J. (2002). *Ecological Assessment of Cyclic Release Patterns (CRP) from Dartmouth Dam to the Mitta Mitta River, Victoria. Environmental Consulting Report No. 27*. Wagga Wagga: Johnstone Centre for Research in Natural Resource Management.

Watts, R.J., Nye, E.R., Thompson, L.A., Ryder, D.S., Burns, A., & Lightfoot, K. (2005). *Environmental Monitoring of the Mitta Mitta River Associated with the Major Transfer of Water Resources from Dartmouth Reservoir to Hume Reservoir 2004/2005. Report to the Murray-Darling Basin Commission. Environmental Consultancy report number 97*. Wagga Wagga, Australia: Charles Sturt University, Johnstone Centre.

Watts, R.J., Ryder, D.S., Burns, A., Wilson, A.L., Nye, E.R., Zander, A. & Dehaan, R. (2006). *Responses of Biofilms to Cyclic Releases During a Low Flow Period in the Mitta Mitta River, Victoria, Australia. Report to the Murray Darling Basin Commission.* Wagga Wagga, NSW: Institute for Land Water and Society Charles Sturt University.

Watts, R.J., Allan, C., Bowmer, K.H., Page, K.J., Ryder, D.S., & Wilson, A.L. (2008a). *Pulsed Flows: A Review of Relative Environmental Costs and Benefits, Summary of Current Practice, and Identification of Prospective Best Practice and Areas Where Future Research Can Contribute.* Draft Report to the National Water Commission.

Watts, R.J., Ryder, D.S., Burns, A., Zander, A, Wilson, A.L., & Dehaan, R. (2008b). *Monitoring of a Pulsed Release in the Mitta Mitta River, Victoria, During the Bulk Water Transfer from Dartmouth Dam to Hume Dam 2007–08.* Report to the Murray Darling Basin Commission. Institute for Land Water & Society Report # 45, Charles Sturt University, Thurgoona, NSW.

Wong, C.M., Williams, C.E., Pittock, J., Collier, U., & Schelle, P. (2007). *World's Top Ten Rivers at Risk.* Gland, Switzerland: WWF International.

Building Trust in a Distrustful World

George H. Stankey

Hardly any aspect of human relationships is more fundamental than trust. Luhmann (1979) writes "trust, in the broadest sense of confidence in one's expectations, is a basic fact of social life." Trust is multi-faceted, involving competency, reliance, and integrity and is the glue that ensures society acts coherently and with purpose. In its absence, conflict and contention reign, with social action dominated by adhocracy and self-interest.

Given its centrality to effective social action, one would expect that understanding of the concept of trust was highly refined. Yet, the literature reveals a notion of complexity, disparate dimensions and meaning. Rousseau et al. (1998, 394) conclude there is "no universally accepted scholarly definition of trust." However, these authors recognize the conditions necessary for trust to arise. First, there must be a condition of risk; trust would not be necessary if actions could be taken with complete certainty. Second, trust requires a state of interdependence; the interests of one party cannot be achieved without reliance upon another. Taken together, these conditions produce definitions such as "undertaking a risky course of action on the confident expectation that all persons involved in the action will act competently and dutifully" (Lewis & Weigert, 1985, p. 971).

In addition to risk and interdependence, other assumptions regarding trust include:

- Trust is dynamic and can move through cycles of building, stability, and dissolution. A state of trust is always tenuous and provisional.
- Trust exists as multiple variables; it can occur as an independent (causal) variable, as a dependent (effect) variable, or as an interaction variable (a moderating condition for a causal relationship).
- Trust occurs at different scales; trust exists among individuals (e.g., citizens and resource managers) as well as at the institutional level (e.g., between citizens and the government agencies). Trust at one level does not necessarily translate to other levels.
- Trust manifests itself in different forms. It can arise from the commonality between individuals or groups that "serve as indicators of membership in a common cultural system" (e.g., race, gender, "good old boys"). It can develop from repeated exchanges over time, perhaps initiated by self-interest

or imposed by external requirements, but which "become overlaid with social expectations that carry strong expectation of trust and abstention from opportunism." Finally, trust can arise from institutions that have become accepted social facts; e.g., we place trust in the presence of professional credentials or in the rules and regulations that government imposes.

How can trust be developed (or, if necessary, restored)? First, it is important to acknowledge that trust cannot be created in a mechanistic manner; restoring trust is not equivalent to restoring riparian conditions. Trust is earned, based on action and outcomes, not rhetoric. It derives from long-term relationships in which there is a continued demonstration of good faith and follow-through. A recurring message in the literature is "do what you say you will do." In their study of partnerships, Wondolleck and Yaffee (2000, p. 149) report "Quite simply, successful partnerships kept their promise to one another in a variety of ways."

Second, trust is a provisional quality of any relationship, requiring constant tending and attention. It is also asymmetric; while the building phase can be lengthy, it can be diminished in a moment. Also, it is not a dichotomous condition (I trust you or I don't). Trust and distrust can exist simultaneously. We must also distinguish between personal trust, grounded in honesty, benevolence, and reciprocity and organizational trust, founded on concerns with fairness and equity. Trust can exist between individuals – e.g., local citizens and the ranger – but if the organization is perceived as untrustworthy, then it will be difficult to fashion productive relationships.

Institutions can make a difference in trust building. For example, they can demonstrate an openness and willingness to engage in self-criticism. They can promote organizational stability and clear role expectations for employees; however, turmoil generated by downsizing and re-engineering act to diminish both. Although regulations provide one means of building shared understanding regarding appropriate and expected behavior, they also undermine trust by substituting formalization for flexible, context-specific management approaches. But the bottom line remains straightforward: organizations that operate openly, transparently, and honestly and that strive to follow through on their promises have an opportunity to foster the trust needed to do their job and to survive politically. Those that don't, won't.

References

Lewis, J. D. and Weigert, A. 1985. Trust as a social reality. Social Forces 63:967–985.
Luhmann, N. 1979. Trust and power. New York: Wiley.
Rousseau, D. M. et al. 1998. Introduction to special topic forum: Not so different after all: A cross-discipline view of trust. Academy of Management Review 23(3):393–404.
Wondolleck, J. M. and Yaffee, S. L. 2000. Making collaboration work: Lessons from innovation in natural resource management. Washington, DC: Island Press.

Chapter 5
Adaptive Management of a Sustainable Wildlife Enterprise Trial in Australia's Barrier Ranges

Peter Ampt, Alex Baumber, and Katrina Gepp

Abstract This project is an example of a participatory research activity that set out, from the outset, to apply adaptive management principles for both improved resource management and enhanced project management. As a result, the entire project exemplifies the application of adaptive management: a complex system with multiple parts where initial interventions are continuously evaluated to determine the next steps in the process. There are multiple actors and theatres of the project, each of which needs to progress before others can progress, and the results of one impacts on the progress of another. A key component of the project is a trial under the adaptive management provisions of the New South Wales Government's Kangaroo Management Program, so it also provides insights on the practical implications of conducting research to meet an institutional requirement for adaptive management.

Introduction

In this chapter we reflect on a participatory research project centred on the Barrier Ranges. To understand the progress of the project it is necessary to have some feel for its context, which is socially, economically and ecologically complex, so we begin with a detailed description of these aspects.

P. Ampt and A. Baumber
Future of Australia's Threatened Ecosystems Program, c/o Institute of Environmental Studies, University of NSW, Sydney, Australia

K. Gepp
Future of Australia's Threatened Ecosystems Program, c/o Western Catchment Management Authority, Broken Hill, NSW, Australia

The Barrier Ranges

The Barrier Ranges area, north of Broken Hill in Western NSW, is a microcosm of much of Australian semi-arid rangelands. It has an average rainfall of between 200–300 mm per year (of moderate to high variability) and is covered with native vegetation such as bluebush, saltbush, grasslands, and sparse woodlands of Mulga (an Acacia) and other small trees. It is crossed by ephemeral streams vegetated with river red gum (Eucalyptus) and associated species. Geomorphically, it consists of alluvial and rolling plains, lowlands, hills and tablelands interspersed with dune fields and sand plains. Prior to European occupation, Wiljakali, Malyankapa and Pandjikali people lived in the area.

Explorer Charles Sturt named the Barrier Ranges in 1841 and pastoralists began settling the area in the 1850s, using the Darling River as their main trade route. The vast shrublands were quickly stocked with sheep over the following decades. Devastating droughts in the 1890s resulted in massive stock losses and land degradation. The area is now under the jurisdiction of the NSW Department of Lands, having being divided into Western Lands Leases overseen by the Western Lands Commissioner.

During the twentieth century, pastoralism continued with a proliferation of bores sunk to extend the areas available to grazing. Crises such as rabbits and droughts occurred, leading to massive soil loss and local extinctions of many species, including small native mammals. This has impacted on the structure and function of the remaining native vegetation and the subsequent productivity of the land for grazing purposes. For the past 8–10 years the area has remained in the grip of drought with only minor reprieves.

Presently, the Barrier Ranges is settled by grazing families on Western Lands Leases who are under considerable pressure on multiple fronts. Traditional enterprises (such as wool growing) are returning marginal incomes. Some landholders are acquiring additional leases to achieve an economically viable area, taking on large areas of land. This leads to extreme labour demands, so traditional enterprises such as wool growing become less feasible, as infrastructure is difficult to maintain under these circumstances with fencing and stock water requiring ongoing attention. Pressure to generate off-farm income is driving some families to separate during the week with partners living in the nearest large town to work and be close to schools. Many families have off-farm investments in property and shares and include ancillary businesses. In some cases this means that generating income from the pastoral enterprise is no longer critical. In addition to wool growing, there is interest in meat sheep breeds and many landholders make a significant income from trapping and selling feral goats. For some families, these enterprises have displaced wool growing because of the increased global demand for sheep and goat meat and because the labour demands are much less than for wool growing.

Since 1990 federal money has been available under a number of schemes for landholders to carry out conservation-orientated works on their properties. These include the historic decade of Landcare (1990–2000), the associated Natural Heritage Trust program, regionalization and the current 'Caring for our Country' initiative.

Regionalization of natural resource management led to the establishment of the Western Catchment Management Authority (WCMA), a key intermediary between individual farmers and federal funding. Like all regional bodies, the WCMA has developed catchment targets for land and vegetation (for example ground cover greater than or equal to 40% to prevent soil erosion) and biodiversity (for example ecological communities of high conservation values adequately protected and 25% of other ecological communities managed for conservation within 25 years).

The state's 2003 Native Vegetation Act and Regulations have put conditions on management that restrict landholders' rights to clear and modify native vegetation. One of the biggest impacts of this in the Barrier Ranges is to require landholders to prepare a Property Vegetation Plan (PVP) before being able to manage the encroachment of invasive native scrub; a contentious issue because while proliferation of native shrub can be classified as native vegetation under the Native Vegetation Act, landholders generally view it as being over-run by woody weeds. Some landholders are also involved in the NSW State Government's Enterprise-based Conservation Scheme. This scheme pays them per hectare to reduce their stock numbers and/or manage for a minimum ground cover target.

In response to these pressures and to the availability of the federal money, the Barrier Area Rangecare Group (BARG) was established by interested landholders in 2002. It is an active, incorporated Landcare group of landholder families with a wide range of ages, property sizes and backgrounds. BARG members have been successful in gaining access to Western CMA funding for a range of activities including goat trapping, invasive native scrub control and improved stock water management. They are clearly committed to maintaining their pastoral, outback station lifestyle despite the pressures described above. As a result they are keen to develop diversified income streams.

Kangaroos in the Rangelands

This vast arid landscape also supports varying populations of four different species of large kangaroos; Reds, Western Greys, Eastern Greys and Wallaroos. Numbers vary according to the seasons, but these species have been very successful despite the dramatic changes in the landscape since Europeans arrived. Pastoralists traditionally view these kangaroos as pests because, apart from shooting the occasional kangaroo for pet food, they obtain no direct material benefit from them. During good seasons kangaroo numbers increase, then as the landscape dries they can move large distances seeking feed in the paths of storms and in washout areas where there is green vegetation. They occasionally descend on properties in large numbers at these times. At other times they are ever present in the landscape. Many landholders are convinced that kangaroos cost them many thousands of dollars through competition with domestic stock and the damage they do to infrastructure.

In Queensland in the nineteenth century, kangaroos were officially considered vermin and bounties were paid. At the same time their commercial potential was being discovered, with a growing skin trade in the late 1800s and into the 1900s.

Kangaroo meat was also used for pet food, and with the collapse of the rabbit industry after the introduction of myxomatosis in the 1950s, it became more valuable. Over the next few decades legislation was introduced into most states to control the harvest. By the 1970s all states had legislation that offered protection to kangaroos as native animals but issued licences to cull kangaroos either for damage mitigation or for commercial use. An industry grew around the cull, supplying skins to tanneries and lean meat to both pet food manufacturers, and to a growing market for human consumption overseas and in Australia. Many of the pioneers of the industry are still in business. They have worked hard to develop domestic and export markets for kangaroo meat, promoting it as a healthy alternative to traditional red meats.

The Barrier Ranges are in the Tibooburra and Broken Hill commercial kangaroo management zones under the management of the NSW Kangaroo Management Program in the NSW Department of Environment and Climate Change (DECC). The goal of this program is to:

Maintain viable populations of *kangaroos* throughout their ranges in accordance with the principles of *ecologically sustainable development* (Department of Environment and Conservation NSW, 2006).

Each year the DECC commissions a population survey which estimates the populations of the four commercial species of kangaroos and sets a quota for harvest which is usually about 15% of the estimated population. Landholders can apply for an 'Occupier's Licence' to harm kangaroos on their properties. The licence involves purchasing royalty tags from the Kangaroo Management Program of DECC and specifies a 'Licenced Trapper' who will undertake the harvest. The trapper fixes a royalty tag to each harvested kangaroo and offers them for sale to a registered fauna dealer.

Two studies have been done recently about the commercial kangaroo industry (Chapman, 2003; Thomsen & Davies, 2007) that came to the following shared conclusions:

- It is rare for landholders to derive direct income from kangaroo harvest.
- Landholders perceive that regulatory regimes are a key disincentive to their participation in the industry.
- Despite many landholders regarding kangaroos as a potential resource, they provide access to harvesters because they derive indirect benefit due to reduction in kangaroo numbers.

The Origins of the Barrier Ranges Sustainable Wildlife Enterprise Trial

For about 2 decades a number of scientists and commentators have called for what has become known as 'sheep replacement therapy' for the rangelands (Grigg, 1987, 1989; Ampt & Baumber, 2006). They argue that kangaroos are superbly adapted to the rangelands, and that a production system based on them would be more

5 Adaptive Management of a Sustainable Wildlife Enterprise Trial 77

sustainable than sheep pastoralism. The reasons why this hasn't happened are multiple. The Future of Australia's Threatened Ecosystems Program (FATE), a small research group based at the University of NSW, is working on conservation through sustainable use (CSU) and common property strategies to improve natural resource management. FATE has picked up on the issue of kangaroos in the rangelands and is active in facilitating change in the kangaroo industry to generate conservation benefits and incomes for landholders through commercial use of kangaroos (Ampt & Baumber, 2006; Ampt & Owen, 2008).

FATE is interested in whether landholder returns from kangaroos can simultaneously improve the viability of rangeland enterprises and create incentives to conserve rangeland habitat. Such conservation outcomes may result from diversifying away from sheep (with a commensurate reduction in grazing pressure) and/or by more effective control of kangaroo grazing pressure through the commercial harvest.

FATE first became involved with BARG in March 2005 when the FATE program manager attended a meeting and discussed the issues around kangaroos in the rangelands. The positive response from the meeting stimulated a preliminary funding proposal which was accepted by the Rural Industries Research and Development Corporation (RIRDC) in October 2005.

FATE then assembled a team and attended a BARG meeting in November 2005, at which ten BARG properties expressed interest in participating in a trial to learn about better ways of managing kangaroos for multiple benefits. In the meeting it was clear from the landholders that the stimulus for their involvement was the belief that:

- Kangaroos made a significant impact on total grazing pressure, especially in dry times when landholders reported influxes of kangaroos onto drought reserve paddocks, flood out areas and in the path of storms where 'green pick' was evident.
- The existing quota setting and tag allocation system was not flexible enough to respond quickly to influxes of kangaroos.
- The existing industry was preventing economic returns to landholders from kangaroos harvested from their property.

FATE then submitted a full proposal to RIRDC in January 2006 which was funded from July 2006 until June 2009. The project synopsis is presented in Fig. 5.1.

Problem Analysis – Is It Fertile Ground for Adaptive Management?

The problem at the centre of this project is declining sustainability of pastoralism in the rangelands and the perceived lack of alternative enterprises. Linked to this is public pressure to manage land for enhanced environmental outcomes. The FATE team viewed landholder involvement in kangaroo management as a management option with potential to improve this situation. In taking on landholder involvement in kangaroo management as a key component, the team immerse itself in a

Barrier Ranges Sustainable Wildlife Enterprise Trial Synopsis

(i) Objectives/aims of the proposed research

The major objective of the project is to investigate whether a Sustainable Wildlife Enterprise (SWE) based on kangaroo harvesting can provide incentives to manage rangelands for biodiversity conservation and landscape rehabilitation. In order to achieve this objective, the project will aim to:

1. Develop a collaborative kangaroo enterprise that provides returns to landholders.
2. Develop a collaborative approach to kangaroo management across the BARG area.
3. Integrate kangaroo management with other enterprises on the participating properties to achieve improved management of total grazing pressure.
4. Establish and undertake community monitoring of landscape function and kangaroo populations to inform adaptive management.
5. Document the process and develop a model for similar initiatives in other locations.

(ii) Outcomes of the proposed research

The chief outcomes of this research will be an improved understanding of the feasibility of collaborative landholder involvement in kangaroo harvesting and its potential benefits for rangeland management. This increased understanding will be reflected in the development of a model for kangaroo harvesting initiatives that is based on the experience of the BARG members but also flexible enough to be adapted to the different economic, social and environmental factors operating in different locations involved in the SWE Program. Ultimately, this trial may lead to increased landholder involvement in kangaroo harvesting across Australia's rangelands and for increased acceptance of sustainable kangaroo harvesting as a viable land-use option.

(iii) Background, relevance and potential benefits

The guiding principle behind the trial is Conservation through Sustainable Use (CSU), whereby the sustainable commercial use of wildlife can provide incentives for land managers to conserve habitats, in this case through improved control of total grazing pressure and decreased reliance on traditional pastoral income. The commercial kangaroo harvesting industry has been operating in Australia for several decades and although it has been shown over that time to be a sustainable use of abundant free-ranging wildlife species, it is not yet a good example of conservation through sustainable use (CSU). In the view of the FATE Program, the missing factor is economic returns to landholders that would create incentives to become more actively involved in the management of kangaroos.

(iv) Research strategies and methodology

The project will follow an active adaptive management framework, devising and testing strategies for:

- How kangaroo harvesting is carried out (e.g., targeting certain areas, species, age and sex classes);
- How landholders will be involved in the industry supply chain; and
- What the level and nature of cooperation between landholders will be.

These strategies will be carried out by the participating BARG members and tested through monitoring of:

- Economic impacts (i.e., returns to landholders vs costs);
- Social impacts (e.g., landholder beliefs, perceptions and attitudes); and
- Environmental impacts (changes in kangaroo populations and landscape function).

(v) Communications/adoption/commercialisation strategy

The trial will allow FATE and the BARG members to assess the commercial viability of an ongoing landholder-driven kangaroo harvesting enterprise, with a commercialisation strategy based on ensuring a smooth transition from a trial phase to a fully commercial phase if desired by the BARG members. The trial will explore the potential for expanded landholder roles in the industry, including barriers to industry entry, investment requirements and sources of investment funds. The trial would also provide a model for other landholder groups to engage in a kangaroo harvesting enterprise, with the results of the trial communicated through a range of activities.

(vi) Time-lines

The project will be carried over three years (2006/07 to 2008/09) and consists of four stages (Team and skill building and proposal development, Implementing the 2007 Adaptive Management Trial, Implementing the 2008 Adaptive Management Trial, and Evaluation and dissemination).

Fig. 5.1 Sustainable wildlife enterprise project synopsis

complex environmental management problem. To describe the dimensions of this complexity we will use the key characteristics of complex environmental management problems as outlined in Chapter 2 of this volume.

Multiple Uses and Multiple Objectives

The Barrier Ranges are used primarily for pastoral activities, but the wider community has an interest in their iconic outback cultural status and in maintaining environmental values. In the same way, kangaroos can be used for meat and skins, to promote local tourism, and as a component of our community well-being – we are happier in the knowledge that they are there in their natural environment. They are also used as a pawn in the political game around animal rights in that they are convenient media target for animal activists. The objectives of the key stakeholders are diverse, sometimes overlapping and sometimes in competition as is described further below.

A Mix of Scales of Interest and Boundaries of Responsibility

Landholders primarily operate at the single property scale except when they are active in a group like BARG; harvesters operate on several properties to spread their risk – kangaroos regularly move across property boundaries so harvesters follow. Full-time harvesters may have up to ten properties on which they harvest regularly while part-timers may have two or three; regulators have a state-wide perspective that in NSW is divided into zones. They assess population and quota at a zone level but apply policy and issue tags at an individual property level. Processors operate across Australian states to ensure they can maintain continuity of supply to large processing plants. Processors may employ area managers to coordinate the harvest effort across localities. They locate field chiller boxes depending on where the harvest is occurring to minimize transport to chillers.

Landholders are primarily responsible for their own property but may recognise the benefit of acting collectively on a number of natural resource management activities. They are also accountable for the impact of their actions off-farm and are restricted in their on-farm actions by legislation, regulation and policy of government departments. They provide access to the kangaroo resource to harvesters through the licence system. They provide this free of charge because they generally perceive it is better that the kangaroo population is controlled, and that if commercial shooters didn't do it they would have to and it would cost them money. If landholders acted collectively, they could choose to exercise power over the harvest by demanding certain conditions be met for access with the ultimate threat of closing down the industry through denying access if those conditions were not met. In reality this is unlikely to happen. Many landholders appreciate the role that shooters play in management and in small local communities.

Harvesters are responsible for ensuring they comply with the licence conditions and for maintaining good relationships with landholders on whom they rely for access to the resource. They are also responsible for the quality of their work, which includes maintaining their equipment, kangaroo selection, marksmanship, field processing and transport to field chiller boxes. A load of kangaroos can and will be rejected at the chiller if they are too small or are unhealthy, if they are not head shots (ensuring instant death), if they have been processed carelessly in the field or if they don't arrive at the chiller in time to be chilled to the required core temperature in the specified time. Harvesters also have to administer the royalty tags correctly and complete accurate harvest returns to the Kangaroo Management Program.

Divergent Needs and Desires of Stakeholder Groups

Landholders wanted better control of grazing pressure due to kangaroos and were curious whether they could generate any income from kangaroos. They were skeptical but had a sufficient level of interest to support FATE in going forward and supported the Steering Committee. There were BARG members who were passively resistant or disinterested in the trial, others that were content to observe its progress and those that volunteered for the Steering Committee and put time into participating in meetings.

Initially most harvesters were suspicious of the trial, and some were strongly antagonistic. At a public meeting in August 2006 FATE personnel and landholders were accused of various degrees of stupidity, opportunism and self-interest. There was widespread skepticism about whether any of the initiatives were worth anything. Views were forcefully expressed that landholders just wanted something for nothing and that the only likely result of the initiative was that harvesters would be squeezed because any income for landholders would come at the harvesters' expense. It became clear during this meeting that what harvesters needed was secure access to the resource, more consistent demand for the product from processors and a fairer and more predictable price at the chiller.

Processors need to be able to manage supply to maintain continuity and to match supply to market demand. They do this by manipulating the price they offer at the chiller and by closing or moving chillers for which the supply is inadequate. They value reliable and efficient harvesters and provide strong incentives to some to keep them loyal. Processors were dismissive about the project. They maintained a consistent line collectively that the industry is functioning fine without landholder input and without FATE's intervention. This position is not surprising as, they retain control over price and supply through regional managers and relationships with key chiller operators and shooters while working hard to maintain markets. A small processor trying to enter the industry expressed a desire to work with landholder groups who could coordinate harvest and maintain quality management to ensure consistent and better than average quality for specific markets.

The regulators (The NSW Kangaroo Management Program of DECC) were skeptical but had the adaptive management provision in their Kangaroo Management Plan (dealt with later in this chapter) so were obliged to engage with the project. They remained clearly focused on the goals of the Kangaroo Management Program and the need to fully comply with their legislative obligations. They expressed the view that the current system was flexible enough to allow landholders to participate more fully, and that the reason they didn't was because of the lack of an adequate profit margin in the industry.

Tight Economic Imperatives Around Ecosystem Exploitation

A continuation of pastoralism requires ongoing maintenance of pastoral infrastructure (fences, yards, stock water, roads, vehicles, silos, sheds) and significant labour associated with stock management (shearing, crutching, drenching, lamb-marking, mustering) all of which come at a considerable cost in terms of time, labour and capital. Commodity prices are uncoupled from this, and until recently, the cost-price squeeze pushed landholders onto bigger and bigger areas to make an economic return. Critical components of productivity are lambing percentages and wool clip. Both rely on maintaining stock numbers and improving genetic lines of stock, a strategy that is not compatible with the extreme year to year variability of feed in the semi-arid rangeland environment. There is a trade-off between production per animal and production per hectare that is mediated by stocking rate, but landholders generally attempt to maintain as high a stocking rate as they can to maximise economic return.

While landholders derive no economic return from kangaroos harvested from their properties any kangaroo is a threat to the profitability of their pastoral enterprise. A complicating factor in this is the increasing reliance of landholders on off-farm income.

Harvesters have to make a significant outlay to get into the business. Once licenced, their big challenge is minimizing the harvest effort. The distance traveled and wear and tear on vehicle are major costs. If Kangaroo density is low the cost per harvest increases. Studies (Hacker et al., 2004) indicate that the commercial industry is not viable at kangaroo densities that might threaten the viability of the commercial species. This indicates that with current cost structures harvesters will cease harvesting long before a critical density is reached.

Reduced Ecosystem Health and Ecosystem Services

The WCMA Catchment Action Plan sets targets aimed at improving ecosystem health and the provision of ecosystem services. These are to some extent in competition with economic imperatives as outlined above. These targets exist despite major rangeland monitoring systems lacking any systematic biodiversity component (Fisher et al., 2008) and reporting little positive or negative change in range condition (Eldridge & Grant, 2004a, b). Catchment action plans emphasise incentives which, in the judgment of the WCMA Board, will move the catchment towards the targets. The targets are precautionary in that they are judged to be sufficient, if achieved, to maintain or improve biodiversity and enhance the provision of ecosystem services. A key ecosystem service is the resilience of the ecosystem and cultural and aesthetic benefits of knowing we are managing ecosystems to maintain and enhance biodiversity.

However, land management remains dominated by the private good need to generate income from pastoralism, and the public good need for improved ecosystem health remains under-resourced. This suggests that strategies that combine private good and public good will be beneficial.

Significant Technical Information on Parts of the System

Sufficient technical information existed on parts of the system such as:

- Land systems in the area
- The colonial history of the rangelands of Western NSW
- Grazing management
- Rangeland ecology
- Past kangaroo harvest data
- Extensive biological, geological and ecological research from the Fowlers Gap Arid Zone Research Station situated in the Barrier Ranges
- Kangaroo behaviour, biology and ecology; kangaroo population survey methodology and results
- Landscape function
- A landholder survey on kangaroo management from Queensland
- Consumer attitude research

An extensive review of this literature revealed to the researchers significant areas where the functioning of the system was far from optimal according to the principles of ecologically sustainable development. As a consequence, there were potential benefits in intervening in the system using an adaptive management framework.

Competing or Open Mandates, with Different Policy Options and System Targets

The Kangaroo Industry Association of Australia aims to significantly increase domestic consumption of kangaroo meat and actively promotes the health and environmental attributes of kangaroo.

Presently landholders provide free access to the resource because they accept the pest status of kangaroos. This is despite significant scientific evidence that kangaroos and sheep only compete when biomass gets below a critical threshold. As a pest control strategy, the commercial industry has limited effectiveness because kangaroo densities that are required to make harvesting profitable are considerably higher than those that landholders perceive to be desirable (Hacker et al., 2004). Also, large influxes of kangaroos rarely are dealt with effectively by commercial harvesters because of difficulties with getting enough harvesters with tags to the influx quickly enough (Landholder survey 2008 unpublished).

While this continues, landholders give away any bargaining power they have in the industry. Increased landholder involvement in the industry has been perceived as a threat by processors, largely because it raises the possibility that landholders will exercise influence to gain commercial benefit from the harvest. Yet landholders increasingly see kangaroos as a potential resource and good relationships between landholders and harvesters are common and mutually beneficial (Thomsen & Davies, 2007). Many consumers also hold a view that landholders are actively involved in some way in bringing kangaroo meat to market (Ampt & Owen, 2008).

The regulators are charged with ensuring harvest is consistent with maintaining sustainable kangaroo populations and a humane harvest according to the principles of ecologically sustainable development. Landholders (Chapman, 2003) and Landholder Survey 2008 (unpublished) report that regulatory arrangements restrict their ability to manage kangaroo component of total grazing pressure especially in times of influx and prevent them from adding value to the industry. Harvesters reluctantly report that it is normal for them to 'work around' the property specific tagging system and use tags issued for one property on another.

There are also problems with the funding of the regulation. KMP is supposed to operate on a cost recovery basis but is currently running at a loss because population surveys and administration are costing more than income from the sale of royalty tags. Processors are not (according to KMP) likely to contribute further, so KMP is anticipating a large increase in cost of royalty tags which currently cost 80c per tag.

Reasons for Taking an Adaptive Management Approach

We had three key reasons for considering using an adaptive approach in this project:

1. The situation was highly complex: It is clear from the previous section that the problem is complex, with various and conflicting values, multiple objectives, and entrenched histories.
2. There was structural support for the use of adaptive management: The RIRDC Program under which we sought funding suggested that adaptive management was a key strategy in developing sustainable wildlife enterprises and the Kangaroo Management Plan also had provision for adaptive management trials.

As researchers we had strong motivation to drive the process and had secure support from UNSW for our work.
3. There was good potential for participation: FATE could see clear applicability for adaptive management cycles to be built into the research process. Despite our open admission that we were not sure how far the project would go, we had a willing group of landholders committed enough to sign up to the project and to join a Steering Committee. Several harvesters volunteered to join the Steering Committee out of their loyalty to the landholders involved and to have a stake in the process. We knew that this support was dependent on progress, and that the research team would be doing most of the work. However it was imperative to gain support for each step and provide feedback and opportunities to influence the directions that we took.

The following points summarise key components in the adaptive management cycle that we needed to include in the process when appropriate:

Learn

- The group needed critical information and understanding key components of the complex systems relevant to kangaroo management.
- We needed to canvass the views and suggestions of the group.
- We needed to negotiate key steps with other parties on behalf of the group without knowing in advance what the outcome would be.
- We needed to provide continual informal access to us to allow opportunities for dialogue.
- We needed to identify gaps in knowledge as they became apparent and seek to fill the gaps.

Describe

- We needed to be able to describe and model key parts of the process and provide opportunities for the group to contribute to the models.
- We needed to provide experts to build and conduct economic and business models for possible strategies.

Predict

- In deciding on the next steps we needed the benefit of the group's and outsider's experience on how our actions would impact others.
- We needed to develop scenarios for an outcome that satisfied the group's motivation and diminished their skepticism.

Do

- We clearly needed action before we could know the next steps. This was particularly true when it came to testing a different regulatory framework.
- The FATE Program was committed to participatory action research.

Five hypotheses were framed for the project (Fig. 5.2). Each of these hypotheses required learning, describing, predicting and doing stages with specific feedback before proceeding to the next stage. What follows is the project synopsis from the full proposal which was subsequently accepted and funded by RIRDC.

The Progress of the Trial

We were committed from the outset to ensuring that the trial itself should be focused on processes that enable the stakeholders to adaptively manage their resources for multiple benefits. We were also committed to use the adaptive management provision in the KMP. It was clear early in the project that there was considerable uncertainty about the likely outcomes of different stages of the process and how they would impact on future directions. As a result, there was no point in being linear and prescriptive in planning the project. As a consequence, we made a deliberate decision that our project management would be adaptive.

What followed was a serious commitment by the FATE team to adaptively manage the group of participating landholders (and kangaroo harvesters) through a series of stages – a process that was still in progress as this chapter was written.

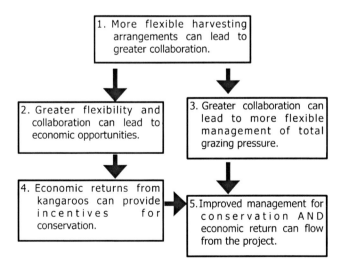

Fig. 5.2 Hypotheses for the Barrier Ranges sustainable wildlife enterprise trial

The FATE team wanted to understand how landholders could add value to kangaroo industry, why landholders weren't involved, and what it would take to get them involved. Learning was involved by all parties, and at the outset it was unclear how different players would react as new information came to light.

In this process FATE played the role of *agent provocateur*, stimulating dialogue by bringing people together regularly and joining BARG events – we kept turning up. We knew that to shift the system, thinking had to move beyond the status quo and this wouldn't happen quickly or without sustained and regular effort. In the spirit of researching *with* people, we were very open about the approach we were taking and were clear about not being certain at the outset about where the project would go. We were emphatic about the need for us to gain feedback from them, and they expressed a willingness to embark on the journey and keep up with what happened.

A critical component was to employ a local research officer who was the face of the project and the voice at the other end of the phone, providing an accessible avenue for expressing views about the project that might not have come out at more formal meetings. The research officer was critical in managing stakeholder expectations, reinforcing our chosen strategies and directions and following people up about commitments. We produced a regular newsletter and maintained good group management practice – agendas, minutes, housekeeping.

FATE recognised that while there was a genuine interest from landholders in being an agent for change in the way kangaroos are managed, even the members of the Steering Committee basically didn't believe that kangaroos could become a significant source of income. It was clear to us that landholder involvement in kangaroo management was our agenda not theirs, and that they were engaging with the process largely because FATE was prepared to provide the resources and do the work. We learnt that we could only make small demands on their time and that if other demands encroached on time set aside for the project, landholder priorities generally lay elsewhere.

As a consequence, a key sign of improvement as a result of the project will be whether the community continues to support the initiative when the project funding comes to an end, and whether a member of the community steps forward to take on the running of the enterprise.

Over time, however, influential harvesters with more moderate views and open minds joined the Steering Committee, and it became obvious, at least to them, that the accusations leveled at the FATE team were unfounded. Participating harvesters became very cooperative and willing to share information and views – one said 'no-one has ever died wondering what I think about this!' One of the most significant shifts has been a deepening level of dialogue and potential collaboration between harvesters and landholders. This was not anticipated when the project was instigated and it is a key point of progress. As a result we made sure that harvesters knew that they were welcome to contribute, and they became more active and vocal in the Steering Committee.

One landholder described kangaroo harvesters as their 'night custodians,' alluding to the value to landholders of trustworthy and reliable shooters. Other landholders expressed frustration that harvesters don't come when landholders want them to. Conversely, harvesters described landholders as very unreliable in determining whether kangaroo numbers are high enough for a successful harvest. Both landholders and harvesters describe the benefits of having a strong trusting relationship – landholders value the extra pair of eyes and that harvesters will sometimes act on the landholder's behalf. Harvesters value landholders that give them accurate reports of harvestable number of kangaroos and that are loyal – being prepared to wait for them rather than to 'give' territory to another harvester. There are examples of positive relationships between landholders and harvesters over several decades.

Harvesters clearly have a vital interest in maintaining a livelihood through kangaroos. As a result they are far more motivated than landholders to push for the sort of improvements advocated by FATE. There are also landholders who shoot kangaroos both on their own places and for other landholders. Some landholders also own and/or operate chillers. The opportunity for ongoing dialogue provided by the regular Steering Committee meetings generated benefits for both – building trust and willingness to work together. This made it possible for a strong group to come forward when the General Licence finally came through.

The researchers elucidated a vision for the system managed for higher standing biomass, higher biodiversity, increased groundcover, improved landscape function, and greater resilience, all of which are consistent with management of total grazing pressure. Overall it was aiming towards a strategy for generating income from a landscape managed for resilience. As part of this, we undertook to train as many landholders as possible in Landscape Function Analysis (Tongway & Hindley, 2004) and through that training to encourage them to incorporate landscape function information into the decision-making process. This is an additional avenue for adaptive management, dealt with in Fig. 5.3 below.

In parallel, FATE also embarked on a project to better understand consumers' beliefs, attitudes and behaviours around the choice and consumption of kangaroo meat. It was through this project that FATE directly engaged processors. The report produced from this project (Ampt & Owen, 2008) became a key component of the thinking in 2008 at the Broken Hill Workshop organized as part of the trial.

We adopted multiple lines and levels of evidence (MLLE) approach as suggested in Chapter 2. See Table 5.1 below for a summary of the types of evidence we have used.

At the time of writing, we were in the process of developing the enterprise and business plan and analysing the results of the Western Division Landholder Survey on kangaroo management. Critical steps were approaching during which landholders choose whether or not to financially support the development of the business plan. The following model had been put forward regarding the nature of the business entity (Fig. 5.4).

A significant aspect that had already emerged was the strong potential for collaboration to generate mutual benefit between a group of landholders and the harvesters that operate on their properties. There had been a significant shift in attitude between landholders and harvesters during the progress of the trial from skepticism and suspicion to openness and a willingness to contribute constructively.

Barrrier Area Rangecare Group (BARG) Proposed Landscape Monitoring and Assessment Program

Reasons for monitoring

- o Provide an additional tool for landholders to use to read their land and understand how it is responding to seasonal and management changes.
- o Provide landholders with benchmarks for important landscape types against which they can compare the condition of their land.
- o Provide evidence over time of changes in landscape condition and landholders' level of land stewardship.
- o Provide reliable and rigorous information on which individual landholders and BARG as a whole can base management and strategic decisions such as domestic stocking rates, control of invasive native scrub, control of introduced pests and kangaroo harvesting strategies.
- o Develop a model that, if successful, can be used in other locations both in WCMA and other areas to provide CMAs with reliable and timely information on resource condition.

Field data collection and analysis

- o Landscape Function Analysis (LFA) will be used as the basis of data collection and analysis. Through LFA, indices for soil stability, water infiltration and nutrient cycling can be generated through a well designed methodology.
- o Important land types will be defined using broad RAP (Rangeland Assessment Program) land types refined to be meaningful across the BARG properties. Three Key Land Types will be selected that best represent the landscape and the objectives of the landholders.
- o Reference Sites will be located on participating properties; at least two for each of the three Key Land Types.
- o Landholders will learn LFA. Data will be collected on whether they have learnt to conduct the LFA consistently during the training.
- o Landholders will establish sites on their properties which correspond as closely as possible to one or more of the Key Land Types and will undertake to monitor regularly (a minimum of twice yearly).
- o Critical locations that have higher levels of growth and thus higher potential herbivore populations (eg 'wash-out' areas or areas that have received water from a localized storm) could be identified and monitored as additional sites.
- o Each landholder can enter their data onto data sheets. They (or the team) can transfer the data onto the LFA software program to generate the 3 indices. The team can add the data to the GIS.

Turning data into information

When new data comes in, the team will enter the data onto the LFA software, do the calculations and provide feedback to the landholder about how to interpret it. The team will develop information via print and the website that will help landholders to understand what their data means using the LFA framework. For example, it might show whether measured values mean that the landscape is functioning well in relation to the Reference Sites and other sites on other properties.

The team will also record and maintain the data, maintaining a balance between confidentiality and making an appropriate level of information available to help make sense of data from individual properties. The team will have access to all data but landholders will be able to de-identify themselves from the data that is viewed by others. The data will appear in GIS format with access to layers managed in accord with the wishes of the landholders and researchers.

Turning information into action plans

The reliable and rigorous information generated from the data can provide evidence on which individual landholders and BARG as a whole can base management and strategic decisions such as domestic stocking rates, control of invasive native scrub, control of introduced pests and kangaroo harvesting strategies. It will also enable BARG to demonstrate any movement towards improved environmental stewardship based on sound and systematic evidence.

Fig. 5.3 BARG landscape monitoring and assessment program

Table 5.1 Multiple lines and levels of evidence

Trial hypothesis	Description of evidence
More flexible harvesting arrangements can lead to greater collaboration	• Meeting agendas and minutes
	• Trial progress as documented in newsletters
	• Documented feedback from participants
	• Western Division Landholder Survey
Greater flexibility and collaboration can lead to economic opportunities	• Documented discussion of economic opportunities
	• Development of an enterprise plan
	• Development of a business plan
	• Establishment of business entity
	• Success of business entity
	• Documented feedback from participants
	• Adoption of business model by other groups
Greater collaboration can lead to more flexible management of total grazing pressure	• Western Division Landholder Survey
	• Document the process of tag distribution
	• Describe any influx events and how the trial responded
	• Contrast trial response with status quo or previous influx events
	• Documented feedback from participants
Economic returns from kangaroos can provide incentives for conservation	• Success of business entity
	• Documented feedback from participants
	• Impact of trial on BARG and other landholders
	• Data from monitoring using Landscape Function Analysis
Improved management for conservation AND economic return can flow from the project	• Adoption of business model by other groups
	• Document examples of collaborative kangaroo enterprises being successfully incorporated into sustainable management

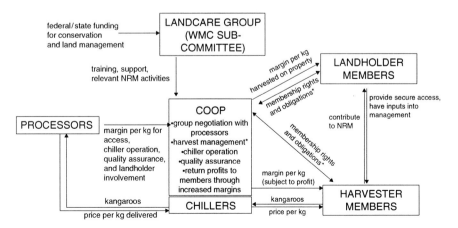

Fig. 5.4 Suggested model for a co-operative for kangaroo harvest management (Cooney et al., 2009)

The Adaptive Management Trial of Group Licencing Under the NSW Commercial Kangaroo Harvest Management Plan (KMP)

The main reason why we sought to have the trial approved as an adaptive management experiment under the NSW KMP was so that a special group harvesting licence could be issued to facilitate collaboration. This negotiation process took almost 2 years before the trial was accepted by DECC under the adaptive management provisions of the KMP (during which the KMP was revised from the 2002–2006 to the 2007–2011 iteration). This delay meant that altered harvesting arrangements didn't start until 2008 instead of 2006.

A key issue for us was that the KMP made provisions for adaptive management trials to improve the program, but there was a lack of clarity about the process for development and approval of the trial. One reason for this may be that the provisions included in the plan were developed largely to provide DECC with latitude in adapting its management actions rather than facilitating adaptive management by other managers of kangaroos, such as landholders, harvesters, processors or researchers such as FATE. It turned out to be costly, time consuming, and contained a strong bias, probably unintended, in favour of maintaining business as usual.

We began the process seeking a period of robust and engaged discussion before settling on a mutually-agreed plan that sought answers to management questions on behalf of the researchers and the regulators. After a short meeting it was suggested that we submit a detailed proposal that would then be assessed. In the absence of real engagement by DECC with this initial process, we set about second guessing what might be possible or acceptable. As a result, the initial proposal was longer than the entire 2002–2006 Kangaroo Management Plan, and much of it was unnecessary because of information that was subsequently revealed to us.

One of the difficulties in arguing for an adaptive management proposal such as ours was that two of our key goals – to make control of kangaroo grazing pressure more efficient and to carve out a sustainable economic role for landholders – are not stated goals of the KMP. This is not to say that control of kangaroo grazing pressure and the distribution of economic returns from the industry are not of concern to DECC, in fact the opposite is clearly true. The KMP has specific management actions designed to assist landholders in their goal of controlling kangaroo grazing pressure, such as the release of 'special quota' when zone quotas are exhausted but grazing pressure persists. It also has a longstanding policy of limiting the number of licensed processors in order to ensure their economic sustainability and has recently introduced a moratorium on new harvester licences for much the same reason, but neither of these policies nor their underlying aims is stated in the KMP.

The reasons why such aims of kangaroo management in NSW are omitted from the KMP are partly political. There has been a deliberate shift in rhetoric away from 'pest control' to 'sustainable use' and there is a need for DECC to portray itself as a manager of protected kangaroo species that is not influenced by the economic goals of industry participants. The lack of explicit goals in these areas also frees DECC from any obligation to monitor whether their desired outcomes (i.e. grazing pressure control and economic viability) are being achieved, or consider what policy changes might be warranted.

However, it also makes it very difficult to argue the case for adaptive management proposals such as ours, which seek to find ways to better deliver these unstated goals. As an example, despite having no stated goals regarding the economic participation of landholders in the industry, no monitoring of such participation and no policy prescriptions to enable such participation, DECC dismissed our initial proposal in August 2006 by stating that it "does not consider that the failure of most landholders to participate in the commercial kangaroo industry beyond providing access for licensed trappers is due to legislative or policy impediments".

We persevered despite these difficulties with a process which stretched well into a second year. It became evident that dealing with our proposal came on top of a significant workload for Kangaroo Management Program staff, suggesting that resourcing the process of adaptive management was deficient.

The relevant parts of the DECC 2002–2006 Kangaroo Management Program are reproduced below indented and in different font. Although brief, the guidelines are consistent with accepted practice in adaptive management. However, the negotiation process we entered revealed differences between accepted practice and what we were required to do to comply. The italicised text is a commentary on the inconsistencies we perceived in the process.

Adaptive Management Uncertainty is inherent in the management of natural systems. This is due to variation in system processes and limits to understanding of system functions. An adaptive management approach makes it possible to acknowledge this uncertainty and improve knowledge through controlled intervention and monitoring of outcomes. Management can then be modified in subsequent management programs.

Active Adaptive Management

Learning can be accelerated through management deliberately intervening in the system and providing contrasts between different management factors or land units. Proposals for active adaptive management experiments will need to be assessed by NPWS following review by the Advisory Panel and must comply with the following criteria:

1. **Awareness of relevant background information.**

    Negotiation revealed what we already knew – that the problem situation was complex and the available information had not generated a level of agreement that provided a basis for shared understanding and consensus in prediction. The result was several rounds of correspondence responding to the trial plan. Disagreements about the likely outcomes of the trial to us obviated the need for the trial. Instead we were challenged to try and provide the very evidence, prior to the trial, that we needed the trial to uncover.

2. **Consideration of alternative models/hypotheses.**

   The trial plan presented an alternative tag allocation model and a series of hypotheses that suggested how the new model might lead to other benefits. KMP expected us to come up with several different models and hypotheses from which we chose the best. After negotiation our model was accepted.

   A key issue was the requirements placed on the AM trial for providing evidence. Negotiation revealed an expectation of a traditional agronomic style experiment with controls. This was later accepted as inappropriate in the context following discussion of the reasons behind our research design, but the expectation suggested a reluctance to accept other research modes such as participatory action research.

3. **A monitoring program must be described.**

   This part was problematic because of the uncertainties inherent in adaptive management generated by the complexity of the situation. Our initial naïve expectation was that we might be able to generate landscape function differences in the life of the trial; we quickly came to realise this was unrealistic. As the behaviour of people was the key aspect in which we anticipated change, our monitoring needed to reflect that. The resolution came through refining our hypotheses to reflect only the change we were implementing and monitoring stakeholder responses to it.

4. **Critical evaluation of the merits of every experiment/proposal including evidence that risk of permanent damage to kangaroo populations is low.**

   We were able to comply with this easily because our intervention did not alter the rate of kangaroo harvest.

5. **Consistent with the Program Goal.**

   There were no issues with this criterion.

6. **As understanding of the system improves, consideration (should be given) of how management may be modified to accommodate the new knowledge gathered from the intervention.**

Dissemination of research findings and the results of this program's monitoring activities will contribute to achieving the program's overall goal of management within the framework of *ecologically sustainable development* principles. Implementation of an adaptive management experiment that affects commercial utilisation would, in addition, need to demonstrate how it provides for reasonable business planning and investment decisions.

This was always incorporated into our plan.

To complicate matters, we prepared our initial proposal under the 2002–2006 Kangaroo Management Plan, but were approved under the 2007–2011 Commercial

Kangaroo Harvest Management Plan. Under the new plan adaptive management is dealt with less thoroughly and the requirements for researchers submitting under the adaptive management trial provisions are more clearly defined but less compatible with accepted views of the nature of adaptive management (Department of Environment and Conservation NSW, 2006).

Evaluating the Project

Regarding the factors listed in Chapter 2 that have been identified as being influential in the success of adaptive management activities, the following comments are relevant to this project:

The presence of a 'champion' for the activity:

FATE has in a sense been the champion for the activity up until now, but the emergence of a champion from within the group of landholders and harvesters is critical to its future success.

The previous history of management or dispute between parties:

A key factor in stimulating landholders to engage with the project was a first hand report from the 'champion' of a previous attempt by a group of Western Division landholders to develop a kangaroo enterprise. The failure of this attempt was attributed to the attitude and behaviour of processors. This generated sufficient anger to motivate the landholders to learn from the experience and devise their own approach.

The political climate, and structure of the participating institutions:

During the life of the project, the potential role of kangaroos in carbon pollution reduction came to prominence due to publicity associated with the release of research into the greenhouse gas benefits that could be generated by a shift towards kangaroos in the rangelands (Wilson & Edwards, 2008). The potential of kangaroos to contribute was picked up by The Garnaut Climate Change Review Final Report 2 (Garnaut, 2008) which generated further publicity and interest.

The RIRDC Rangelands and Wildlife Program was the ideal funding vehicle for the trial, and the support provided to the FATE Program by the Faculty of Science at UNSW was critical. The leadership of BARG was open and inclusive. The responsiveness of the WCMA to the project resulted in that organization administering and supporting the employment of the local research officer. The level of resourcing of the Kangaroo Management Program at DECC and the lack of clarity in their guidelines for developing a trial under their adaptive management provisions caused difficulties that were, with goodwill on both sides, ultimately overcome. However, a key lesson learnt is the barrier to adaptive management that can be posed by a lack of clearly articulated management goals, particularly if political concerns prevent goals being stated upfront.

Extreme natural events such as tropical storms, droughts or floods:

The entire research area has been in the grip of drought throughout the period of the research. This has increased the level of stress on landholders, reducing their availability to attend meetings and engage with a peripheral issues compared to their survival as landholders. It has also caused a significant drop in kangaroo numbers resulting in a reduced harvest.

The occurrence of a significant kangaroo influx on a participating property would greatly increase the amount of evidence that can be collected, but clearly we have no control over this.

While the funded part of the project will conclude in June 2009, the ultimate measure of success will be whether the landholder and harvester group maintains the group licencing trial and proceeds with setting up a collaborative business. Along the way, it is clear that the adaptive management framework adopted for the overall project has contributed significantly to maintaining progress on the journey.

References

Ampt, P. & Baumber, A. (2006). Building connections between kangaroos, commerce and conservation in the rangelands. *Australian Zoologist, 33*(3), 398–409.

Ampt, P. & Owen, K. (2008). *Consumer attitudes to kangaroo meat products*. Barton, ACT: Rural Industries Research and Development Corporation.

Chapman, M. (2003). Kangaroos and feral goats as economic resources for graziers: Some views from south-west Queensland. *Rangeland Journal, 25*(1), 20–23.

Cooney, R., Baumber, A., Ampt, P. & Wilson, G. (2009). Sharing skippy: Models for involving landholders in kangaroo management in Australia. *Rangeland Journal*, 31(3).

Department of Environment and Conservation (NSW) (2006). *New South Wales Commercial Kangaroo Harvest Management Plan 2007–2011*. Sydney: Department of Environment and Conservation (NSW).

Eldridge, D. & Grant, R. (2004a). *Rangeland Change in the Hard Red Range-Type [Brochure]*. Sydney, NSW: Department of Infrastructure, Planning and Natural Resources.

Eldridge, D. & Grant, R. (2004b). *Rangeland Change in the Western Riverina Saltbush Range-Type*. Sydney, NSW: Department of Infrastructure, Planning and Natural Resources.

Fisher, A. & Eyre, T. (2008). Developing a collaborative biodiversity monitoring program for Australian rangelands. *A Climate of Change in the Rangelands: 15th Conference of the Australian Rangelands Society*. Charters Towers, Queensland: Conference Organising Committee.

Garnaut, R. (2008). The Garnaut climate change review: Final report. Available from http://www.garnautreview.org.au.

Grigg, G. (1987). Kangaroos: A better economic base for our marginal grazing lands? *Australian Zoologist, 24*(1), 73–80.

Grigg, G. (1989). Kangaroo harvesting and the conservation of arid and semi-arid rangelands. *Conservation Biology, 3*(2), 194–198.

Hacker, R. B., McLeod, S., Druhan, J., Tenhumberg, B., & Pradhan, U. (2004). *Kangaroo Management Options in the Murray-Darling Basin*. Canberra, ACT: Murray-Darling Basin Commission and NSW Agriculture.

Thomsen, D. & Davies, J. (2007). *People and the Kangaroo Harvest in the South Australian Rangelands: Social and Institutional Considerations for Kangaroo Management and the Kangaroo Industry*. Barton, ACT: Rural Industries Research and Development Corporation.

Tongway, D.J. & Hindley, N.L. (2004). *Landscape Function Analysis Manual: Procedures for Monitoring and Assessing Landscapes with Special Reference to Minesites and Rangelands*. Canberra, ACT: CSIRO Sustainable Ecosystems.

Wilson, G.R. & Edwards, M.J. (2008). Native wildlife on rangelands to minimize methane and produce lower-emission meat: Kangaroos versus livestock. *Conservation Letters, 1*, 119–128.

Chapter 6
Learning About the Social Elements of Adaptive Management in the South Island Tussock Grasslands of New Zealand

Will Allen and Chris Jacobson

Abstract Adaptive management initiatives are frequently used in multi-stakeholder situations. The more immediate barriers to success in these cases are proving to be organizational and social. We use a case study set in the South Island tussock grasslands of New Zealand to reflect on some of the social elements required to support ongoing collaborative monitoring and adaptive management. We begin by siting the case study within its wider policy context to show how this influences the choice and application of scientific inquiry. The next section concentrates particularly on the processes by which information and knowledge are shared across the different stakeholder groups involved. Finally, we expand on some specific lessons that emerge as important for sharing information and knowledge in adaptive management, including tools to support dialogue and improved tools for evaluation.

Introduction

Although adaptive management approaches have been advocated for environmental management for around 40 years (Holling, 1978; Walters & Hilborn, 1978), their success in practice has been less than spectacular. There is a growing appreciation that the more immediate barriers are organisational and social, rather than technical, given the multi-stakeholder nature of most environmental situations (McLain & Lee, 1996; Dovers & Mobbs, 1997; Gregory et al. 2006). These barriers include a tendency to discount non-scientific forms of knowledge, institutional cultures within research and policymaking that work against genuinely participatory approaches, and a failure to provide appropriate processes to promote the development of shared understandings among diverse stakeholders (e.g. Campbell, 1995; Pretty, 1998; Stankey et al., 2005; Feldman, 2008).

W. Allen
Landcare Research, P.O. Box 40, Lincoln 7640, New Zealand

C. Jacobson
School of Natural and Rural Systems Management, The University of Queensland, Gatton, Queensland, Australia

Another problem we face as we try to develop the next generation of adaptive management programmes is that these initiatives are often multifaceted, unfolding over timescales that are longer than a single project or programme cycle. Unfortunately this characteristic means it is difficult to easily evaluate the success of the programmes. Adaptive management programmes generally include a number of learn–describe–predict–act cycles that should unfold over the 5–15 years of a policy cycle (Raadgever et al., 2008). In some cases, adaptive management programmes may only progress some of the way through these steps before the next policy or management issue overtakes them, and the original programme fades into obscurity. Alternatively, we find ourselves learning lessons from adaptive management programmes that are artificially squeezed into too short a time frame. In the latter, the language and steps inherent in adaptive management are often put in place, but the essence of reflective and scientifically robust discussion and adaptation is missed (Gregory et al., 2006).

We begin this chapter by outlining the social context of tussock grassland management in the South Island, New Zealand. Some sense of the major framings of high country issues over the past three decades are provided – in particular the ongoing emphasis on sustainability, along with an interest in monitoring during the 1990s and tenure review in more recent years. Activities that occurred between 1994 and 2000 were targeted to support adaptive management and increase understanding about the potential outcomes of alternative management strategies for this area of New Zealand. Using these experiences, we reflect on some of the social elements required to support an ongoing collaborative monitoring and adaptive management programme. We concentrate particularly on activities related to sharing information and knowledge across the different stakeholder groups involved. Finally we expand on some specific lessons that emerge as important for sharing information and knowledge in adaptive management.

Case Study Context

Policy Setting

Agriculture represents an important interface between people and their environment. The tussock grasslands of the South Island of New Zealand run up the eastern slopes of the Southern Alps, and are commonly referred to as the South Island "high country". These grasslands are renowned for producing high quality meat and wool for export. At the same time they represent a microcosm of the major resource management issues surrounding extensively grazed ecosystems worldwide. These lands have been used for extensive pastoral management since European settlement in the mid-1800s under leasehold tenure. As O'Connor (2003) points out, high country sheep runs (properties) remained as Crown pastoral leases for a variety of reasons, including climatic, topographic and politico-economic value. In the early

1990s around 350 pastoral leases existed, covering about 2.4 million hectares of land (Walker et al., 2006).

Over the past three decades the high country has shared the worldwide trend of moving towards a more holistic, multi-use, multi-value view of such extensively grazed grasslands (Allen, 2001). Grazing by sheep has increasingly become a variable component, or even been abandoned in some areas. This change highlights the diverse management values that grasslands are now expected to serve. In New Zealand these not only encompass traditional pastoral considerations but extend to national aspirations concerning issues such as indigenous Māori land rights, preservation of biodiversity and natural landscapes, sustainable management, tourism, and recreation.

As these values have gained recognition, high country resource use has been characterized by tensions between different interest groups (Allen, 1997).While changing the social worldview which underlies land use practices and management may appear a daunting task, Bawden (1991) reminds us that we should recognise that it is something that happens quite regularly in response to different societal concerns and aspirations. So marked are these changes in many rural areas in countries such as Australia and New Zealand that he suggests we can identify several different perspectives of rural land management since European settlement. These different perspectives are outlined in Fig 6.1 as they relate to the South Island high country since World War II. However, as Bawden (1991) points out, these issues are more complicated than they appear because each emerging perspective (or world view) complements rather than replaces its predecessors, making for increased complexity in resource management.

The first worldview apparent in contemporary high country management was about *production*. Until the 1980s, those working in the high country were at least confident in the knowledge they were dealing with what everyone knew was a

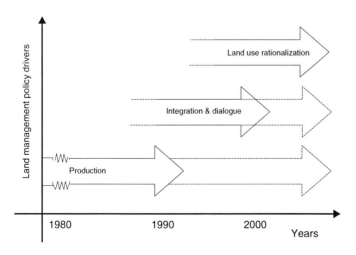

Fig. 6.1 Different worldviews that drive change in the high country

largely extensive pastoral system (Allen, 1997). This manifested itself as a production system, with a science input oriented towards improving that production. The removal of agricultural sector subsidies by the New Zealand government in the mid-1980s encouraged farmers and policy-makers to rethink the viability of extensive pastoralism as the main land-use.

The second predominant worldview we draw attention to manifests itself in *integration and dialogue*, and stems from a growing interest in sustainability. In the 1990s a number of changes happened which encouraged an emphasis on sustainability issues. New Zealand redesigned its environmental legislation. The Resource Management Act (1991) replaced around 50 previous statutes. It was notable in being one of the earliest pieces of legislation to explicitly incorporate 'sustainable management' as the purpose placed at the heart of the regulatory frameworks for resource management (Harris, 1993).

Alongside this change was a national-policy-level focus on the high country, particularly the semi-arid regions, which raised questions around economic and ecological sustainability. Concerns included land degradation, weeds (particularly *Hieracium* spp., an introduced forb), pests (particularly rabbits) and the ability of farmers to manage for market and climatic variability (Martin et al., 1994). The Rabbit and Land Management Programme (RLMP) of 1986–1996 was established to address problems of the semi-arid high country regions, while landcare groups arose in the early 1990s to promote sustainable management of rural communities through environmental, economic and social reforms (Mark, 2004). The RLMP took advantage of this to support high country families to work in landcare groups. Each group took an approach most suited to their local environment, but more importantly one based on their interpretation of a range of information (Ricketts, 2001).

In more recent years, policy efforts have been focused on *land use rationalization*. As a result of tenure review processes, a return to Crown management has occurred for lands where significant inherent values (predominantly indigenous biodiversity) exist, while tenure on the remaining pastoral leasehold land has been freed up for economic use (McFarlane, 2008). While this conversation doesn't preclude consideration of multiple uses, it has tended to be more focused on positions than interests. However, this approach tends to support negotiations that are based around positional bargaining and support compromise around existing uses, rather than encourage joint exploration of new ways forward (Walkerden, 2006).

Developing an Adaptive Management Approach

The Semi-Arid Lands (SAL) research programme (MAF, 1996) was developed within the Rabbit and Land Management Programme. Most of the work was carried out between 1994 and 2000, and was designed to support the integration and dialogue worldview that was driving high country discussions of the time. The SAL research team comprised around five to seven scientists representing disciplines covering plant, landscape and wildlife ecology, and including social systems. Both authors have been involved in this research, and the lead author in this chapter was the primary social researcher in the SAL team.

Key to the SAL approach was recognition that the development of sustainable management (e.g. grazing) strategies requires an emphasis on experimental rather than descriptive ecology, and this required learning from large-scale management (experiments) by farmers, in addition to more detailed research experiments carried out by scientists. A report by the Parliamentary Commissioner for the Environment (1995) stated that ongoing monitoring by land managers was essential to increase the understanding of issues affecting tussock grasslands. The same report also stressed that decision makers and land managers needed to promote and adopt management approaches that were based on both research and monitoring. In 1994 the High Country Committee of Federated Farmers put together a farmer resource kit that detailed various monitoring methods farming families could use on their properties.

In response to these calls, one component of the SAL programme, the Hieracium Management Programme (HMP), emphasised an adaptive management process to more closely link research with management and policy. The wider benefits were seen as increased information sharing and dialogue among the different sector groups (e.g. farmers, scientists, policy managers) that collectively contributed to high country decision-making. More specifically, the Hieracium Management Programme (HMP) was initiated to encourage adaptive management as an approach to addressing an invasive weed, and improving understanding of the tussock grasslands in the high country. The programme had two main strands: the first brought together and integrated existing local and scientific knowledge, and the second involved development of a monitoring programme that could be used to learn from farmer experience.

The first strand included activities around accessing existing farmer and science information through the use of interviews and questionnaires, synthesising this information, and then holding workshops (or community dialogue processes) that would more actively involve farmers and researchers in developing the structure and content of a first-version decision support system that made use of this information. This activity relates to the information sharing component of adaptive managemnet described in Chapter 2 of this volume. The second strand ran concurrently and focused on development of a farmer-friendly monitoring system for use in the tussock grasslands. A project linking researchers and farmers in the development of condition assessment models for measuring (monitoring) and interpreting vegetation change was developed (Bosch et al., 1996a, b; Gibson & Bosch, 1996). With the outputs of these two strands – an integrated knowledge system and user-friendly monitoring tools – the research team (perhaps naively) thought that the hardest work of establishing the conditions for a community-based adaptive management programme, which would enable the use of local knowledge and the adoption of a continual enhancement process to information management, had been achieved. Instead, the search for ways to support such a programme continues today. Exploration of the social and institutional issues involved in the SAL/HMP project has provided ample grounds upon which to reflect on the practice of adaptive management. Rather than guiding readers through a traditional case study that describes the adaptive management process at each step, this case study focuses on the roles of information sharing, engagement and dialogue in supporting adaptive management.

Integrated Systems for Knowledge Management (ISKM)

The Integrated Systems for Knowledge Management framework (ISKM) (Bosch et al., 1996c, 2003; Allen et al., 1998a, 2001) was used in this case study. ISKM (Fig 6.2) focuses on strengthening participation and self-help in natural resource management projects. As such, it is not a new project type or innovative development concept, but rather a specific approach that emphasises a number of key steps applicable to developing the knowledge and action needed to address problem situations in a constructive way.

The ISKM framework is designed around the steps of adaptive management. Two phases are involved: the first supports finding out about a situation and the second aims to take action to improve the situation. Activities associated with the first phase involve establishing a climate for change with the different parties involved, setting goals and objectives (including joint problem framing), searching for information, developing a shared understanding and creating action plans to address the issue at hand. Monitoring plans also need to be developed to monitor progress and help check that the action plans remain on-track. The final activity in this first phase of ISKM involves the development of a management information system that captures decision-making information for the benefit of the wider community of stakeholders. Computer technology is often relevant at this stage as it offers a way of

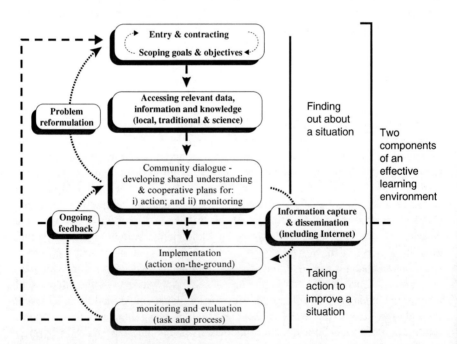

Fig. 6.2 ISKM – a participatory research framework to facilitate the identification and introduction of more sustainable resource management practices. The two phases interact to create an effective learning environment (Allen, 2001)

organising information in ways that make it easily accessible to a range of audiences. The second phase of ISKM stresses the need to develop feedback loops to maximise the benefits from monitoring and evaluation and develop a collaborative-learning/ self-improving environment.

The skills required for managing the process involved in ISKM will naturally vary according to the specifics of the initiative. For instance, there is a substantial difference between pursuing a collaborative approach within an already well-functioning situation and trying to initiate collaboration in a social environment characterized by existing conflict. In the latter, the need for effective facilitation and expert mediation of conflicts is greater.

ISKM Phase 1: Finding Out About the Situation

Entry and Contracting

This first phase of ISKM includes making sure the right people are involved and establishing the ground rules for working together. However, while open access to a collaborative effort is important symbolically, making sure that key stakeholder representatives are involved is critical (Wondolleck & Yaffee, 2000). This is because some people are more suited to, and interested in, participating in a collaborative inquiry than others. As Bunning (1995) points out, the reality of that because of current global pressures (e.g. production squeeze, downsizing, reduction in organisational levels, increased accountability) there are higher levels of stress and pressure around than ever before. While it is precisely those symptoms that indicate that change and development is needed, if people are not provided with the capacity to participate successful change is unlikely to be developed. Thus more will be learnt by a few genuinely committed co-researchers dedicated to exploring change within a smaller case study approach, than may be gained by engaging with a larger number of less willing participants in a bigger inquiry (Allen, 2001) (see also Chapter 18, Fazey and Schultz, this volume).

It is important to cultivate relationships that make it easy for people to talk about their needs, share information, and work together. Previous experience is one of the most important influences on attitudes to collaboration. People may be extremely reluctant to enter into a further participatory process if they have been involved in an unsuccessful one in the past – "we've already tried that and look what happened!" The emotional part of the conflict (which often forms a hidden barrier to uncovering the real issues) may have to be dealt with first. Equally, as people begin to work together successfully in new systems, so trust manifests itself as an emergent property of the new way of working. Rules for working together should not be imposed from outside, but should be developed in conjunction with the people involved, as described in the box on trust in Chapter 4 of this volume.

The SAL/HMP began its adaptive management initiative in the tussock grasslands by convening a steering committee including scientists and farmers.

Much of the initial contact with farmers was done through the farmers on the committee, who effectively provided the researchers with personal introductions. Conversely, when the funding by the Ministry of Agriculture finished and subsequent funding from the Ministry of Environment redirected the focus towards conservation issues (with a new title, the Tussocks Grasslands programme) the research team became acutely aware that existing relationships built with the farming and local government communities would have to be extended. This necessitated finding the time to develop new relationships with agencies such as the Department of Conservation (DOC) who have responsibility for managing public conservation lands within the high country. Ensuring that enough time is given to developing the right relationships was one of the most important lessons from this exercise.

Accessing Information

Changing land management systems requires the many parties involved to change the way they work with the land and with each other. It is important to acknowledge the validity of different worldviews and concentrate on showing how individuals with different worldviews can work together. Information often remains fragmented because we do not have the mechanisms to collect it. However, as Allen and Kilvington (2005) point out, strong emotions associated with information can also create a barrier to its availability. Among science researchers, much personal self-worth and commercial worth is linked to the information generated. Fear over misrepresentation affects the willingness of researchers to offer their information for use in situations over which they have no future control. Other stakeholders may have similar fears that their information might be used inappropriately, or against them, if released publicly. Consequently, the exchange of information between different levels and groups in society is often inadequate.

Years of experimentation with different sheep stocking rates and other management regimes have provided individual high country managers with much knowledge about local land-use and environmental systems. Unfortunately, this knowledge resides in the heads' of farmers, and is seldom available to the wider community on a collective basis. Similarly, much of the valuable knowledge accumulated by scientists was fragmented, held in different databases and, consequently, was not readily available, even to other scientists. The need to develop protocols that safeguard information use, and protect against its misuse emerged as a way to address these concerns. Farmer interviews were undertaken with clear expectations on both sides on how the information would be used, and safeguarded. How farmer derived monitoring information might be used was also a topic for discussion. Table 6.1 provides an example of a protocol that was developed by a Landcare group to agree on how they would use their monitoring information.

Concerns similar to those of farmers were raised regarding the use of information from research sites on private land. In one case during the programme, access to sites in one farm cluster was denied, largely because farmers were unsure about what use would be made of the subsequent research findings. However, because

Table 6.1 Draft protocol for monitoring information sharing (Allen et al., 2001)

To specify data ownership:
Information stored on central database is the property of the group and individual owner, and to be controlled by the land management group or its agent.

To protect individual privacy:
The site data and property identification are to be coded to retain anonymity and are not to be divulged to third parties without the property owner's consent.

To enable the benefits of sharing data within the group:
However, unless otherwise specified by the individual, pooled results can be released in summary form.

To provide for working in with other parties (e.g. local government):
Where joint/collaborative arrangements with third parties exist, then third parties share ownership and access to the results for the sole purpose of that specified in the arrangement.

the project process was prepared to address this conflict, with appropriate skills for conflict resolution, the situation was able to be resolved (Allen & Kilvington, 2005). A subsequent conflict management exercise resulted in the establishment of information management protocols that enabled the research to proceed. These protocols protected the rights of landowners to be advised of research results prior to their release to third parties, and provided for discussions of the implications of research results for different stakeholders involved, before publication.

Community Dialogue

Enormous gains can be made by promoting an understanding of what different stakeholders and other groups, such as local land managers or indigenous people, have to offer to the resolution of complex environmental problems (Bosch et al., 2003). However, there is often an understandable reluctance on the part of agency and research staff to bring together factions where there is a risk, or perceived risk, of conflict. For example, staff in most, if not all, of the high country research initiatives that preceded this case study tended to work separately with government conservation management staff and local farming families (who collectively manage all the tussock grasslands), or solely with one or other group, largely to avoid having to deal with possible conflict (Allen, 1997). Given that one of the main land-use debates revolves around determining trade-offs and synergies between conservation and pastoralism, there is little doubt that both groups would have been better served by science had they been provided with more, well-facilitated opportunities to come together and discuss the implications of emerging research findings.

Information may have different meanings and hence values in different situations. Making sense of information has two principal components. First, all stakeholders must agree and clearly understand the intended use of the information. This may, for example, be to resolve a particular environmental problem or to attain a particular resource management goal. Second, the context within which the information was originally collected is a key to its strengths and weaknesses. Addressing this requires clarifying issues such as why the information was collected and by whom; its source (e.g.

practical experience, observations, science research etc.); whether the information relates to a specific situation or site and whether it can be extrapolated to other situations. Skilled facilitation is needed to ensure that all participants and stakeholders share a common understanding of these components of new information.

For the tussock grasslands Allen and Bosch (1996) point out that scientists concentrated on determining the effects of grazing on *Hieracium* (describing and accounting for some phenomenon). In contrast, farmers asked more focused questions such as the effects of different grazing regimes (rotational grazing vs set stocking, different grazing intensities and frequencies), and were concerned with applying the answers to their own context. Similarly conservators often place a high priority on protecting individual species – such as a rare lizard. On the other hand, farmers are unlikely to identify the same species if asked to list conservation issues in order of importance (Allen & Bosch, 1996). As part of this programme, a number of workshops were held that brought together different people and groups that provided their information, knowledge and experience gained in the tussock grasslands. The collective discussions that ensued helped groups make sense of others' contributed information, and enabled it to be understood in the context in which it was generated. Ensuring that information is understood "in context" is a main reason that scientists and others are often reluctant to share their data until they are confident these have been understood and interpreted.

Information Capture and Dissemination

Using collaborative approaches provides all those directly involved with an environment in which information is synthesized through a participatory process (Allen & Bosch, 1996). At the workshops, participants clarified management questions, sorted information on the basis of its applicability to addressing these, and identified the starting points for stakeholder-specific information needs. Essentially, this provided a way of understanding information relevant to the entire high country and the management of multiple, sometimes competing values. It was then possible to develop a management information system (MIS) that served to integrate collected information and organize it in a way that matched the questions asked by land managers, so that it could benefit others who have not had the opportunity to be directly involved in the ISKM process. The resulting Internet-based Tussock Grasslands Management Information System – TGMIS (2000) provided background ecological knowledge and best practice guidelines for managing different vegetation states. It was designed as an open-ended system that could be continually updated as new information became available through research and monitoring (Bosch et al., 1999). It drew on farmer, conservation manager and science knowledge that had been discussed at forums with representatives of these different groups.

Underlying the need to develop the MIS system is the need to look beyond a presenting symptom (in this case an exotic weed) to presenting information about the management of the wider system in which it is embedded (Bosch et al., 2003). Farmers do not manage for *Hieracium* alone, but are primarily concerned with managing for increased stock production or available forage supply, without degrading the system. Accordingly, the TGMIS provides information not only on *Hieracium*, but also on a whole range of inter-related management issues such as conservation, grazing management, burning and water quality. It brings information from many different sources together into one place for easy access by land managers, policy-makers, researchers and other interest groups. Importantly, dissenting opinions are not dismissed, but are included with a descriptor of the variety of existing perspectives and, where appropriate, acknowledgement that they are a minority opinion.

ISKM Phase 2: Taking Action to Improve the Situation

Development of the MIS provided a link between the two phases of ISKM: finding out about a situation and taking action to improve it. The ability to access the MIS was one way of supporting the ability of land managers to take action to improve the management of their grasslands. Support for ongoing farmer-based monitoring was also provided through a concurrent research project involving scientists and farmers in the development of Condition Assessment Models for measuring (monitoring) and interpreting vegetation change in the different ecological areas within the tussock grasslands (Gibson & Bosch, 1996). This information was contained in a user-friendly computer tool (REDIS) that enables land managers to interpret the results of monitoring by indicating where a particular site is situated along a condition gradient (Gibson & Bosch, 1999). These models were subsequently made available to individual land managers through Landcare groups in the high country. Training was provided to help land managers identify key indicator plant species and to use the software package. REDIS was also made available through the Tussock Grassland Management Information System TGMIS (2000).

Funding for the SAL/HMP project finished in late 2000. The TGMIS was subsequently evaluated in 2001 and 2002. Copies of these reports are available online (Jacobson, 2001, 2002). At that stage, indications were that the website was being used by Department of Conservation staff, and to a lesser extent by local government agency staff. Although participants noted that much of the background information was not new to those with a history of involvement in the high country, they valued MIS' potential to provide research summaries, including summaries of new research. Farming participants noted that the website would be of particular use when changing a management regime, when diversifying practices, applying for resource consents (to undertake different land uses on pastoral lease land), or when unusual observations were made. Publicly available website statistics have been maintained for the front page of the website since its launch in September 2000. Since that date more than 36,000 visits have been made to this page of the website. The site receives substantially more visits from New Zealand Internet-users, than it receives from any other country. On average, the MIS has been visited more than 260 times each month over the past eight years. The highest monthly number of visits was 672 in November 2004, more than three years after the site launch.

Broader Lessons on Sharing Information and Tracking Progress

This case study has highlighted some aspects of collaboration in adaptive management that are not commonly discussed. Two lessons were particularly evident: (1) the need to use good facilitation tools and processes to help people share information, and (2) the need to move from the term 'monitoring' to 'evaluation'.

Models and Pictures for Information Sharing

Information gathering and sharing is not just a matter of asking people what they know and then passing the information on. Frameworks, pictures and representations are powerful aids to help people unlock and discuss the information and experience they have with others (Heemskerk et al., 2003). This process is best described as a form of participatory modelling (Heemskerk et al., 2003; Lynam et al., 2007). By using modelling processes, we begin to expand the use and richness of the word 'model' in the adaptive management literature beyond that of quantitative systems modelling (e.g. Walters, 1986) or even that of Bayesian predictive modelling (e.g. Johnson & Williams, 1999) to one of helping people sort out and represent different forms of knowledge.

Importance of Pictures

The idea of bringing people together to develop a common understanding of issues and what an appropriate set of responses might be sounds easy enough. In practice, one of the main challenges turned out to be the development of a common language. To illustrate this with a simple, but crucial, example, it became apparent at the workshops previously described that everyone had their own idea of what different 'states' of tussock grassland were. Some people regarded short tussock grassland as being up to the top of their work-boots, while others regarded it as being more akin to the height of gumboots. Finally, a successful – but unplanned – solution was developed when one of the ecologists sketched out the diagram shown in Fig. 6.3.

Jointly developing models helps participants clarify the system boundaries, formulate questions, and reveal assumptions of the different people involved. The most difficult communication gaps to bridge are those between science disciplines. Similarly, it can be extraordinarily difficult to get managers to set out the underlying knowledge behind their practices. Many of these practices are highly contextual, and it is necessary to find ways to help them express this. During the workshops described earlier, a decision tree approach was used to unlock and structure existing knowledge (Bosch et al., 1999). An example illustrating a completed version of

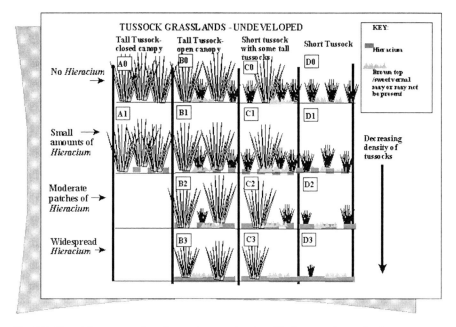

Fig. 6.3 Pictorial representation of tussock states developed in the tussock grassland management workshops (Allen, 2001)

such a decision tree is given in Fig 6.4. This example also illustrates the holistic way in which the programme sought to deal with the exotic weed *Hieracium* spp. That is, farmers do not manage for the weed alone, but rather address it as one problem within a wider goal – in this case as part of the management of a tussock grassland community.

The session began by defining the management goals and targets. These are written on the left-hand side, and participants are asked how they would achieve these goals (from their own experience and knowledge). The various options and best management practices are listed on the right-hand side as participants supply them. Once this is done, the facilitator returns to the top of the options/actions list, and initiates a second round of discussions among participants with a question such as, "To achieve goal x, could you use this option or strategy under all circumstances and conditions?". This process is repeated for all options or actions for each management goal initially identified. The decision trees, additional information, and question marks form the basis for further refinement with knowledge from scientists and other experts, the identification of questions and research gaps, and for easy processing into manuals or computerised information systems. Participants (end-users) are able to see their inputs in the design and content of the final information system. An important principle, however, is never to summarly dismiss any piece of information given by an individual, even if most participants disagree on its applicability (Bosch et al., 2003).

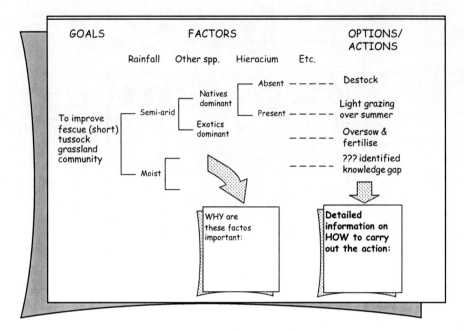

Fig. 6.4 Example of a decision tree being developed during a workshop, indicating the points at which "why" and "how" questions could unlock information in a structured way (Allen, 2001)

Looking for Measures of Success

All too often we think of agricultural and natural resource management projects and programmes as being vehicles for the provision of a particular on-the-ground outcome. However, what became very clear in this case study was that there are many stakeholders, all with different perspectives, involved in an endeavour that is unfolding over time. Accordingly we found that we needed to develop new evaluation approaches that recognized this.

Multistakeholder situations like the high country challenge the common perception of what a "programme" is. A multistakeholder perspective clearly recognizes that each group of participants has its own viewpoint on an issue, and its own reasons for becoming involved in a project. As Schwedersky and Karkoschka (1994) point out, it is traditional to observe programmes within an operational cycle. However, to take into account the various perspectives and interests of the participants, it is necessary to look beyond this cycle. Inevitably, "the programme" can be regarded as a number of sub-projects, each of which is "steered" by a different group of participants in accordance with their values and aspirations, as was the case in the RLMP, which involved farmers, conservation managers, local government and other central agency staff, scientists and other interest groups.

While researchers and policymakers tend to concentrate on the environmental outcomes sought, it is easy to forget that much of the challenge of implementing integrated management within these wider situations lies in promoting change in the behaviour of the different user-groups, departments and even wider communities. Collaborative multistakeholder situations inevitably involve integration of multiple perspectives on the most significant values inherent in a landscape (e.g. production, conservation) and require a more holistic view of the problem as an interconnected system. Other changes may still occur (e.g. building capacity of communities) that are equally important if not the original intent of management programmes.

To evaluate programmes we need to go beyond judging success by primary outcome measurement, and look to evaluation frameworks that raise awareness of processes that contribute to them. Evaluation frameworks also need to help us evaluate over an appropriate timescale. One such approach for grouping the outcomes of an integrated governance initiative is known as the Orders of Outcomes model (Olsen, 2003). It highlights the importance of changes in state (such as better environmental or social outcomes), but recognizes that for each change in state, there are correlated changes in the behaviour of key human actors. Importantly, the model helps plan activities in sequence so they build on each other over time (Fig. 6.5).

First-order outcomes are the organisational conditions that must be present when we begin any programme to bring about a change such as those proposed by cross-theme policy frameworks. Together these form the "enabling conditions" that are required if policy frameworks are to be implemented successfully. *Second-order* outcomes are evidence of the successful implementation of a behaviour-change programme. They mark changes in the behaviour of individuals and organizational groups, and include evidence such as new forms of collaborative action among stakeholder groups, investments in infrastructure, and the behavioural changes of actors in response to policy, regulations, and by voluntary actions. *Third-order* outcomes are the socio-economic, structural, and environmental results that define the ultimate success or failure of the programme. These must be defined in unambiguous terms early on in any management process, although this is often not an easy task. Such terms

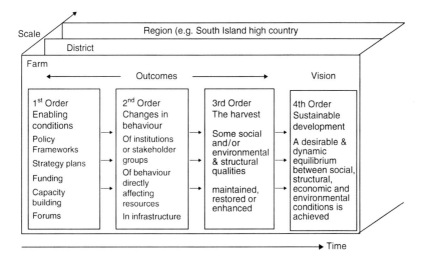

Fig. 6.5 Orders of outcome model approach to monitoring and evaluation (Adapted from Olsen, 2003)

could be, for example with reference to biodiversity issues, in the form of goals that specify the percentage of area we require in a specific tussock state. This long-term goal of sustainable high country development is recognized in the model as one of the *fourth-order* outcomes. Rather than being seen as an externally designed goal to be achieved, sustainability is better viewed as a desirable and dynamic relationship between environmental, social, and economic aspects.

In adaptive management, a focus on monitoring and analysis of that data results in a focus on the ecological processes, and judgements about "success" are often based on whether the results were used as a basis for management adaptation. While the "orders of change" approach to evaluation was not applied in the high country, it might have provided a more comprehensive way of evaluating a range of outcomes of the programme if it had been. In the high country, the SAL/HMP programme was only just beginning to develop the relationships and partnerships to address land management change in a constructive manner before the overarching policy environment changed dramatically with the instigation of tenure review. Today the perceived need to manage for multiple land use goals on single properties that characterised the integration and dialogue era (Fig. 6.1), has been sidelined as efforts to rationalize land-use through tenure review take centre stage.

However, a focus only on third-order land-use change misses the many other achievements that the SAL/HMP programme supported both within the high country, and in other areas. The most significant of the programme's high country successes revolve around capacity building and information sharing, and represent a mix of first- and second-order outcomes. For example the program clearly supported improvements in relationships between conservation managers and farming interests resulting from conflict management exercises (Allen et al., 1998b). In the same exercise new ground was broken by the community inviting a scientist to play a mediating role in supporting better communication and relationships. The Tussock Grasslands Management Information System represents one of the first Internet-based systems to link local and science knowledge (Allen et al., 2001a).

Outside of the high country, the programme can also point to other areas where the Integrated System for Knowledge Management (ISKM) approach has been used to support community-based learning initiatives. These areas include pest management in New Zealand (Allen et al., 2001b), learning about issues related to oil and gas in British Columbia, Canada (Booth et al., 2004), and understanding the links between land use practices and livelihoods around Lake Victoria in Africa (Albinus et al., 2008). The ISKM approach has also been used as an evaluation framework to look at an environmental health surveillance system in California (Abinader, 2004).

Concluding Comments

The goal this programme was addressing (i.e. changing land use patterns in recognition of the need for sustainable development) was quite substantial, even though the programme time period was substantially short. Changing perspectives on land use were evident over the course of the project. For example, diversifying land uses are apparent in the information requests received during an evaluation of the MIS. While the website originally included ecological management information (e.g. effects or fire, grazing and fertilizer on biodiversity), information on tourism impacts, viticulture and statutory processes, regarded as changes in land management practice, were sought (Jacobson, 2001). In this case study we can clearly point to the development of conditions that enable behaviour change. Many of these serve to build capacity, and so leave the wider communities and agency partners with more skills and relationships that can be used to take the process on across a range of fronts.

Collaborative adaptive approaches should be flexible, and designed to grow. It may be appropriate to defer involvement of reluctant stakeholders in the beginning, and new stakeholders may be identified along the way. It is always important to consider the timing for bringing groups together and, as mentioned previously, it may be more culturally appropriate and progressive to work separately with some groups at the commencement of a project, with a view to building collaboration or participation as the project evolves. Overall the process must be able to change to accommodate this growth. Community involvement helps create ownership and a feeling of accomplishment in working together to solve a problem. This group dynamic will encourage others from the community and government agencies to participate and provide and manage the information required for making decisions about sustainable resource use.

What is important is that skilled facilitation is used in adaptive management processes such as that described here. As Reed (2008) points out, highly skilled facilitation is particularly important for natural resource management given the high likelihood of dealing with conflict. To take up these challenges, interdisciplinary science approaches need to include personnel with complementary skills in the management of participation and conflict, and the integration of biophysical and social aspects of collaborative learning.

In combination, the lessons drawn from this case study have highlighted that in cases such as the high country where the management goal is long-term, adaptive management won't "solve" a problem. Funding for programmes such as that presented in this case study should be seen as part of policy directives that represent a changing interaction between society and the resources required to support it. In this sense, the programme contributed to raising awareness about sustainability, and highlighted two key issues needed for it to succeed: that informational integration and capacity of managers and scientists to work together are essential. Perhaps, most importantly, programmes such as those described here leave capacities that can and are used by communities and policy makers as building blocks to support the greater success of future adaptive management initiatives.

Acknowledgments Participatory adaptive management endeavours such as described here are not possible without the support and goodwill of all those involved, and we record our appreciation for the efforts of all those who have put their time and effort into the projects described here. Work on this chapter has been supported by the following New Zealand FRST-funded research programmes: Integrated Water Resource Management in Complex Catchments (ICM–Motueka) – CO9X0214, and Building Capacity for Sustainable Development – CO9X0310. We thank Grant Hunter and

Chrys Horn for their helpful and perceptive comments on early drafts of this paper. We would also like to acknowledge conversations had with Don Ross, Ken Hughey, Ockie Bosch, John Aspinall, Bruce Allan, Grant Norbury and Jean McFarlane to reflect on the changing perspectives in the high county and the legacy of the case study. Responsibility for the merging of these views into the paper is the authors' alone and we are not claiming that these people share our views.

References

Abinader, S. and Associates (2004) Evaluation report of the Senate Bill 702 Expert Working Group process and initial outcomes. Available on-line http://www.catracking.com/resources/ewg/sb702_evaluation_report.pdf (Downloaded 2 December 2008)

Albinus, M.P., Makalle, J.O., & Yazidhi B. (2008) Effects of land use practices on livelihoods in the transboundary sub-catchments of the Lake Victoria Basin. African Journal of Environmental Science and Technology 2(10): 309–317. Available on-line http://www.academicjournals.org/AJEST/PDF/pdf%202008/Oct/Albinus%20et%20al.pdf

Allen, W.J. (1997) Towards improving the role of evaluation within natural resource management *Development Studies* XVIII, Special Issue: 625–638

Allen, W.J. (2001) Working together for environmental management: the role of information sharing and collaborative learning. Ph.D. (Development Studies), Massey University. Available on-line athttp://learningforsustainability.net/research/thesis/thesis_contents.php

Allen, W.J. & Bosch, O.J.H. (1996) Shared experiences: the basis for a cooperative approach to identifying and implementing more sustainable land management practices. pp. 1–10 in Proceedings of Symposium "Resource management: Issues, visions, practice" Lincoln University, New Zealand, 5–8 July

Allen W.J. & Kilvington M.J. (2005) "Getting technical environmental information into watershed decision making." Chapter 3 in Ed. J.L. Hatfield *"The Farmers' Decision: Balancing Economic Successful Agriculture Production with Environmental Quality"* Publisher: Soil and Water Conservation Society. pp. 45–61. Available from http://www.learningforsustainability.net/pubs/AllenKilvington2005

Allen, W.J., Bosch, O.J.H., Gibson, R.G., & Jopp, A.J. (1998a) Co-learning our way to sustainability: An integrated and community-based research approach to support natural resource management decision-making. In: *Multiple objective decision making for land, water and environmental management*. pp. 51–59 (Eds: El-Swaify, S.A. & Yakowitz, D.S.). Boston, MA, USA: Lewis Publishers

Allen, W., Brown, K., Gloag, T., Morris, J., Simpson, K., Thomas, J., & Young, R. (1998b) Building partnerships for conservation in the Waitaki/Mackenzie basins. Landcare Research Contract Report LC9899/033, Lincoln, *New Zealand*. http://www.landcareresearch.co.nz/research/sustainablesoc/social/partnerships.asp

Allen, W.J., Bosch, O.J.H., Kilvington, M.J., Harley, D., & Brown I. (2001a) Monitoring and adaptive management: addressing social and organisational issues to improve information sharing. *Natural Resources Forum* 25(3): 225–233

Allen, W., Bosch, O., Kilvington, M., Oliver, J., & Gilbert, M. (2001b) Benefits of collaborative learning for environmental management: Applying the Integrated Systems for Knowledge Management approach to support animal pest control. *Environmental Management* 27(2): 215–223

Bawden, R.J. (1991) Towards action researching systems. In: *Action research for change and development*. pp. 21–51 (Ed.: Zuber-Skerritt, O.). Brisbane, Australia: Centre for the Advancement of Learning and Teaching, Griffith University

Booth, J., Layard, N., & Dale, N. (2004) A strategy for a community information, knowledge and learning system. Prepared for The University of Northern British Columbia's Northern Land Use institute, Northern Coastal and Research Programme.

Bosch, O.J.H., Allen, W.J., & Gibson, R.S. (1996a) Monitoring as an integral part of management and policy-making. In: *Proceedings of Symposium "Resource Management: Issues, Visions, Practice."* Lincoln University, New Zealand, 5–8 July. pp. 12–21

Bosch, O.J.H., Allen, W.J., & Gibson, R.S. (1996b) Monitoring as an integral part of management and policy making. In: *Proceedings of Symposium "Resource Management: Issues, Visions, Practice."* Lincoln University, New Zealand, 5–8 July. pp. 12–21. http://www.landcareresearch.co.nz/research/sustainablesoc/social/monpaper.asp

Bosch, O.J.H., Allen, W.J., Williams, J.M., & Ensor, A. (1996c) An integrated system for maximising community knowledge: Integrating community-based monitoring into the adaptive management process in the New Zealand high country. *The Rangeland Journal* 18(1): 23–32

Bosch, O., Allen, W., McGleish, W., & Knights, G. (1999) Integrating research and practice through information management and collaborative learning. In: *Proceedings 2nd International Conference on "Multiple Objective Decision Support Systems for Land, Water and Environmental Management (MODSS'99)."* Brisbane, Australia, August 1999

Bosch, O.J.H., Ross, A.H., & Beeton, R.S.J. (2003) Integrating science and management through collaborative learning and better information management. *Systems Research and Behavioural Science* 20: 107–118

Bunning, C. (1995) Professional development using action research. Action Learning, Action Research and Process Management Internet Conference, Bradford, England, MCB University Press

Campbell, A.C. (1995) Landcare: Participative Australian approaches to inquiry and learning for sustainability. *Journal of Soil and Water Conservation* 50: 125–131

Dovers, S.R. & Mobbs, C.D. (1997) An alluring prospect? Ecology, and the requirements of adaptive management. Chapter 4 in Frontiers in ecology: Building the links. In: *Proceedings, Conference of the "Ecological Society of Australia 1–3 October 1997."* Charles Sturt University, Elsevier, Oxford, UK

Feldman, D.L. (2008) Barriers to Adaptive Management: Lessons from the Apalachicola-Chattahoochee-Flint Compact. *Society and Natural Resources* 21(6): 512–525

Gibson, R.S. & Bosch, O.J.H. (1996) Indicator species for the interpretation of vegetation condition in the St. Bathans area, Central Otago, New Zealand. *New Zealand Journal of Ecology* 20(2): 163–172

Gibson, R.S. & Bosch, O.J.H. (1999) Resource and Environmental Data Interpretation System (REDIS) <http://redis.landcareresearch.co.nz/> (Accessed 20 November 2008)

Gregory, R., Failing, L., & Higgins, P. (2006) Adaptive management and environmental decision making: A case study application to water use planning. *Ecological Economics* 58: 434–447

Harris, B.V. (1993) Sustainable management as an express purpose of environmental legislation: the New Zealand attempt - *Otago Law Review* 8: 51–76

Heemskerk, M., Wilson, K., & Pavao-Zuckerman, M. (2003) Conceptual models as tools for communication across disciplines. *Conservation Ecology* 7(3): 8.

Holling, C.S. (Ed.) (1978) Adaptive environmental assessment and management. New York: Wiley.

Jacobson, C. (2001) *Meeting End User Needs: Extension and Evaluation of the Tussock Grasslands Management Information System.* Wildlife Management Series, Department of Zoology, University of Otago, New Zealand.

Jacobson, C. (2002) Tussock grassland MIS evaluation – April 2002. Tussock Grassland MIS http://www.tussocks.net.nz/evaluation2.html

Johnson, F. & Williams, K. (1999) Protocol and practice in the adaptive management of waterfowl harvests. *Conservation Ecology* 3(1), 8

Lynam, T., de Jong, W., Sheil, D., Kusumanto, T., & Evans, K. (2007) A review of tools for incorporating community knowledge, preferences, and values into decision making in natural resources management. *Ecology and Society* 12(1): 5.

MAF (1996) Semi-Arid Lands research: To develop co-operative, community based, and integrated research approaches for application to sustainable agriculture issues. Internet page in MAF Operational Research Results for 1995/96 – http://www.maf.govt.nz/mafnet/rural-nz/research-and-development/research-results/1995–1996/9596sc18.htm

Mark, A. (2004) Our golden landscapes: An historical perspective on the ecology and management of our tussock grasslands and associated mountain lands. Hocken Lecture, Botany Department, University of Otago. Available on-line at http://www.botany.otago.ac.nz/staff/markhocken.html

Martin, G., Garden, P., Meister, A., Penno, W., Sheath, G., Stephenson, G., Urquart, R., Mulcock, C., & Lough, R. (1994) *South Island high country review. Final report of the working party on sustainable land management*, South Island High Country Review Working Party, Wellington

McFarlane, J. (2008) The social construction of ecological sustainability in the South Island, New Zealand, high country. Paper presented at the International Symposium on Society and Natural Resources.

McLain, R. & Lee, R. (1996) Adaptive management: promises and pitfalls *Journal of Environmental Management* 20: 437–448

O'Connor, K.F. (2003) Conflicting innovations: A problem for sustainable development of New Zealand high country grasslands. *Mountain Research and Development* 23(2): 104–109

Olsen, S.B. (2003) Frameworks and indicators for assessing progress in integrated coastal management initiatives. *Ocean & Coastal Management* 46: 347–361

Parliamentary Commissioner for the Environment (1995) *A review of the government system for managing the South Island tussock grasslands: with particular reference to tussock burning*. Published report of the Office of the Parliamentary Commissioner of the Environment, Wellington, New Zealand

Pretty, J. (1998) Participatory learning for integrated farming. Available on-line at http://www.orphanmissions.com/documents/integratedfarming.doc (Accessed 20 November 2008)

Raadgever, G.T., Mostert, E., Kranz, N., Interwies, E., & Timmerman, J.G. (2008) Assessing management regimes in transboundary river basins: do they support adaptive management? *Ecology and Society* 13(1): 14

Reed, M. (2008) Stakeholder participation for environmental management: A literature review. *Biological Conservation* 141: 2417–2431

Ricketts, H. (2001) Sustainable Land Management Through Community Involvement: The NZ Landcare Trust Experience. Paper presented to the International Community Development Conference, Rotorua (April 2001) Available http://www.iacdglobal.org/files/ricketts.pdf

Schwedersky, T. & Karkoschka, O. (1994) Process Monitoring (ProM): Work document for project staff, Eschborn, Deutsche Gesellschaft fr Technische Zusammenarbeit (GTZ) GmbH

Stankey, G.H., Clark, R.N., & Bormann, B.T. (2005) *Adaptive management of natural resources: theory, concepts, and management institutions*. Gen. Tech. Rep. PNW-GTR-654. Portland, OR: U.S. Department of Agriculture, Forest Service, Pacific Northwest Research Station. 73 p.

Tussock Grasslands Management Information System (2000) http://www.tussocks.net.nz (downloaded 26 November 2008)

Walker, S., Price, R., & Stephens, R.T. (2006) An index of risk as a measure of biodiversity conservation achieved through land reform. *Conservation Biology* 22(1): 48–59

Walkerden, G. (2006) Adaptive management planning projects as conflict resolution processes. *Ecology and Society* 11(1): 48

Walters, C. (1986) *Adaptive Management of Renewable Resources*. New York: McMillan.

Walters, C.J. & Hilborn, R. (1978) Ecological optimization and adaptive management. *Annual Review of Ecology and Systematics* 9: 157–188

Wondolleck, Julia M., & Steven L. Yaffee (2000) Making Collaboration Work. Washington, D.C.: Island Press

Participation

Catherine Allan

'Participation' has entered the environmental management lexicon because coping with high levels of complexity and uncertainty requires the involvement of multiple players from many disciplines, and possessing many forms of knowledge, in some form of collaboration (Ludwig, 2001). Reasons cited for promoting broad participation in research and management are both ethical (providing voice and empowerment for a greater range of people than would otherwise be recognised) and pragmatic (better information and processes, as well as avoiding legal and political challenges). Conley and Moote (2003) describe an 'idealised narrative' of collaborative natural resource management which emphasises reduced conflict among stakeholders (justice, fairness), increased social capital (capacity building) and better decision making.

Closely linked with providing justice/fairness and voice, especially to those previously denied these, is the idea of building the 'capacity' of individuals and communities. Capacity building is more than just 'training', as it involves exploration and sharing of values, knowledges and goals. 'Social learning' is one term used to describe this experiential, participatory and transformational learning that developed in response to dissatisfaction with linear models of technology and information "transfer". Muro and Jeffrey (2008) provide an excellent review of the genesis and current practice of social learning with reference to a number of different theories of learning.

Participation is a process that can be used for different ends, including research, evaluation, planning, management, information sharing and social network development and maintenance, so participatory methods such as facilitated workshops may be employed to achieve very different objectives.

From the earliest emergence of participatory development there have been critics, many of whom have questioned the power relationships within collaborative arrangements. Rahnema (1997) was particularly sceptical of the enthusiastic promotion of 'participation' by powerful governments and world organisations. He suggested that participation was quickly coopted so that, rather than being subversive, it legitimised and enabled mainstream development processes: i.e. that participation has been co-opted and its value reduced.

References

Conley, A. & Moote, M. A. (2003). Evaluating collaborative natural resource management. *Society & Natural Resources, 16*(5), 371.

Ludwig, D. (2001). The era of management is over. *Ecosystems, 4*(8), 758–764.

Muro, M. & Jeffrey, P. (2008). A critical review of the theory and application of social learning in participatory natural resource management processes. *Journal of Environmental Planning & Management, 51*(3), 325–344.

Rahnema, M. (1997). Participation. In W. Sachs (Ed.), *The Development Dictionary: A Guide to Knowledge as Power*. Hyderabad: Orient Longman.

Chapter 7
Kuka Kanyini, Australian Indigenous Adaptive Management

George Wilson and Margaret Woodrow

Abstract In some of the remotest regions of central Australia, Anangu Pitjantjatjara are better managing their land and wildlife resources using adaptive management plans. The plans are based on *Kuka Kanyini*, which means *looking after game animals*. Kuka Kanyini draws on traditional land management practices and sets out priorities for scientists to work with Indigenous communities to help them manage their lands themselves. Using these plans as a basis, in this chapter we present a Regional Wildlife Adaptive Management Plan template, RWAMP that can be used to guide other Indigenous communities through an adaptive management planning process. To show how the plan works in practice, we review the progress against Angas Downs' adaptive management plan as a case study. The RWAMP plan describes strategies and actions that could be used in a 'predict, do, learn, describe' Adaptive Management (AM) cycle. The plan contains science-based proactive wildlife management and supports Indigenous law and culture, and the desire to care for the land. It also helps conserve biodiversity and generate new enterprises such as sales of bushtucker and tourism. Importantly, it has wider implications for helping to close the gap on Indigenous disadvantage by providing a focus for training and employment, and improving self esteem and health.

Remote Indigenous Communities

Social Context

Large investments are being made to address health, social and legal issues affecting Indigenous Australians; reflecting the Australian Government's commitment to closing the gap between Indigenous and non-Indigenous Australia. The 2007 Australian Election policy documents commit to closing the gap in literacy, numeracy, infant

---

G. Wilson and M. Woodrow
Australian Wildlife Services, Canberra, ACT

mortality, health outcomes and overall life expectancy (Macklin, 2008). The Australian Government is also committed to working with Indigenous Australians to ensure they are able to fully participate – both socially and economically – in the life of the nation. This includes providing access to high quality education, health services generally and addressing alcohol, violence and homelessness in those communities where these threaten the safety and wellbeing of individuals and families.

Land management has a high priority in Indigenous eyes and presents opportunities for greater employment and a methodology for addressing Government Indigenous policy. These employment opportunities extend to the sustainable use of resources found on Indigenous land by supplying products for both market based enterprises and subsistence through enterprises such as tourism and new industries (e.g. carbon economy). Unfortunately, support for sustainable use of land and wildlife is rarely given the priority by funding programs that reflects the central focus that land and wildlife has for Indigenous communities. This chapter presents an adaptive management (AM) template of how these opportunities and priorities might be delivered. It discusses case studies of Indigenous communities in central Australia managing their land and wildlife resources in a process that is AM. It demonstrates how scientists, wildlife managers and agencies experienced in land and wildlife management can use the AM process to help address the urgent community health and employment challenges facing Indigenous Australians in remote communities. The names given to wildlife resources and the Indigenous language used in this chapter are those of the Anangu Pitjantjatjara Yankunytjatjara people whose lands lie to the southwest of Alice Springs in central Australia. The principles and examples have a wider reference and potential application as a template in development of plans of management on Indigenous Land.

A History of Indigenous Adaptive Management

Indigenous Australians have been using a form of AM in land and resource management for millennia through a process of trial and error experimentation to achieve desired outcomes. Their hunting practice and land management have been described by many authors (Altman, 1987; Berndt & Berndt, 1988; Bomford & Caughley, 1996). They learnt from their experiences and then modified their behaviour in light of what worked and what didn't. They also developed prescribed mosaic or patch burning practices as management of the landscapes to maximize their value as food and game animal producers (Latz, 1995).

Elsewhere in the world indigenous communities have used AM likewise. In a report detailing the AM of natural resources, Stankey et al. (2005) describe the AM processes used for generations by the Yap of Micronesia. The Yap carried out environmental management activities, observed and recorded results through stories and songs and codified practices through rituals and taboos. Using these AM processes they created and maintained coastal mangrove depressions and seagrass meadows to support fishing. Stankey et al. describe these indigenous AM examples as traditional incrementalism because they lack the purposeful experimentation attributed to contemporary AM. The modern day Indigenous AM we describe is tied to a formally-agreed upon set of objectives, and trialling and testing deliberate strategies and activities with scientific support, and can be described as purposeful.

Alienation of Indigenous land from traditional owners and its use in Western market-based economic production means that today there are few parts of Australia where traditional land management practices are conducted. This chapter proposes techniques for using western science, and in particular the AM processes, to support traditional Indigenous practice and maintain culture. We introduce the concept of Regional Wildlife Adaptive Management Plans (RWAMP).

Whitehead et al. (2003) propose mechanisms for wider application of Indigenous patch burning prescriptions in tropical landscapes to meet a range of land management objectives. This view is supported by the current West Arnhem Land Fire Abatement Project (WALFA) project where Indigenous fire managers in Australia's fire- prone tropical savannas work with support from ConocoPhillips Australasia and scientists to implement strategic patch fire management across 28,000 km^2 of Western Arnhem Land. Patch fires give off less greenhouse gases than larger wildfires, and the project offsets some of the greenhouse gas emissions from the Liquefied Natural Gas plant at Wickham Point in the city of Darwin. The project also protects culture and biodiversity on country, and brings social and economic benefits to their communities (Tropical Savannas CRC, 2008).

Why Are Indigenous Land and Wildlife Management Plans Important?

Land and wildlife are central to Indigenous culture and so management plans for these are more than just descriptions of physical program activities. Senior Indigenous spokespeople regard caring for country as integral to their cultural well-being.

Current Australian Government policy acknowledges that natural resource management is an important way to build Indigenous communities. The Working on Country program objectives include:

- Support Indigenous aspirations in caring for country
- Protect and manage Australia's environmental and heritage values by providing paid employment for Indigenous people to undertake environmental work on country and
- Provide nationally accredited training and career pathways for Indigenous people in land and sea management, in partnership with industry and others (Department of the Environment, 2008)

These objectives could have a stronger emphasis on productive use of land to supply Indigenous communities with food and resources. Our proposals for RWAMP describe the connection between land management and cultural aspirations, give motivation and self-esteem and deliver enterprise options and the supply of food resources.

Management Plans

Kuka Kanyini, a Regional Wildlife Management Plan

In 2005, *Kuka Kanyini-looking after game animals*, a Regional Wildlife Management Plan was prepared by Australian Wildlife Services (2005) for the Anangu Pitjantjatjara Yankunytjatjara (APY) Lands of north-western South Australia. The APY plan, which was prepared with the support of the Australian Government's Natural Heritage Trust, draws extensively on the surveys conducted by Robinson et al. (2003), presents key features of the wildlife resources in the area, and offers culturally appropriate options for sustainable use and enterprise development of the land, fauna and flora.

An important part of APY *Kuka Kanyini* is passage of traditional knowledge from the elders to the younger generation. Maintenance of the *Tjukurpa* – law and culture is the highest priority for the Anangu, and regenerating and caring for the land, wildlife resources, flora and fauna forms a vital part of maintaining culture.

The plan provides strategies and activities for wildlife and land management, including documenting Indigenous traditional ecological knowledge to the extent that it can be made publicly or semi-publicly available, and the identification of aspects that can be supported by western science.

Angas Downs Indigenous Protected Area Management Plan

Elements of the APY management plan were also applied to the Angas Downs Indigenous Protected Area Plan of Management (Wilson et al., 2005). Angas Downs is a pastoral lease held by the Pitjantjatjara/Luritja community at Imanpa, midway between Alice Springs and Uluru National Park (Ayres Rock). It has significant conservation value and tourism potential and, using the management plan as a guide, in 2009 was declared an Indigenous Protected Area (IPA) that will support both biodiversity conservation and human communities. The IPA plan outlines the natural and cultural resource base, land management operations, sustainable development opportunities, training, education and collaborative relationships and partnerships.

An Adaptive Approach to Indigenous Land and WildLife Management

The approach taken to develop the APY and IPA management plans requires science and traditional practice to inform and complement one another to help maintain culture, contribute to wider biodiversity conservation on Indigenous lands on behalf of all Australians while also providing food. The plans propose management techniques driven by local needs and aspirations. Better management of wildlife and bushtucker can lead to maintenance of culture, and better health and employment opportunities within their local community.

7 Kuka Kanyini, Australian Indigenous Adaptive Management   121

The plan aims to increase the numbers of wildlife preferred as bushtucker by Anangu so as to improve health and well being of communities while maintaining culture.

> Ara nyangatja mukuringanyi pulkanytjaku mai putitja, Anangu Pitjantjatjara-ku Yankunytjatjara-ku tjutangku wirura palyantjaku munu palyanyku kanyintjaku nganampa ngura munu Wapar/Tjukurpa witintjaku.

The key strategy is blending Anangu customary knowledge – the Tjukurpa (law) with Piranypa (non-Anangu) scientific knowledge to improve wildlife habitat, enhance landscapes, and harvest species on a sustainable basis.

> Ara nyangaku tjungurni Anangu-ku ara (Tjukurpa/Wapar) munu piran-ku (scientific) ara wirura Malu; Kalaya; Tinka munu Tjulpu tjuta-ku ngura, palyanyku atunymankunytjaku, nganampa ngura munu mai ngaranyangka uranma.

The process involves balancing the Tjukurpa with scientific information in a spirit of Ngapartji-ngapartji – 'give and take'.

Both management plans drew on requirements of the Indigenous Protected Area Program (Smyth & Sutherland, 1996), which recognised the need for a formal planning instrument or Plan of Management to guide the management of an IPA for a specified period (generally 5 years). IPA planning guidelines also include the other aspects of AM such as major management objectives and strategies for the IPA, resources, monitoring criteria, and timing and processes to review the plan. IPA plans are developed in consultation with the landholding group and other interested parties, usually including government conservation agencies.

Role of Traditional Knowledge and Culture

Tjukurpa

Culture and religion link the people, their land and nature through ancestral beings from pre-existence. Laws of behaviour and ceremony are laid down in the Tjukurpa or law from the anthropomorphic gods who created all things. Anangu reinforces these rules of life through Inma ceremonies which involve dance, song, body painting and storytelling. Concurrently, 75% of the population are nominated as practicing Christians.

The Tjukurpa comes from the creation period when ancestral beings, Tjukaritja, created the world as it is. The world was once a featureless place. None of the places we now know existed until Anangu ancestors, in the forms of people, plants and animals, travelled widely across the land. Then, in a process of creation and destruction they formed the world as we know it today. Many exploits of Tjukurpa involve ancestral beings going underground. Thus Anangu land is inhabited by dozens of ancestral beings. Their journeys and activities are recorded at sites linked by Iwara (paths or tracks) which link places that are sometimes hundreds of kilometres apart and beyond Anangu country.

Land is 'mapped' through the events of Tjukurpa, and is therefore full of meaning. When Anangu travel across the land they do so with the knowledge of the exploits of their ancestral beings. Their knowledge of the land, and the behaviour and distribution of plants and animals is based on their knowledge of Tjukurpa. They recount, maintain and pass on this knowledge through ceremony, song, dance and art. The Tjukurpa provides Anangu with a system of beliefs and morality with which they judge right and wrong. Tjukurpa guides daily life through a series of symbolic stories and metaphors. The stories are not simple stories, but represent technically complex explanations of the origins and structure of the universe, and the place and behaviour of all elements within it.

Understanding of such stories increases throughout their lives. For a child, a story may be a moral tale about greed, while for an adult it may provide complex explanations of ethical behaviour.

The knowledge of how the relationship between people, plants, animals and the physical features of the land came to be, what they mean and how they work must be maintained. Where Anangu are born, where they live and where they die are of great significance to them.

Land Is Central

Thus the land is the Tjukurpa, providing instruction on everything, defining codes of behaviour and sanctions for inappropriate activity. In trying to gain an appreciation of Indigenous relationship to land, little makes sense to a non-Indigenous person unless there is an understanding of the centrality of:

- Culture = Land = Law

It is not possible to have any one without the others, they cannot be compartmentalised. The land has a spiritual significance such that the people belong to the land rather than the other way around. Indigenous peoples take their teaching about culture from the land, and take their law from the land. It is for this reason that dispossession or separation from land can have such a powerfully destructive impact, leaving people lost and without a framework for living.

Confusion arises when non-Aboriginal observers imply that the relationship between Aboriginal people and land is that of ownership. The concept of custodianship or "caring for country" seems much closer to the reality, but even this implies a one way relationship that ignores the fact that the land also looks after the people.

Alongside custodianship (which brings with it responsibility to care for the spiritual as well as the physical characteristics of the land) rights of access also exist. These give people authority to visit certain sites and country which they are not necessarily responsible for maintaining.

The plans seek to emphasise the importance of western science supporting land management to the wider agenda of health welfare, self esteem employment and training. Addressing these issues independently of one another or independently of land management is unlikely to be either efficient or effective.

Tjukurpa Places Are Often Refuges from Hunting

Certain places in the Aboriginal landscape have profound importance for the continuity of people, as well as plant and animal species. The sites may need to be protected, and their spiritual characteristics maintained through ritual. The sites may be kept secret from the opposite sex and from outsiders to the group. There are also sites that are significant for the protection and proliferation of particular species, which need to be maintained if the species are to prosper.

Sites with sacred significance are well known to the traditional occupants but if they become known to others, it can cause concern. The land has the power to produce signs and omens and to kill people who do not have traditional rights to visit those places. Certain places in the landscape pose particular dangers for outsiders and people of the opposite gender to the business of that place. The responsibilities of traditional owners include the appropriate introduction of strangers to country, ensuring their safety and protecting the integrity of country. Likewise, disrespect for sacred or significant sites can impact adversely upon the spiritual health of people who hold the responsibility for maintaining country.

Indigenous knowledge about places may either be communally held or known only to elders with the appropriate rights to hold the story of a place or physical feature in the landscape. These individuals have the right to speak for country or sanction others to do so. Care of the sites (which may involve singing, dancing and other ritual practice) is likely to be designated to particular individuals, who then have a responsibility to pass their exclusive knowledge on to selected people before they pass away. Disruption of this chain of cultural knowledge and responsibility is highly damaging to the well being of the clan or tribe, and possibly to others, because the sites may lose their spiritual power.

The plan seeks to identify some of these special places notwithstanding the issues discussed above. It does so because access constraints can benefit the species which live within them by limiting hunting pressure and disturbance. The areas are effectively refuges. Where they are also sites of biological importance such as containing better quality soil and water they have added conservation significance.

Endangered Species and Biodiversity Hotspots

Anangu ecological and local knowledge, in addition to adept tracking skills, have greatly enhanced the quantity and quality of ecological knowledge collected in the APY. The most significant of these surveys was the ten years of work by the AP Biological Survey 2003 (Robinson et al., 2003).

Anangu local knowledge assisted in locating survey sites that would maximize the number and variety of habitats that could be studied. For some animal species, traditional ecological knowledge (TEK) was gathered on diet, shelter, breeding, habitat, seasonal movement, response to fire and flood events, predator response, and current and historic distribution and abundance. TEK was especially important in locating sites that supported rare and endangered flora or fauna.

Consultative Planning Process

The need to engage the Indigenous community in any planning process that will affect their community is widely acknowledged. Activities must be pursued in ways that culturally match the way in which different remote communities of livelihoods function (Rea & Messner, 2008). Walsh and Mitchell (2002) describe a participatory planning and action process, which was used to develop the APY and Angas Downs plans. The following provides examples of how these consultative processes were used to develop the APY plan:

- Cultural mapping to identify as much information as possible from the published literature, to engage with traditional owners and identify wildlife refuge areas.
- Books and papers in the library of the Institute of Aboriginal and Torres Strait Islander Studies were summarised.
- A questionnaire was developed and trialled with traditional owners, with extensive discussion undertaken with Tjilpis and Myinkmaku (senior Indigenous elders in the Pitjantjatjara language) and with anthropologists to ensure appropriateness and sensitivity of the issues being addressed.
- Discussions took place with the custodians of a number of areas.
- Surveys of Traditional Ecological Knowledge were developed.
- Sixteen to eighteen interviews were conducted with one to three participants at a time lasting 45 minute to an hour.
- Participants were selected where possible so as to represent Traditional Owners or knowledge holders of as many areas of the APY Lands as possible.
- A laminated picture book version of the Kuka Kanyini concepts was prepared to use during consultations and a Yankunyatjatjara language version of the mission statement drafted.

Stankey et al. (2006) emphasise the need to align with organisational goals, have a shared language and involvement and commitment by all stakeholders. These come about through a complex planning process, and help achieve the 'will to act' in the AM cycle. The APY and Angas Downs participatory planning processes achieve these AM prerequisites. In our AM template we have concentrated on a core set of strategies and activities that can help guide Indigenous Communities wishing to plan and make decisions on how best to use and care for their land. We propose a model that must be adapted to meet local needs as part of the consultation process advocated by Stankey et al.

A Regional Wildlife Adaptive Management Plan for Indigenous Communities

Regional Wildlife Adaptive Management Plan

Based on the APY and IPA plans, we have developed a RWAMP for Indigenous communities. The Anangu experience, *Kuka kanyini* and Angas Downs, represent trials of the principles underpinning the RWAMP. To acknowledge the Anangu's contribution, Anangu examples and significant Pitjantjatjara terms are presented as examples in the RWAMP. The examples are particularly relevant to management of arid zones, but could be applied to coastal areas with modification.

The RWAMP strategies and actions support Indigenous law and culture and their desire to care for the land, whilst helping to conserve biodiversity. Strategies become a series of AM projects in which feed back from participants and scientists will modify direction.

RWAMP covers a range of issues. It incorporates the AM cycle of what to "do" and how to "monitor" the outcomes against each issue. How to "review" the results and adjust the plan as required are not spelled out. Like the complex consultative processes, these are left to the existing management processes of the communities.

Four Steps in the Regional Wildlife Adaptive Management Plan Process

1. As for AM generally, the basic steps for the RWAMP (based on the Stankey et al. model and Greening Australia (2007)) are **Plan:**
   - Identify management issues (e.g. camel damage to waterholes)
   - Identify management goals (e.g. camel impact managed)
   - Determine management strategies available (e.g. capture and remove camels for profit; erect exclusion fences around waterholes)
   - Select appropriate management action (e.g. erect exclusion fences around waterholes to protect waterholes and deny camels' access to water)
   - Determine what will be monitored and how (e.g. bi-annual survey of camel numbers; photographic record of waterhole condition; measure waterhole water quality)
   - Determine how change and success will be evaluated (e.g. waterhole restored; absence of camels)
2. **Act:** Carry out actions (e.g. erect exclusion fence around waterhole)
3. **Monitor:** Monitor results (e.g. take site photo before exclusion fence is erected and six months after erection; undertake bi-annual camel survey)
4. **Evaluate:** Assess management strategy and modify if necessary (e.g. supplement with trapping or shooting camels at problem sites on a needs basis) (Fig. 7.1)

Central to all of this is the need for a RWAMP to be based on what is important to Indigenous Australians. For example, weed management seeks to maintain and enhance the production of locally evolved species and reduce exotics. But, weeds can have value. Buffel grass from South Africa can be a major weed - dominating local species and becoming a fire hazard, although it is considered a valuable food source for cattle and was imported to assist cattle production in northern Australia.

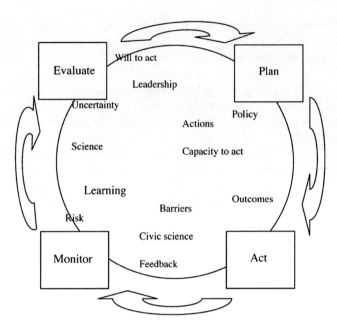

Fig. 7.1 Plan act monitor evaluate adaptive management cycle (Stankey et al., 2006)

To this end the RWAMP is not prescriptive and only acts as a guide or checklist to help the consultative and plan production processes.

The RWAMP provides the structure in the 'predict, do, learn, describe' AM cycle, as described in Chapter 2 of this volume. It identifies strategies and actions, which are listed in Tables 7.1–7.4 covering 1. Cultural and Natural Resources; 2. Resource Management Operations; 3. Enterprise Development; 4. Training.

Monitoring is built into the strategies and actions. Evaluation and feedback are built into reporting processes, including regular Board meetings and comprehensive annual reports.

The RWAMP could act as a starting point to assist other Indigenous communities carry out AM. The management objectives, strategies and actions listed here were developed for the environmental management plan used for the Indigenous Community on Anangu Pitjantjatjara Yankunytjatjara Lands. They provide a basis for other Indigenous communities wanting to engage in AM. It should be noted that the following needs to be adjusted to meet the needs and aspirations of the Indigenous communities involved: to relate priorities for the environment and reflect the natural and cultural values of the area. Put simply, to succeed, the program needs to be based on what is important to each community. Describing the need for Government to engage Indigenous people in the design and delivery of programs, Jenny Macklin MP (2008), made this point clearly saying:

> Solutions developed on the ground must be driven by the communities that will ultimately determine their success or failure.

7 Kuka Kanyini, Australian Indigenous Adaptive Management

Table 7.1 Describing the cultural and natural resources – objectives, strategies and actions

	Cultural and natural resources
Objective	To describe the resource base, cultural and biodiversity values, in order to define management targets and enable monitoring.
Strategies	Actions
1.1 Land features and resources	
• Blend local customary knowledge or law (Tjukurpa) and practices with scientific knowledge and practices to improve wildlife habitat, enhance landscapes, and harvest species on a sustainable basis	• Identify special places and biodiversity hotspots • Record traditional knowledge and customary practice to blend with scientific information • Discuss and document restrictions on hunting and consumption with traditional owners who 'speak for country' • Incorporate Indigenous elements into Regional Wildlife Plans and other relevant documents
• Incorporate local knowledge on water, fire and vegetation into maps and data bases on land condition	• Use remote sensing to provide objective monitoring of land condition • Establish monitoring sites and reference points
1.2 Biological surveys	
• Conduct broad scale surveys of common species of plants and animals, both preferred and pest species	• Gain information from people travelling from place to place, and consider using it (with discretion) to target areas for monitoring or inspection • Conduct biannual aerial surveys for the collation of broad scale data on feral animals and native fauna and frequent surveys in hunting areas to assist intensive management of the native fauna resources
• Define the status of endangered species • Document the destruction to the land and resources by feral and introduced animals, and the impact competition has on native species	• Organise tracking and spotting expeditions with Indigenous involvement • Use tourists on Indigenous-led ecotours to help collect data
1.3 Mapping and information collation	
• Present data as clearly as possible with many opportunities for the Indigenous communities to view it	• Use spatial mapping tools such as Geographic Information Systems (GIS) to analyse and present trends and results

(continued)

Table 7.1 (continued)

	Cultural and natural resources
	• Introduce CyberTracker (software for data collection) to the Indigenous group to be used to gather detailed information (i.e., location of kuka in the landscape)
• Ensure maximum use of the information gathered	
	• Provide digital cameras to assist with the recording of visual data
• Conduct long-term monitoring to identify changes and trends in the condition of natural resources that can then be used to Better manage the land and its assets	• Create hardcopy/plasticised maps of the results and GIS data

Table 7.2 Describing Resource Management operations - objectives, strategies and actions

	Resource management operations
Objective	To identify techniques of proactive management of preferred native wildlife species and consider how they might increase numbers and contribute to a biodiversity conservation strategy.
Strategies	Actions

2.1 Land use and protected area management zones

• Nominate zones for prescribed management in support of Kuka Kanyini (looking after game animals) objectives and to help maintain the Tjukurpa	• Identify areas for zones as:
	○ Places with special significance for cultural purposes
	○ Sites of significance for biodiversity conservation
	○ Areas for sustainable resource use
	○ Tourism
• Apply Indigenous Protected Area (IPA) principles to the zones and use international guidelines from the International Union for Conservation of Nature (IUCN)	• Discuss the possibility of declaring parts of the Lands as an IPA

2.2 Fire Management

• Plan fire control and response to minimise uncontrolled wildfires	• Burn as soon as possible after rain to ensure fires are small to prevent wildfire on a larger scale
• Burn in patches to increase spatial heterogeneity and enhance diversity	• Grade around settlements and development areas to be protected and slash the Regrowth when necessary
	• Use roads as fire breaks
	• Engage professional support to create fire breaks around special areas

(continued)

Table 7.2 (continued)

	Resource management operations
	• Review the use of back burns to control wildfire and other fire management procedures as they impact on wildlife abundance • Develop a fire strategy covering small areas rather than larger areas • Select and train a group of younger people in all aspects of fire management, with a view to employing them to implement intervention plans as decided by Traditional Owners and any Indigenous Land Management group
2.3 Weeds • Eradicate isolated patches of weeds • Identify sleeper weeds, where they are known to occur and where they are suspected to occur	• Establish a weed distribution database • Identify weed species known on the Lands being managed
	• Prepare a list of potential weeds to eradicate • Develop a control program for feral grasses, which will integrate eradication or sustainable control with the reinstatement of Indigenous species including:
• Prevent the spread of weeds	○ Remove feral grasses around communities and modify road grading practices to minimise its spread along roadsides ○ Manage feral grasses at key biodiversity sites, to reduce fire hazard and improve habitat quality ○ Examine options for managing feral grasses at key biodiversity sites, to reduce fire hazard and improve habitat quality ○ Control feral grasses along arterial roads where appropriate to minimise spread into key biodiversity areas
2.4 Water supplies • Keep rockholes full for as long as possible to ensure animal dispersal • Maintain bores and sink new bores to ensure distribution of animals and grazing that is sustainable • Prevent unwanted species from obtaining water	• Clean rockholes and soaks as well as use natural catchments to maximise runoff and take • Carry out ground inspections and map the distribution, availability and condition of rockholes, soaks and bores • Identify suitable sites for bore location and test quality of water available • Record both the quantity and quality of water from bores to assist with long-term sustainability of livestock enterprises and wildlife management • Construct exclusion fencing around significant rockholes, soaks and bores installed for wildlife to prevent feral camels, horses, cattle and donkeys gaining access • Use innovative fencing and electrified fences to enable selective access for preferred native species

(continued)

Table 7.2 (continued)

Resource management operations	
2.5 Reintroductions	
• Conserve the natural range, distribution and diversity of native animals while respecting the rights and interests of the Indigenous people, including those related to hunting and gathering on the lands	Produce an all encompassing plan for each animal species being reintroduced (examples could include Wayuta/Brushtail possums, Waru/Black-footed Rock wallaby, Mala/Rufous Hare wallaby, Tjalku/Bilby and Itjaritjari/ Southern Marsupial mole): ○ Assess the possibility of establishing a captive breeding and reintroduction population ○ Provide an appropriate habitat, including a buffer area to keep predators out
• Establish predator-free facilities to breed animals and promote their survival and subsequent release	○ Reduce predator activity by species specific management techniques and monitoring ○ Ensure predator management does not impact on other species
• Partner with private foundations and contract with endangered species breeding organisations to develop reintroduction programs and deliver expertise rather than use valuable Indigenous resources to undertake resource intensive and specialized reintroduction activities	○ Undertake field survey of key localities from museum records to access local abundance and factors affecting species' security ○ Where possible, consider using grey water irrigated woodlots as breeding, refuge and release areas
	• Develop a management plan, including estimates of populations and current levels of hunting for common native species currently harvested or over exploited (examples could include Malu/Red kangaroo, Kalaya/Emu, Kipara/Bustard and Ngintaka/Goanna,) ○ Conduct initial research to determine options for the establishment of a breeding facility ○ Establish refuge areas as a means of conserving species based on sites where access is limited for cultural reasons and sites of biological importance ○ Estimate numbers currently hunted, relate these figures to populations and natural rates of increase ○ Calculate estimates of sustainable harvest and discuss the results with Indigenous communities ○ Recommend that hunting should only occur on a needs basis ○ Maintain water points, both bores and rockholes beyond any livestock areas ○ Address predator control in line with the section in this table on predator management

(continued)

7 Kuka Kanyini, Australian Indigenous Adaptive Management

Table 7.2 (continued)

	Resource management operations
	○ Identify communities and individuals interested in hand rearing and release, or catch and breed programs for species
	○ Monitor progress with research projects to determine the survival rate of animals released
	• Approach private foundations for support and contract with endangered species breeding organisations to develop reintroduction programs
• Foster preferred vegetation in selected sites by sowing seeds for trees, plants and bush foods (mirka) in their natural habitats	• Establish a local business, involving the traditional owners
	• Survey and use GIS technology to map rare plant and bushfood species and weeds
	• Collect the seed from the most favoured trees, set up a nursery for germination of seedlings, plant an orchard and harvest the fruit
• Encouraging cultivation in the wild of any preferred plant species	
	• Protect existing species (e.g. Quandong trees) with camel and rabbit proof fencing
	• Encourage cultivation in the wild species e.g. wattle seed; bush tomatoes with a view to potential commercial cultivation
	• Offer training in field horticulture and similar enterprises growing bushfoods for the commercial market
2.6 Feral Animals	
• Develop a predator control strategy that considers benefits and costs of baiting and trapping programs.	• Produce a plan for each animal species being controlled (examples could include: Rapita/Rabbits, Kamula/Camels, Nyantju /Horses and Tangki/Donkeys)
	• Improve coordination and communication between regional agencies to benefit monitoring and control of pest animals
• Develop coordinated programs of mustering, poisoning, warren ripping and trapping in water exclusion zones	
	• Establish monitoring programs to track changes in predators and prey subject to control programs
	• Maintain information about water points (domestic, stock, soakage, rockhole); patterns of predator distribution and preferred feed sites in a database
	• Map and monitor plant health and the impact of ferals at selected sites
	• Conduct research into the relationship between predators on the lands
	• Prioritise water points requiring protection

(continued)

Table 7.2 (continued)

	Resource management operations
	• Bait feral predators in a coordinated baiting program targeting high priority areas • Review new baiting and trapping technologies, and consider instigating them • Map locations of major rabbit warren aggregations with a GPS from the air • Adopt a rabbit warren ripping program as needed • Map treated warrens and revisit one or two months later to determine efficacy of fumigations • Muster, trap or shoot ferals at problem sites on a needs basis • Where appropriate, incorporate capture and removal for profit • Establish a relationship with Industry Associations (e.g. for camels) to conduct an aerial survey of the populations to determine numbers and trends • Prepare appropriate training and education programs for implementation of predator management by Indigenous
2.7 Regional cooperation • Enable regional cooperation and strategic approach for management beyond the Indigenous land	• Seek special funding to address cross border issues and regional management and information exchange

Table 7.3 Describing Enterprise Development – objectives, strategies and actions

	Enterprise development
Objective	To outline options for sustainable use of the wildlife resources in a culturally appropriate manner so that species provide for the needs and aspirations of the Indigenous community
Strategies	Actions
3.1 Sustainable food production • Establish wildlife harvesting programs that supply food and other needs to local communities through trials and pilot operations in which communities express interest	• Establish microcredit facilities for assisting family based businesses • Incorporate wildlife management and tourism operations in to the Community Development Employment Projects (CDEP) program and underpin salary resources during the establishment phase • Establish links to the Australian bush food industry and research opportunities

(continued)

Table 7.3 (continued)

	Enterprise development	
	• Develop joint venture harvesting operations that build on industry technologies to deliver quality products and satisfy Indigenous cultural requirements	• Get professional advice from locals involved in bush food harvesting and increase skills in harvesting and to maximise product quality
		• Identify most important preferred species of plant and animal habitats on Indigenous Lands and manage to produce numbers
		• Collect information on the 'catch per unit effort' by local hunters in terms of hours and distance travelled for animals shot
	• Aim for less reliance on exotic species such as cattle or camels and processed food from off the lands	
		• Ban hunting if population numbers are unsustainable until they recover
		• Implement a breed and release program
		• Estimate sustainable harvests once numbers have recovered
		• Work with relevant Livestock Boards to establish a monitoring program to support decision- making on livestock numbers
		• Set upper stock limits at watering points particularly in drier periods rather than set stocking rates due to the dynamic nature of the production and decay cycles of the pasture
		• Ensure transparency in payment of any agistment fees
		• Investigate the possibility of selling animals such as camel meat to be used by the pet food market
		• Monitor developments in new markets, such as export markets, and determine whether it is feasible to enter
		• Conduct surveys of availability of guns, their calibre and serviceability and activity by hunters
		• Provide training in rifle maintenance, marksmanship and processing of animals to hygienic standards
3.2	**Supplying local demand and wider marketing**	
	• Support regional stores policy that seeks to provide clean food with good nutrition and quality	• Obtain support from Community Councils, managers of stores and officers responsible for implementing the regional Stores Policy
		• Determine the needs and the size of the opportunity for locally produced food in stores in the region. Integrate kuka and bush plant retailing with other foods
	• Establish wholesale facilities to purchase bushfoods and wildlife Produce and distribute to community stores	

(continued)

Table 7.3 (continued)

	Enterprise development
	• Educate Indigenous on the importance of a good diet
	• Ensure community stores are providing healthy food
	• Encourage Indigenous to eat bush tucker
	• Attempt to resolve any issue of complexities and cultural constraints surrounding the sale of products hunted in the Indigenous Lands by community stores through discussion
	• Ensure the continuation, support and improvement of Outback Pride and the Indigenous Australian Foods Ltd.
3.3 Tourism joint ventures	
• Develop the experiential tourism industry so that tourists visit the area to enjoy and experience its alluring mixture of history, culture, environment, music, art, archaeology, astronomy and even gastronomy	• Get state, territory and local government tourism bodies to work in partnership with Indigenous operators to maximise the tourism opportunities
	• Develop concepts and seek expressions of interest from potential collaborators in ecotourism or nature-base tourism
	• Identify potential partners in tented accommodation ventures
• Partner with nature based tour operators and accommodation to deliver ecotourism, art and heritage experiences	
3.4 Sustainable arts and crafts	
• Ensure harvesting rates of materials used in arts and crafts are sustainable and linked to the wildlife management plan	• Integrate the taking of timber and other raw materials for arts and handcrafts with management of habitats and biodiversity conservation
	• Ensure wood harvesting for punu (wood carving) is sustainable
	• Get an estimate of sustainable harvest rates as harvesting can have an impact on suitable trees, particularly close to communities
3.5 Mining	
• Mining might play a significant role in the future in regional and national biodiversity programs, and given the correct degree of planning and stakeholder input, exploration agreements may be considered	• Approach the mining industry to support biodiversity programs
3.6 Carbon Offsets/Alternative energy	
• Explore opportunities to provide carbon offset products under a carbon trading scheme	• Approach industry to support carbon offset programs
	• Work with researchers to help guide land management practices such as tree clearing/planting, grazing and cropping to store carbon

(continued)

Table 7.3 (continued)

Enterprise development	
• Partner with interested businesses such as mining or energy providers to provide carbon offsets	• Investigate the role of reinstating patch burning to improve fire management and provide carbon offsets
	• Investigate opportunities to build businesses around alternative energy sources, such as sunflowers
• Develop experimental businesses around alternative energy sources	
3.7 Corporate identity	
• Encourage wider use of any corporate logo to inspire a corporate identity and inspire loyalty to training and other programs	• Focus on a strong badging campaign for training and other programs

Table 7.4 Training – objectives, strategies and actions

	Training
Objective	To train Indigenous in wildlife management that enables wildlife-based enterprise opportunities to deliver benefits for cultural maintenance, employment and biodiversity conservation
Strategies	Actions
4.1 Opportunities in schools	
• Develop a stronger emphasis on land and wildlife topics in schools to enable the passage of land and wildlife information to the younger generation and assist in increasing school retention rates	• Examine current biology syllabus and look for opportunities to link to Kuka Kanyini – Wildlife Management concepts. Involve community members in school field trips, and encourage the passing on of traditional knowledge
	• Enable schools to get children involved with wildlife surveys, land management and feral animal control programs. Link any education programs to Kuka Kanyini
4.2 Transition programs	
• Use the attractiveness of wildlife management as an incentive for ensuring continuity between school and tertiary education programs	• Establish school-to-work transition programs which use wildlife management as a means of making academic study and vocational training more relevant for students within Indigenous communities
	• Support leadership programs which bring wildlife management into wider life-skill training
4.3 Adult training	
• Expand the Vocational Educational Training (VET) initiatives to include wildlife management and land management	• Offer the opportunity to enrol in a Certificate course such as Certificate II in Conservation and Land Management
	• Train Indigenous as wildlife rangers as part of the VET program

(continued)

Table 7.4 (continued)

	Training
• Support training in financial management enterprise development and business skills	• Identify current level of firearm ownership in communities and offer a firearms training course to improve the skill of hunters • Make the computer program Money Story® available to assist managing and communicating financial information • Encourage Indigenous to enrol in Tertiary Education Programs
4.4 Cultural maintenance	
• Search out young people to become more involved in land management by drawing on both contemporary and traditional skills • Balance Tjukurpa/traditional law, which emanates from the creation stories with scientific information in the spirit of Ngapartji-ngapartji/give and take'	• Support programs that assist elders to take young people on camps to homeland properties
4.5 Health and welfare	
• Encourage physical aspects of wildlife management operations	• Promote health benefits of kuka and mirka (bushfood) instead of processed western food
• Connect health improvement and education campaigns to the consumption of fresh local produce and Kuka Kanyini	
	• Enable sales of bush tucker in community • Develop new outdoor competitions based on aspects of wildlife management operations that could become competitive sports

Status of the Case Studies (Where Do Things Stand)

Although some projects in the APY plan have been funded, the totality of the concept is yet to be implemented, primarily due to lack of financial support and clarity over responsibilities for land and wildlife amongst agencies in the APY Lands. On the other hand, Angas Downs has been declared an IPA and funding has been received from the Department of the Environment, Water Heritage and the Arts Caring for our Country 2008–2009 program to implement significant land and wildlife management components of the plan.

Progress on Angas Downs Case Study

The Angas Downs IPA Plan of Management was prepared consultatively with the Indigenous and key government and local bodies to guide the management of Angas Downs. It draws on traditional land management practices and sets out priorities for scientists and wildlife managers to work with Indigenous to increase wildlife populations and estimate hunting yields, to identify refuge areas, to restore patch burning practices and waterhole cleaning, to control feral animals, and to exchange information across the region. It details environment restoration and development of a wildlife sanctuary and breeding facility and a tourist facility. It restricts cattle to a 250 km^2 zone to protect other more fragile and significant regions of Angas Downs.

As is the case for many remote Indigenous Communities, the Imanpa Community faces significant health, employment and educational challenges. The IPA plan was prepared with members of the community to help solve some of these challenges. It details how to restore the station environment, but is also designed to improve the self esteem and motivation of the Indigenous people by appealing to their aspiration to care for their country, and provide opportunities for training, employment and economic development.

It is an ambitious program, employing rangers with the Caring for our Country funding.

Some of the current works build on projects that had already been successfully undertaken with other government funding; closing the AM loop. *Envirofund* 2005/2006 and *IPA* funding in 2006/2007 were used to erect fences to protect a badly eroded and culturally significant waterhole, Wilpiya. As a result, camel damage has ceased at Wilpiya and reeds and other wetland values have come back, with wider biodiversity benefits. (See Figs. 7.2 and 7.3).

Protecting waterholes from camels provides a simple example of effective AM in practice. Initial attempts to fence off the waterhole using a standard stock fence failed, indicating that a stronger fence was required. With support from Envirofund,

Fig. 7.2 Wilpiya soak before camel exclusion

Fig. 7.3 Wilpiya soak after camel exclusion

a stronger steel cable and concrete footing fence was installed that excludes camels while allowing smaller native animals like kangaroos and emus access to drink unimpeded. The waterhole was restored and the lesson learned has since been incorporated into the Caring for our Country support to protect another eight waterholes on Angas Downs. In this example, the initial experimental activity did not succeed; the group then adapted the technique in light of the result and tried again.

The Caring for our decision-makers and the Imanpa community were able to see visible progress and results from the AM process.

Monitoring the water quality at Wilpiya and results of animal surveys will further inform the success of the project and determine whether other activities are needed to fully restore the culturally significant site.

Issues

Reporting and Evaluation

The AM process involves improving management policies and practices by learning from the outcomes of previous actions. Besides finding funding to support the work, perhaps the most difficult aspect of applying the AM process to Indigenous planning relates to the need for ongoing rigorous monitoring and evaluation of outcomes. To account for the funding received by Angas Downs, a prescriptive monitoring and reporting process is being implemented, supported by scientific and technical expertise. This should ensure the ongoing collation, review and evaluation of results and data and application of the full AM cycle.

As the plan is implemented regular reports are made to the Indigenous Company directors, who in turn are responsible to the Imanpa Community Council. Meetings review progress and consider changes to strategies and activities based on results to date and any new directions the community wish to take.

Reports for Caring for our will include:

- Photo records of facilities constructed and water points that have been restored to show evidence of vegetation regeneration and wildlife being given access to a better grade of water and ferals excluded.
- Water quality monitoring results of salt levels.
- Results of animal surveys (native fauna as well as feral animals), showing sightings, density and distribution.
- Progress report for weed control based on hectares of weeds.
- Progress report on fire management giving areas patch burned.
- Reports on numbers Anangu Rangers completing training.

Leadership and Community Commitment

Strong leadership is also essential to successful implementation of AM; it establishes direction, contributes resources, and aligns, motivates and inspires people (Stankey et al., 2006). Support for leadership development and mentoring is available from a number of sources – including Indigenous Community Volunteers, and Australian Indigenous Leadership Centre. Support from outside sources assists the considerable 'other work' associated with the Angas Downs IPA plan that is not included in the current *Caring for our Country* funding. This other work includes opportunities to develop market economies with development of a wildlife sanctuary and breeding enclosure, tourism facilities, and carbon trading.

Conclusion

Based on the experiences and concepts in the APY and IPA plans, we have developed the RWAMP as a template of strategies and activities that can be used as a starting point to guide Indigenous communities through an AM process to manage their land and wildlife resources. It can act as a checklist for planners and scientists contributing to resource planning for Indigenous communities.

The RWAMP is not meant to be prescriptive. Rather it is a guide to be adapted to meet local needs and local economies and aspirations to support Indigenous to manage their land themselves.

The process will help them manage their land and deliver land wildlife management whilst also helping to address the urgent community health and employment challenges facing Indigenous Australians in remote communities. It sits well with their traditional links to the land and historic use of AM.

Support for science-based land management is helping improve Indigenous self esteem, motivation and enterprise. Traditional Ecological Knowledge was vital in collating information for the surveys which underpinned the planning processes. It is also helping health, welfare, and employment. Dealing with these issues independently is neither efficient nor effective and is inconsistent with the national agenda on support for Indigenous communities.

Although not yet fully implemented, the employment of Indigenous Rangers and restoration of waterholes in the Angas Downs case study provides evidence that the AM process can deliver desired outcomes of improved land management along with a more prosperous community.

References

Altman, J. (1987). *Hunter Gatherers Today and Aboriginal Economy in Northern Australia*. Canberra: Australian Institute of Aboriginal Studies.
Australian Wildlife Services (2005). *Anangu Pitjantjatjara Yankunytjatjara Regional Wildlife Management Plan*. Canberra: Anangu Pitjantjatjara Land Management http://www.awt.com.au/content.htm.
Berndt, C., & Berndt, R. (1988). *Bewilder the First Australians*. Canberra: Australian Institute of Aboriginal and Torres Strait Islander studies.
Bomford, M., & Caughley, J. (Eds.) (1996). *Sustainable Use of Wildlife by Aboriginal People and Torres Strait Islanders*. Canberra: AGPA.
Department of the Environment, Water, Heritage and the Arts. (2008). *Working on Country*. Retrieved 19 November 2008 from www.environment.gov.au/indigenous/workingoncountry/index.html.
Greening Australia. (2007). *Native Vegetation Property Management*. Retrieved 19 November 2008 from http://live.greeningaustralia.org.au/nativevegetation/pages/page201.html.
Latz, P. (1995). *Bushfires and Bushtucker: Aboriginal Plant Use in Central Asutralia*. Alice Springs: IAD Press.
Macklin, J. (2008). *Budget, Closing the Gap Between Indigenous and Non-Indigenous Australians*. Canberra: Attorney General's Department.
Rea, N., & Messner, J. (2008). Constructing Aboriginal NRM livelihoods: Anmatyerr employment in water management. *The Rangeland Journal, 30*(1), 85–93.
Robinson, A., Copley, P., Canty, C., Baker, L., & Nesbitt, B. (2003). *Biological survey of the Anangu Pitjantjatjara Lands of South Australia, 1991 - 2001*. Adelaide: Department of Environment and Heritage.
Smyth, D., & Sutherland, J. (1996). *Indigenous Protected Areas: Conservation Partnerships with Indigenous Landholders*. Canberra: Environment Australia.
Stankey, G. H., Clark, R. N., & Bormann, B. T. (2005). *Adaptive Management of Natural Resources: Theory, Concepts, and Management Institutions* (General Report Number PNW-GTR-654). Portland, OR: U.S. Department of Agriculture.
Stankey, G. H., Clark, R. N., & Bormann, B. T. (Eds.) (2006). *Learning to Manage a Complex Ecosystem: Adaptive Management and the Northwest Forest Plan*. Portland, OR: U.S > Department of Agriculture.
Tropical Savannas CRC. (2008). *The West Arnhem Land Fire Abatement Project (WALFA)*. Retrieved 15 September 2008 from http://savanna.ntu.edu.au/information/arnhem_fire_project.html.

Walsh, F., & Mitchell, P. (Eds.) (2002). *Planning for Country*. Alice Springs: Jukurrpa Books.
Whitehead, P.J., Bowman, D.M.J.S., Preece, N., Fraser, F., & Cooke, P. (2003). Customary use of fire by indigenous peoples in northern Australia: its contemporary role in savanna management. *International Journal of Wildland Fire, 12*(4), 415–425.
Wilson, G., Pickering, M., & Kay, G. (2005). *Angas Downs Indigenous Protected Area Plan of Management*. Australian Wildlife Services http://www.awt.com.au/content.htm.

Chapter 8
Crisis as a Positive Role in Implementing Adaptive Management After the Biscuit Fire, Pacific Northwest, U.S.A.

Bernard T. Bormann and George H. Stankey

Abstract Environmental crises develop from undesired, generally large, and unexpected events. They often result in a crisis response by natural-resource managers, especially in the absence of advance planning. These crises, however, have the potential to be harnessed to help overcome typical barriers to adaptive management, including little-noticed uncertainty, societal and scientific polarization, and institutional inertia, aversion to risk, and limited resources. Crisis can ripple across polarized groups, getting them to better tolerate others' views – and ripple through institutions, getting them to question what is known, frame bigger questions, and take risks by employing new strategies. If crisis responses can be harnessed to help formalize adaptive management, new understandings are likely to emerge that support decisions that can help avoid or better respond to future events. We look for evidence of this theory in adaptive management generally unfolding under the Northwest Forest Plan, responding to the spotted-owl injunction in the Pacific Northwest states, U.S.; and specifically under the post-fire management plan, responding to the 200,000-ha Biscuit fire in southwestern Oregon in 2002.

Introduction

A large, poorly controlled wildfire results in a major environmental crisis for many people – local citizens whose lives and well-being are threatened and resource management specialists concerned with the loss of important wildland values as well as the costs associated with control and restoration. Such a perception of wildfire has helped promote the widespread sense that such events are perilous and that steps need to be taken to avoid their occurrence. Yet, a contrary, but evolving perspective is

---

B.T. Bormann
U.S. Department of Agriculture, Forest Service, Pacific Northwest Research Station, Corvallis, Oregon, USA

G.H. Stankey
Social Scientist, Private consultant, Seal Rock, Oregon, USA (Retired research social scientist), Pacific Northwest Research Station, USDA Forest Service, Corvallis, Oregon, USA

that such events can play a critical role in reframing conventional thinking and behavior; indeed, such crises are a necessary prerequisite to systemic reform and change. In this paper, we examine the various positive and negative effects of such a crisis event – the 200,000 ha Biscuit fire in southwest Oregon, U.S. – and its impacts on ongoing efforts to implement an adaptive management program on federal lands in the region.

The dangers of high-intensity, large-scale fires are well known to rural communities and firefighters worldwide. Given dry conditions and sufficient fuels, these fires can make their own local weather, spread at alarming rates, and often become nearly unstoppable. The monetary and human costs of fighting such fires – loss of property, timber, wildlife habitat, water quality, carbon stocks, and other resource values, and remediation expenses – can be substantial (Neuenschwander et al., 2000; Dombeck, 2001; Lynch, 2004). An analysis of the effects of such events must begin with a more nuanced view of how people approach the problem, given different worldviews. Typically, such large scale wildfires are considered catastrophic events; by local residents who have had to flee the fires, firefighters and land managers responsible for protecting and recovering resource values, local governments concerned about employment, and in some cases, politicians seeking to make the crises a political issue upon which they could capitalize. However, some environmentalists and forest researchers prefer to characterize events in more positive terms, because of their belief that wildfires can act as a critical component of natural processes that regulate fuel accumulation and successional patterns.

The short and long-term costs and benefits of fire also shape how people debate this issue. Fire attack in the U.S. now consumes nearly half of the entire Forest Service budget. For example, the direct cost of fighting wildland fires in the U.S. in 2002 was $US 1.6 billion (GAO, 2004). If burned trees are harvested quickly after fire, they can contribute to the bottom-line of forest industries and the local tax base. However, this benefit can become a cost when timber supply is reduced while the forest grows back, especially if the forest grows back more slowly (Bormann et al., 2008). Critical habitat for rare and endangered species can be destroyed outright or its development set back many decades. For fire-dependent species, however, critical habitat – in short supply with fire suppression – is created. The news media can attract viewership by using the drama of wildfire to play on our fears – there are, unfortunately, few incentives for the media to devote the kind of detailed attention to such topics that they warrant. However, the media can play an important role in raising public consciousness about issues such as fire management, particularly by heightening public concern and fostering readiness for action (Yankelovich, 1991).

To the extent that data can influence people's views, what specific data are considered and how they are displayed becomes important. We include an example of this idea that possibly extends to all applications of adaptive management. Wildfires in U.S. Pacific Northwest forests are common (Agee, 1993), and the area burned annually has varied over the last 90 years (Fig. 8.1). Many people have chosen to focus on only part of the historical record. For example, the recent increase in wildfire after 1980 is widely thought to depart from the historical norm, largely as a result of fuel accumulation caused by fire exclusion. This conclusion only can be sustained if data further back in time is overlooked, ignored, or somehow otherwise justified, and when other evidence is not sought out. When the historical fire record is extended from 1954 back to 1916 and broader evidence is sought, new conclusions emerge, such as that recent wildfire-burned hectares are actually less than that observed from 1916 to 1945. The decline in burned area after 1930 is attributed mostly to more effective fire fighting focused on putting fires out before they could reach large size and high intensity. More recent changes in fire fighting strategies (e.g., early direct attack) may be involved in recent increases in fires (GAO, 2004). Changes in regional climate, as indicated by the decadal oscillation index (Fig. 8.1, upper right), appear to strongly correspond to area burned in wildfire, posing an alternative explanation for at least some of the recent fire increase (McKenzie et al., 2004; Gedalof et al., 2005; Westerling et al., 2006).

One general conclusion that emerges from this experience is that a broad and open exploration of the evidence more often than not provides insights into the difficulty of associating change with specific management actions. This, in turn, can reveal that underlying uncertainties can arise from numerous interacting factors, statistical interpretations, and temporal perspectives. Acknowledging the nature and extent of uncertainty helps frame the hypotheses around which adaptive management implementation efforts take place.

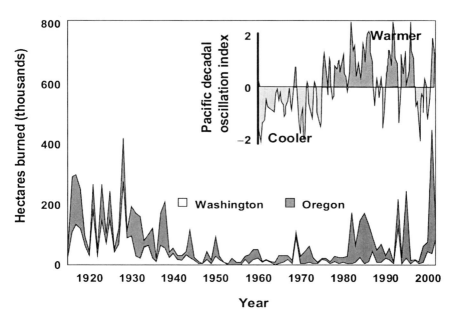

Fig. 8.1 Annual area burned by wildfire in Oregon and Washington on Forest Service and Bureau of Land Management (BLM) lands, showing increased area after 1980 (area graph, Bormann et al., 2006), and corresponding changes in ocean temperatures (*upper right*)

High uncertainty sets the stage for catastrophic events driven by unexpected events with large societal consequences. Holling (1995) has observed that in response to increasingly severe ecological and environmental problems, many people seek increasingly precise data upon which to act, particularly to avoid litigation. Rather than seeking to increase understanding of these perplexing problems, both public and private decision-makers are seemingly interested only in information that protects them against legal challenge, a particular problem in the U.S. However, as Holling (1995, p. 5) continues, "the issue should not be seen as a lack of certainty and precision of data or of predictions. Rather, there is a fundamental loss of certitude – loss in the belief that any of the ground rules work anymore." Unfortunately, the dominant tendency has been to seek what Gunderson (1999) has called "spurious certitude;" i.e., codified, rule-based planning systems where the principal concern is compliance with the rules. This stands in contrast to a view where the principal strategy is one of building understanding and modifying subsequent behaviour in light of that learning.

Ironically, such understanding can arise from the ashes of a catastrophe. Light et al. (1995) offer three postulates regarding how dramatic change can precipitate systemic shifts in how society copes with such events. First, crises are inevitable; they can arise from either unanticipated or unforeseen changes in the underlying natural system or as the result of previous human interventions whose consequences take time to be revealed. Second, such crises can precipitate a relatively quick response or restructuring that leads to a new sense of appropriate strategies. However, because these reconfigurations can conflict with tradition and conventional wisdom, they are often met with skepticism and disbelief. Third, once some new understanding does emerge, a new set of conventions, practices, and operating premises come to be adopted (law, policies, institutions, beliefs). These new approaches then dominate practice until the next (and inevitable) crises emerges, resulting in yet another reformation.

Thus, crises play two key roles. They help redefine the underlying scientific paradigm and belief systems that lead us to question and challenge convention and encourage a search for new (perhaps radical) change. Second, major environmental crises – from the major fires that have plagued Australia, the U.S., and Europe in recent years to a 1,000 km algal bloom on the Murray River – can help awaken and activate public awareness and the possibility of mobilizing that awareness and concern to trigger needed political action.

We propose a general relationship between perceptions of the intensity of a crisis and adaptive-management effectiveness. The more an event is thought to have catastrophic consequences, the more people will support a serious response (and redirect resources) to solve the problem (Fig. 8.2). Working against this trend is the extent that people are willing or interested in setting up and exploring the polarized views needed to form competing hypotheses that could be transformed into comparative approaches.

Local groups focused on activities with small or limited-area effects can more easily facilitate agreement among traditionally polarized groups. Although building trust among individuals is important, the lack of resources and the limited inference of these efforts make them less than optimal adaptive management efforts. When whole communities are threatened, as Hurricane Katrina did for the citizens of

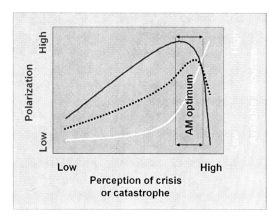

Fig. 8.2 To the extent that adaptive management benefits from having a variety of strongly held views on how to proceed and available resources, an optimum may be found when people see the problem as severe, but not fully catastrophic

New Orleans, resources become non-limiting, but a social tipping point is achieved above which polarization and alternative views disappear. These situations are also non-optimal for adaptive management.

Role of Disturbance in the Plan

The Northwest Forest Plan (Record of Decision, 1994) was an ambitious attempt at regional-scale management involving 10 million hectares of U.S. federal forest estate in the Pacific Northwest. The Plan – requested by then president Clinton – responded to discordant interest groups, an increased appreciation of larger-spatial-scale ecological processes (such as the nesting, roosting, and feeding habitat for old-growth-dependent species, including the spotted owl), and the application of this knowledge in legal strategies by environmental groups, who successfully argued for an injunction curtailing all forest harvesting. (Judge Dwyer's 1992 decision, as described in Yaffee, 1994).

Within the 10 million hectare region, nearly 30% was already protected as Wilderness and National Parks by federal law. Another 30% was newly designated as "late-successional" reserves, where the management emphasis was on protecting or restoring late-successional, old-growth conditions. Eleven percent was in "riparian reserves," areas along streams, wetlands, and lakes where conservation of aquatic and riparian-dependent terrestrial resources would receive primary emphasis. Another seven percent was already administratively withdrawn for other purposes (e.g., recreation, visual quality). In total, nearly 80% of the planning region was designated as some type of reserve, reflecting in large part the high levels of uncertainty about appropriate management prescriptions and policies.

Although the principal focus was on late-successional habitat, architects of the Plan recognized that these forests are dynamic and subject to disturbance events. The Plan increased the total area designated as late-successional reserves assuming that a portion would be lost to disturbance events – fire, windthrow, insects and disease – each year. About 2.5% of reserved areas were estimated to be subject to such disturbances per decade (FEMAT, 1993, pp. IV-55). In response, the Plan directed managers to reduce fire hazards by reducing fuels in these reserves.

The unresolved debate about causes of increased wildfire, concerns about immediate effects of fuel reduction on the viability of the Northern Spotted Owl population (a key indicator species in the Plan), shifting social priorities regarding forest management, and unaddressed economic issues all contributed to a considerable shortfall in planned fuel-reduction activities. In retrospect, the Plan put forth conflicting directives by limiting thinning in older forests, to avoid degradation to old-growth habitat, but at the same time, called for fuel reduction to protect them. Without net revenue from thinning, fuel reduction rarely happens on Forest Service lands because of the small budget available to support such activity. The specter of increasing large wildfires during the Plan's first decade also shifted the priority for fuel-reduction activities to areas near towns, rather than in the reserves.

Actual losses of owl habitat during the first decade over the entire region were close to that predicted, but fire losses were concentrated in the dry forests in the southern and eastern portions of the region to an alarming extent. The Klamath Province in Oregon lost 7% of its owl habitat in a single fire, the Biscuit Complex Fire. Subsequent fires in other dry forests have claimed an even higher proportion of owl habitat. Continued high losses to wildfire, or even higher losses with projected climate changes, recently have prompted managers to look for ways to reduce fire hazard beyond what the Plan directed.

Not unexpectedly, arguments quickly erupted over the suggested direction in the Plan regarding harvesting in burned reserves. Scientists involved in the original preparation of the Plan have disagreed over what they agreed to at the time. Some believed that limited harvest of burned trees within reserves was needed to meet Plan timber targets; others were equally convinced that harvesting was not allowed because the argument that such harvests would help meet habitat goals could not be proven. Disagreement over Plan intentions and directives boiled over into positions on the proposed study discussed in this paper. During peer review of the study proposal, scientists saw considerable value in the study, but afterward a noted scientist (J.F. Franklin) actively opposed the study, asserting that while the study objectives and purposes were appropriate, it was highly unlikely that local forest managers would cover the substantial monitoring costs, thereby making the study a pre-destined failure.

The Biscuit Fire as a Case Study

The Biscuit Fire started from a series of lightning strikes on 13 July, 2002 on the Rogue River-Siskiyou National Forest in southwestern Oregon. Accompanied by low humidity and dry east winds, the fire grew to nearly 202,000 ha before it was declared contained on 5 September, 2002, and until it was finally extinguished by rain on 8 November, 2002. At times, it burned as an intense crown fire with rates of spread up to 2.4 km h^{-1}. It was the largest wildland fire in Oregon recorded history and one of the largest ever on National Forest land. It burned primarily on the Siskiyou National Forest, including nearly all of the Kalmiopsis Wilderness Area (Fig. 8.3). About $US150 million was spent fighting the fire (Sessions et al., 2004).

Societal and Organizational Responses to the Fire

Debate about the role of fire in the Plan was played out in deciding what to do with late-successional reserves burned in the Biscuit wildfire. On one side, reserve-centric constituents and one set of scientists argued that the Plan fully accounted for wildfire and that natural processes were essential to recreate habitat over whatever timeframe it would take. For example, woody debris from burned snags was recognized as important for a variety of processes that give old-growth forests their essential quality. Contrary to this view, constituents representing local governments, timber industry interests, and another set of scientists argued that extensive post-wildfire harvesting was needed in the reserves to capture revenues to support local government, schools, and management costs, especially given the long-term losses in timber production caused by the fire. To make matters worse, the constituent divide fed into the 2004 presidential politics, forcing politicians and constituents to take polar-opposite sides. If one accepts the notion that a true crisis breaks down polarization among societal groups and fosters a sense of community and mutual interest – the Biscuit fire did not constitute such an event.

Forest Service decision-makers found themselves in a familiar position between two polarized groups, trying to meet a diverse set of management objectives. The decision to proceed with post-fire harvest in reserves was both a complex and contentious one, but was driven in part by the lack of funding to meet high post-fire management costs; e.g., mandated post-fire planning, repair of damaged roads and trails, and managing to recreate habitat and meet other objectives, including support of the local economy. The only available revenue was from harvesting and selling dead trees.

Remote sensing suggested that vegetation mortality was low in 45% of the burn area and moderate to severe in the remaining 55% of the area (Harma & Morrison, 2002). Somewhat surprisingly, studies of forest inventory plots suggested that 99% of all stands had ground fire, even though, as noted above, nearly half of the area sustained low tree mortality (Campbell et al., 2007). Another study of burned research plots intensively sampled before and after the fire suggested remarkable losses of mineral soil, soil C, and soil N in hotly burned areas (Bormann et al., 2008). Unexpectedly, less severely burned areas with low mortality had up to 60% of the soil losses found in more severely burned areas. Nutrient losses received little attention in the post-fire management plan – changes in productivity focused more on retaining woody debris rather than replacing nutrients lost in the fire.

Adaptive management became a major part of the post-fire management plan for a variety of reasons. Certainly leadership from many different quarters and from

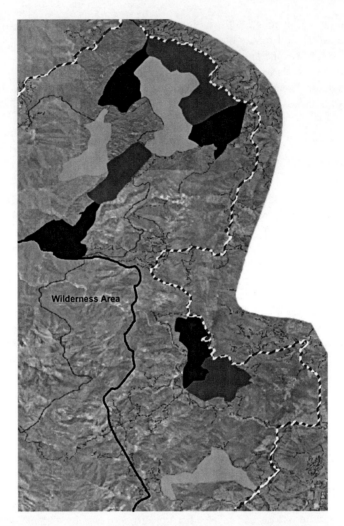

Fig. 8.3 The Biscuit landscape management strategy. Inside the Biscuit fire perimeter (striped line), in the northeast quadrant of the fire, three management strategies (grey areas) were applied to approximately 1,200-ha units in each of four larger areas (blocks) using a randomized block design. The study is on late-successional reserves and the objective is to compare different ways of restoring late-successional habitat. The three strategies – all allowed under the NW Plan – are: (1) to harvest dead trees where economically possible, plant to Douglas-fir, and tend plantations to produce large conifers as quickly as possible (light grey); (2) to harvest dead trees as above, but plant more fire-resistant pines, and then tend fuels through frequent low-intensity prescribed fire, especially in areas underburned by the wildfire (dark grey); and (3) to promote natural recovery without any harvesting or planting, but with fuel treatments focused along the perimeter (mid grey). The first two strategies are close to the positions of the polarized constituents, and were supported by different groups of scientists. The underburning strategy was an idea of one of the agency specialists. Monitoring of the complex landscape involves repeated lidar-based remote sensing to examine development of late-successional habitat and fuel changes. The uncertainties are high because no one has tried these strategies before at this scale and future fires are unpredictable. The Pacific Northwest federal agencies are now looking to apply this design to new wildfires

different geographic scales was necessary. The Forest Supervisor (the principal executive officer at the Forest level) sought special input from the science community and adopted a model for adaptive management, presented by scientists, based on a project on a different Forest (Bormann & Kiester 2004). The model became part of the plan because scientists worked directly with the field management specialists who wrote the plan. Because such a model had not been used often, continued support from the Forest Supervisor was needed to convince the specialists to remain committed and involved. Clear support for adding a major adaptive-management element also came from the regional office of the Forest Service and the Research Station, and even the undersecretaries of the Departments of Agriculture and Interior.

A key element underlying successful implementation of the adaptive management effort lay in specifying a clear learning objective in the decision documents. Requirements of U.S. public-land decision documents typically stem from the legal mandate to keep agencies from making "arbitrary and capricious" decisions by explaining to the public the rationale behind a specific decision (APA, 1947). Environmental laws, based on this rationale, dictate that documents have clear objectives, examine environmental evidence, and discuss alternative actions (NEPA, 1969). Although federal-land management objectives typically are focused on resources and the goods and services they provide, learning also can be included as an objective. This is what happened in the response to the Biscuit fire. By doing so, a formal process (i.e., an environmental impact statement) was invoked to analyze the effects of learning and to set a benchmark against which progress could be assessed. In other words, a key element of adaptive management – learning – became institutionalized, grounded in a familiar, acknowledged part of traditional planning processes.

Adopting adaptive management as a major element of the response had other roots as well. There was a lack of management experience upon which to base new objectives regarding post-fire management in the Plan. Similarly, and somewhat surprisingly to many, there was little unanimity within the scientific community as to what was the most appropriate course of action. Following fires in the region, so-called salvage logging had taken place for decades, often entailing extensive harvesting

---

as a way to replicate these findings in areas with different conditions, thereby extending inferences across the entire region. Remote sensing suggested that vegetation mortality was low in 45% of the burn area and moderate to severe in the remaining 55% of the area (Harma and Morrison, 2002). Somewhat surprisingly, studies of forest inventory plots suggested that 99% of all stands had ground fire, even though, as noted above, nearly half of the area sustained low tree mortality (Campbell et al., 2007). Another study of burned research plots intensively sampled before and after the fire suggested remarkable losses of mineral soil, soil C, and soil N in hotly burned areas burned (Bormann et al., 2008). Unexpectedly, less severely burned areas with low mortality had up to 60% of the soil losses found in more severely burned areas. Nutrient losses received little attention in the post-fire management plan – changes in productivity focused more on retaining woody debris rather than replacing nutrients lost in the fire

and replanting of burned and partly burned forest, with the main objective of restoring timber stands. The term "salvage" had also been a source of debate and conflict in the environmental debate in the Pacific Northwest. National legislation had been passed in an effort to resolve conflict over salvage harvest, but became little more than script for political theater to counter harvest reductions in the Plan by allowing harvest of older stands by companies associated with existing timber sales. Few of these timber sales ever were sold because of prolonged legal battles. As a result, the management concept of "salvage" came to have considerable negative baggage associated with it.

Because the Northwest Plan established an entirely different objective of restoring late-successional habitat on large parts of the Biscuit fire area, managers considered limiting harvest to large patches of dead trees, less intensive replanting, and placing an emphasis on stream and fuel management. However, little experience and no experimental studies existed to assess these new ideas. This helped make the case that the uncertainties surrounding any proposed prescription were very high. It also set the stage for a large scale field application of an adaptive management approach, where policies and practices would enable managers to reduce those high uncertainties.

When Forest specialists convened to discuss implementing a landscape-scale replicated experiment as part of their decision, choosing treatments was made easy by the ongoing debate. Rather than making a decision of choosing one or the other of competing alternative management options, decision-makers realized it would be both simpler and wiser to implement *both* of the polar-opposite management strategies – let nature restore itself *and* aggressively intervene to recreate conifer stands as quickly as possible (with post-fire harvesting to help pay for it and to support local communities). The specialists came up with a third treatment of their own, one not widely debated – recreate late-successional habitat through post-fire harvesting, planting more pines rather than Douglas-fir, and reintroduce low intensity fire to keep fire hazards low over time. In effect, the public, interest groups, and others had already done the ground work and assembled the evidence in support of each strategy.

The Biscuit landscape management study (Fig. 8.3) was included in the decision documents, in three of the seven alternatives in the recovery plan (FEIS, 2004), including the alternative eventually chosen. The recovery plan was challenged vigorously in court from many legal directions. In one ruling, the court recognized competing scientific perspectives and chose not to decide which group was right (Hogan, 2004). By including considerable discussion about uncertainties and possible responses to them, the study appears to have played a role in these decisions.

The study was implemented fully on the ground between 2004 and 2006, but resources to do the monitoring described in the study plan have proved difficult to find. The Forest did not receive as many receipts for harvested timber as they had hoped because of legal delays in timber sales that allowed deterioration of standing timber (GAO, 2006). Further, bids for timber might have been lower because of projected costs associated with anticipated civil disobedience. The Forest Supervisor, facing continuing decline in resources, sought to shift the monitoring costs to the Regional Office and the Research Station. Although some resources were found to begin monitoring in 2008, this appears to be another case of a lessening of enthusiasm as the crisis wanes. But, perhaps enthusiasm would have dropped more dramatically with a lesser crisis. Yet to occur is an interpretive step that could change how future decisions are made, including how to manage after fire and how to learn after fire.

Evidence of "interim" learning includes progress in the regional framework for adaptive management where interpretation might be more appropriate (next section) and a richer debate takes place, including: fire-attack strategies in mixed severity regimes; more consideration of prescribed fire to reduce fuels; research on regeneration, carbon loss, soil effects, and fire patterns in relation to pre-fire conditions. This debate has involved participants from many sectors, including the management and scientific communities, interest groups, and local citizens.

Although many people have played important roles in promoting the management study, a few were disproportionately responsible. Inside the agencies, the Forest Supervisor Scott Conroy, pushed for the study even when some members of his interdisciplinary team started to waiver. Conroy's leadership was also essential in getting the study implemented on the ground. As discussed later, Conroy's difficulty in fully funding for monitoring has led some to question his motives.

Outside the agencies dogged support came from a seemingly unlikely source. A retired sawmill executive and small-community advocate, Wayne Giesy, used his connections with the Undersecretary of Agriculture in charge of the Forest Service, Mark Rey, to pressure the agencies to begin monitoring the study. Wayne has strongly advocated the need for quality evidence about both the good and bad of post-fire harvesting. Historically the timber industry has been wary of research that often seemed to underpin their opponents' arguments. The resistance to the study expressed by some members of the environmental community – Dominick Delasalla, of the World Wildlife Fund office in Ashland Oregon, in particular – seems to indicate a change in the power structure. Environmental groups, sensing a more dominant position, are starting to show some of the resistance to new information that might endanger their positions.

Westley (1995) discusses the key role of "visionary-led collaborations;" she argues that visionaries have a strong facility for creating and manipulating symbolic language that helps build a bridge between communicators (in this case, scientists and specialists) and audiences. They rely on face-to-face exchanges and help foster intensive interaction within the organizations. They are particularly adept at mobilizing organizational levels that have become alienated by planning processes by giving legitimacy to the idea that those closest to the action are empowered to act (i.e., they can act without being told to do so). Visionaries often appear in a time of crisis and can help forge new alliances between knowledge and actions where previous paradigms and models have proved inadequate to enable effective management of ecosystems.

What Was Learned from the Biscuit Fire?

Implementation of an adaptive management program in the area burned by the Biscuit fire remains a study in progress, and the efficacy of the three treatments or of the adaptive management process itself has yet to be assessed. However, we can offer some preliminary observations about the process and particularly how various factors have contributed to, or constrained, implementation efforts.

On the positive side, a regionally-organized effort to implement adaptive management was enhanced by:

- The public's attention to a seemingly rare situation of a huge fire, and availability of national resources for traditional research on the fire. As Yankelovich (1991) has suggested, media coverage helped raise awareness among citizens of the scope, intensity, and causative factors associated with such a large scale fire.
- Debate within the science community helped frame alternative hypotheses that provided a basis for on-the-ground policies that could be tested in an adaptive management framework. Despite differing, strongly held convictions, scientists contributed to the impetus to test their ideas on the ground, provided guidance for specific aspects of the management policies, and for the most part helped encourage efforts to make the adaptive management effort work.
- Agency leadership, at multiple organizational levels, helped ensure support for adopting learning objectives in decision documents and implementing a management study at an unprecedented scale.
- Despite long-term enmities between them, in the end, quiet industry and environmentalist support helped provide the social license necessary for the project to be undertaken. Although initially opposed to the study because of reduced harvest resulting from the study, industry eventually came to recognize it as a mechanism that would likely result in at least some harvesting. Some environmentalists agreed that learning about post-fire management had value, but others remain convinced that the study was little more than an excuse for logging.

At the same time, certain factors encouraged, while others constrained, the adaptive management effort. These included:

- There persisted a strong motivation to take the debate to a national constituency in the long-standing war between industry and environmentalists (both sides made significant efforts to gain public support for their legal position). In particular, the environmental community was concerned that the gains they had attained in the original terms of the Plan (e.g., 80% of the region in a reserve status) might be jeopardized by the results of the adaptive management program, which could have resulted in allowing additional harvest (primarily for thinning and fire reduction purposes). When the results of an adaptive management program have potentially adverse consequences for one's interests, resistance can quickly materialize.
- Crises have the potential to mobilize rapid, strong support as they unfold, but such gains can rapidly decay once the crisis passes (or is at least perceived to be over). Among other things, this can lead to rapid changes in organizational priorities. In the case of the Biscuit fire, for example, once the study was implemented, there was a diminution in the willingness of local management authorities to invest resources – time, money – in monitoring. In part, this stemmed from the not uncommon desire to see someone else meet what can

become substantial ongoing costs. However, it also reflected a question that once the current crisis passed, was there a continuing need to invest organizational resources in long-term monitoring, given the uncertainty as to whether that data would, in fact, be needed. Unfortunately, this reflects a pervasive lack of long-term thinking; e.g., given the likely implications of global warming, the likelihood of continuing large-scale wildfires makes ongoing monitoring of such issues as fuel accumulation critical information for informed decisionmaking. Confronting polarization through an adaptive approach nonetheless has the potential – in the Biscuit fire example as well as elsewhere – to be a productive implementation strategy. Perhaps near-crisis is the optimum condition, where polarized groups are still motivated to develop ideas on the right thing to do, but where moderates can find room to promote learning. A platform built around the idea of adaptive management might represent the kind of "forum for working through" advocated by Yankelovich (1991) for fostering thoughtful, deliberative thinking about complex issues facing society (Stankey & Shindler, 1997).

- An idea long associated with adaptive management is that diversifying management alternatives represents a way in which we "hedge" our bets. This is especially critical under conditions of uncertainty. In many cases, the underpinnings of scientific knowledge is sparse, many random and unpredictable disturbance and climate change risks are at play, and management strategies are being proposed that have never been tried before. Diversification responds to these uncertainties by not putting "all our eggs in one basket." For example, if another wildfire got started in the area, one strategy is likely to perform better than another. In addition, increased landscape heterogeneity might slow fire progression. Although diversification was employed in the Biscuit fire response, the concept and potential benefit was not described in the decision documents. Perhaps managers and others failed to grasp fully the uncertainties involved or there was a sense that this particular crisis was a unique event, unlikely to occur again.
- In theory, managers comparing alternative strategies should desire the outcome of no significant difference between some or all strategies (Bormann et al., 1999). If differences prove unimportant, then evidence has been provided to support a wider range of approaches that then might be used to prioritize strategies to meet a wider set of objectives. It is unclear whether managers understand this potential benefit or whether they have adopted adaptive management for other reasons.

In 2008, the Forest Service Regional Office and Research Station allocated $US50,000 to begin monitoring the effects of the strategies, but future resources remain uncertain. Key advocates, in response, have sought funding from a non-profit organization to meet five objectives: further demonstrate the values of question-focused monitoring, develop inexpensive landscape-scale monitoring techniques involving lidar, develop a civil science educational webpage, replicate this study on future fires, and institutionalize a network of landscape-scale management studies in the Pacific Northwest.

A Regional Framework for Adaptive Management

Implementing adaptive management is not easy. The Northwest Plan had a major investment in the adaptive management areas, allocating 6% of the 10 million hectares planning region to foster on-the-ground application of adaptive management. More importantly, the concept of adaptive management was seen as an essential element for the long-term successful implementation of the Plan. In many ways, given the extraordinary high levels of uncertainty and complexity revealed during preparation of the FEMAT report in 1993, it became apparent to many that an adaptive approach would be the only successful mechanism whereby effective implementation could ever occur. In this sense, it was the "engine" that would drive successful implementation, making adaptive management more than a tactical strategy, but the crucial cornerstone (Pipkin, 1998). But in time, a host of political, institutional, and leadership issues intervened to significantly undermine this critical role.

The authors of this chapter participated in a major evaluation study to assess the performance of adaptive management over the first decade following release of the Plan (Stankey et al., 2006). That assessment suggested a considerable shortfall in efforts to make adaptive management a clear and decisive element of the Plan. It identified a number of requisite attributes of any potential strategy to improve implementation: a closer alignment of adaptive management with organisational goals, a demonstrated organisational commitment and will to act in an adaptive fashion, increased capacity (skills, resources), a clear, shared set of terminology; agreement on expectations within and outside the management organisations, a reasonable likelihood of continuity to ensure the process had a fair chance to succeed, clear performance benchmarks, and formal and explicit documentation protocols.

Given the extensive, systemic shifts needed, it is not unreasonable to assess the likelihood of achieving success in implementing adaptive management, or at least improvement, as problematic. Nonetheless, the concept has remained attractive to many throughout the region, and to a certain extent the genie is now out of the bottle.

Notable progress in the Plan's attempt to implement adaptive management has made by institutionalizing regional monitoring and publishing a periodic interpretive report. Various monitoring reports were completed after the first decade of the Plan and a 10-year interpretive report (Haynes et al., 2006), was published, based on a synthesis of regional monitoring and ongoing. Eight monitoring modules were created to address a range of issues (Table 8.1). The choice of issues, delays in starting some modules, and apparent imbalance in allocation of funds reflect the difficulty of starting a program from scratch with entrenched interests. The focus on owl monitoring continued through 2008, in part because some think all management is in legal jeopardy (based on the Endangered Species Act) without it, while others think that it drains resources from nearly all other learning activities, including efforts like the Biscuit landscape management study.

A further problem uncovered in the interpretive report was that no formal questions were documented by decision-makers, and only one monitoring module established quantitative expectations that would facilitate a more useful interpretation.

8 Crisis as a Positive Role in Implementing Adaptive Management

Table 8.1 Regional monitoring under the Northwest Forest Plan ($, US)

Module	1994	1995	1996	1997	1998	1999	2000	2001	2002	2003	2004	2005	Total
						Thousand dollars							
Spotted owl	1,840	1,840	1,840	1,740	1,626	2,291	2,117	2,363	2,553	2,369	2,548	2,612	25,774
Marbled murrelet						1,490	854	1,139	987	767	814	738	6,789
Olderforests						752	446	411	486	777	551	433	3,856
Watersheds						422	450	1,426	1,053	1,007	1,252	1,223	6,833
Implementation			234	252	200	250	200	239	263	280	225	216	2,359
Socioeconomics						17	25	140	200	383	400	395	1,560
Biodiversity						75	75	35	58	47	47	27	364
Tribal issues								10	40	58	105	76	289
Program Management						225	80	165	582	523	315	455	2,345
Total	1,840	1,840	2,074	1,992	1,826	5,522	4,247	5,928	6,222	6,211	6,257	6,175	50,134

These problems were analyzed in a section of the interpretive report focused on adaptive management, and course corrections were proposed. An issue yet to be addressed by researchers is how to collectively answer more questions, or the same question in more areas, by making individual monitoring efforts more efficient. A new monitoring model is needed that seeks out the least expensive way of framing and answering priority, durable, quantitatively defined questions.

One strategy that worked well was what became know as "the handshake model," adopted for writing the interpretive report (Bormann et al., 2007). Authors of the interpretive report agreed to work with, and learn from, decision-makers while preparing the report. In turn, decision-makers agreed to provide resources and formally accept and respond to the report. The time spent together increased the likelihood that the report interpretations would be quickly used in policy with fewer losses in translation. It helped foster mutual learning, sensitized all parties to the perspectives, concerns, and realities of one another, and created a great appreciation of the constraints and barriers one another faced. This interpretive forum appeared to be important, and the concept likely would be applicable to other situations.

A key outcome of the interpretive report was the decision to alter the Plan approach by adopting a new regional-scale adaptive-management framework (Fig. 8.4). The framework is driven by a formal decision on priority questions specified by regional decision-makers. Selecting the problems upon which attention and resources will be devoted is a critical first step in framing a strategic, effective

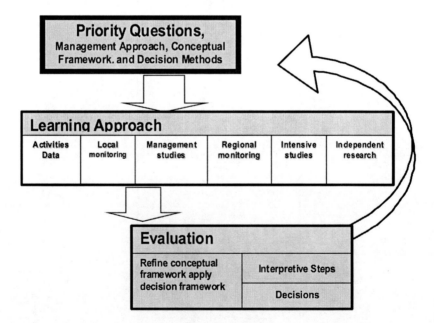

Fig. 8.4 The Plan adaptive management framework adopted July 2007. The Biscuit fire project (Fig. 8.3) was chosen as the first management study

adaptive management program. Although it does require the direct involvement of decision-makers who are responsible for allocating people, money, and time to get the work done, it also requires the perspectives and insights of people throughout the organization (Clark & Stankey, 2006). In addition to selecting the most important problems, it is also necessary that these problems be *framed* appropriately (Bardwell, 1991). Problem framing is an important but difficult phase. Its main purposes are to generate a representation of the problem that is an expression of more than the sum of individual perspectives, to foster mutual understanding and learning among participants (e.g., citizens, scientists and managers), and to identify existing knowledge as well as gaps in understanding. Framing also must consider how long it will take to address a question, and the likelihood that the question will remain relevant to future decisionmakers. In short, decisionmakers need to strive to frame *durable* questions. A workshop helped managers discuss and prioritize their questions and then choose a learning approach best suited to address them. How to manage after wildfire was chosen as a priority question, a management studies learning approach was chosen to address this question (Fig. 8.3), and the study is currently underway. The framework also specifies a formal evaluation step where interpretations feed back to alter the next priority questions. Formality in evaluation is needed to continue and strengthen learning through time.

The regional adaptive-management framework adopted by agency executives in 2007 partially recognizes the importance of *mini-feedback loops* in the adaptive management cycle. As noted in Chapter 2 of this volume, the adaptive management cycle is often depicted simplistically – plan, act, monitor, evaluate, and moving through this cycle is often cited as a measure of success. However, for broad adaptive-management programs, cycling can be seen at different rates and in different paths across time and space. Under the Plan, we can see that the adaptive management cycle became much more complex than previously assumed (Fig. 8.5). Local actions began to change before the full cycle was completed. External forces such as civil disobedience, lawsuits, and funding declines fed some of these changes (such as halting planned harvest of old-growth trees), but local adaptive-management activities and research also had a major impact. We have come to call these latter changes "adaptive mini-loops," occurring in the context of a larger regional adaptive framework. Recognizing these mini-loops, and acknowledging their shorter temporal scale in the regional framework has lead to a much more realistic depiction of the adaptive management cycle.

During the first decade of the Plan, mini-loops emerged at the local scale. A notable effort focused on examining the value of different strategies of managing existing plantations of Douglas-fir to achieve late-successional habitat objectives. The first landscape-scale management-study prototype was developed (Bormann & Kiester, 2004) to examine this question. Results from several traditional research projects also helped make the case that thinning late-successional reserves could enhance the capacity of the Plan to achieve habitat objectives. One unanticipated consequence of this was that timber harvests from these thinning sales began to account for a major portion of the overall regional harvest. The Biscuit fire study is a direct product of the new regionally-sanctioned management study and mini-loop

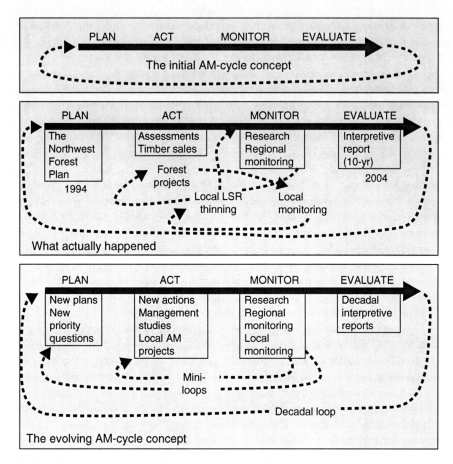

Fig. 8.5 Unfolding of the simplistic plan, act, monitor, evaluate adaptive-management cycle under the Northwest Forest Plan. Local feedback started to inform regional policy before a full cycle was completed, suggesting that mini-loops could be institutionalized to speed adaptation

regarding the priority question of how to manage forests following wildfire. The hope is that the thought given to the design and initial findings will begin to shape post-fire management before the Plan needs to be formally revised.

In sum, the Biscuit fire, seen as a catastrophe, had several major effects on how adaptive management was conceived in the region. First, it made the issue of uncertainty (both in terms of causation as well as appropriate intervention) much more evident and revealed the significant debates that existed among the scientific community. It is likely this kind of conflict and debate will continue, both in regards to fire management as well as to many other complex resource management issues. Second, much of the motivation for the Biscuit fire study arose at the local level, but it also served to impress upon regional management executives the benefits of a regional-approach to learning and the value of a regionally-recognized adaptive management program. Third, the high levels of uncertainty that virtually all parties acknowledged helped precipitate increased support – including funding – for a strengthened traditional research program focused on fire. Research papers and results from this work have attracted extensive coverage not only within scientific circles but in the wider media as well. Fourth, despite the best efforts of many parties, the search for more thoughtful public discourse and debate regarding fire management remains elusive. The seemingly natural tendency to frame debate of ongoing research results in polarized, "either-or," "right-wrong" terms largely continues to dominate discussion. But, in the final analysis, the Biscuit fire study does represent an example of how a polarized debate could help facilitate the design of an adaptive management strategy.

Conclusions

As noted earlier, the adaptive management experiment in the Biscuit fire remains a study in progress. The effort has been able to seize upon the catastrophic event as the stimulus for applying adaptive management and as a means of examining alternative courses of action, rather than either letting paralysis set in or selecting one "best" answer even when it is clear that the underlying uncertainties are large. Overcoming deep-seated traditions and beliefs has been a major challenge; attempting to move to a culture grounded in learning, tolerant of risk, and a resistance to single-solution, rule-based management has been difficult. Existing institutional structures and processes have been major stumbling blocks to innovation and change. The decline in timber revenues (an issue that derives from more than this particular fire) has seriously handicapped efforts to fund both the adaptive management program as well as activities in other resource sectors (wildlife, recreation). An organizational culture dominated by a short-term perspective can act to limit programs, such as long-term monitoring, that are essential components of adaptive management.

Yet, the Biscuit fire study also reveals important insight about the adaptive management process and these insights, in turn, provide an increased appreciation of the types of structures and processes needed to make an adaptive program effective.

The fire did result in the kind of studied change anticipated in the face of a crisis. By encouraging debate about appropriate management strategies following the fire, it helped facilitate the pursuit of alternatives, rather than simply relying on some standardized, one-size-fits-all strategy. It has thus begun the kind of reformulation and rethinking that Holling (1995) has called for. Even so, resistance to change remains and only time will reveal the extent to which substantive and fundamental shifts in fire policy will occur.

Several important conclusions can be drawn from the Biscuit fire case study. We organize our closing comments around five major topic areas. These areas are not mutually exclusive; just as with ecological systems, the social, administrative, and political systems within which an adaptive management strategy might be employed are inter-connected and linked in many ways. First, as Walters(1997) concluded, many of the issues are fundamentally institutional in character. **Institutions** include the array of formal and informal ways in which society organizes to achieve its objectives; it includes laws, administrative structures and processes, educational curricula, and a host of informal ways in which we behave to achieve our aims. Second, the **risk and uncertainty** that gives rise to much of the interest in adaptive management also represents one of the major sources of barriers to implementing effective strategies. Many institutions are inherently risk-averse and such a prevailing belief system, augmented by structures and processes to promote risk aversion, act to suppress an adaptive approach. Third, thinking and acting in adaptive ways requires a particular suite of skills and resources. Thus, **organizational capacity** becomes a major challenge facing those who seek to promote adaptive management. Fourth, traditional resource decision-making has been dominated by a technical-rational model that relies heavily on expert opinion and rule-based models. **Effective decision-making** in adaptive management will likely require new structures, models, and approaches. Finally, all of the above dimensions interact in ways that create a host of **barriers** that challenge the ability of natural resource organisations to act adaptively. However, by explicitly acknowledging these barriers, the likelihood of identifying effective and appropriate strategies is enhanced.

Innovative Institutional Structures and Processes are Called for to Think and Act Adaptively

- There are significant institutional challenges facing adaptive management in terms of addressing operations that occur at multiple spatial and temporal scales, across multiple resource sectors and tenures, and involve multiple actors and interests. This reaffirms the continued search for innovative institutional structures and processes as well as strong, effective leadership.
- A new monitoring model is needed. Questions and methods developed by researchers tend to be far too expensive (for example owl monitoring, Table 8.1). Questions and monitoring by managers tends to be unfocused and usually lack quantitative hypotheses. A new blend of these perspectives is needed to develop a cost-effective monitoring approaches focused on durable, priority questions.
- There is a need for time, forums, and leadership to promote effective assessment and evaluation of data garnered during adaptive management. The heavy focus on monitoring (which is essential) must be matched by a similar investment of organizational resources that facilitates processing of these data to create information and understanding across a wide range of players, both internal and external.
- In the case of the Northwest Forest Plan, a key interpretive forum was developed that became known as a "handshake." The idea was to mutually agree beforehand on the roles of interpreters and decision-makers. Interpreters worked with and learned from decision-makers; decision-makers provided resources and formally accepted and responded to the report. The time spent

together better assured that the report interpretations would be quickly used in policy with fewer losses in translation.
- Adaptive management can be thought of as a strategy for coping with the "law of unintended consequences." The increasing recognition and appreciation of complexity and uncertainty that faces many resource management sectors requires a more flexible, adaptive management capacity, as outcomes become more difficult to identify in unequivocal terms. However, it is also important to resist approaches that simply "make it up as we go." This likely means there is a need for new organizational structures and processes; the exact nature of these needed changes remains an issue requiring attention.

Risk and Uncertainty Must Be Acknowledged Explicitly and Seen as the Basis for Framing Adaptive Management Strategies and Policies

- Implementing adaptive management is challenged by the pervasive reluctance of some local and regional decision-makers, to admit to uncertainties, either with regard to understanding the complex causative processes that shape and influence events such as catastrophic fire as well as the efficacy of alternative interventions. Many factors likely underlie this, including an unwillingness to admit to the lack of any single "simple" solution.
- Because they are exposed to complexities of the natural world, field staff more often recognize local variation and worry about the uncertainties. As you go up the chain of command, people start looking for savings by doing the same thing every where. High level decision-makers are also the ones that defend the strategy (and tend to downplay arguments that can be used against them).

New Structures and Processes of Decision-Making Are Needed to Frame Appropriate Adaptive Management Policies and Actions

- Effective implementation of adaptive management must take place in the face of a comprehensive process for framing important, enduring (durable) questions to guide subsequent inquiry. This process, by its fundamentally political nature, must involve a broad spectrum of interests, including various technical experts, citizens, and others with an interest in any outcome. It is also important that the question-framing process involve individuals at multiple organizational levels.
- Effective problem-framing and construction of questions not only helps better focus the adaptive management process by targeting key issues and information

needs, but it also can help create a more realistic set of expectations, across multiple parties, as to what is likely (and unlikely) to result from the process.
- As efforts to implement adaptive management unfolded, there was a growing recognition of the need to acknowledge the multiple mini-feedback loops that operated within the broader adaptive management cycle. These feedbacks are often ignored or not recognized, yet they provide for the ongoing refinement that is a key characteristic and quality of effective adaptive management. They also have implications for the time the adaptive management process will take, costs, and can contribute to a sense, both within professional ranks and outside, that there is a lack of clarity of purpose and objective; however, this is an issue that professionals must engage forthrightly and more effectively than they have in the past.
- Adaptive management has a potential strength to redress the typical "win-lose," "either-or" mentality that has dominated much resource management. It can do so by treating alternative perspectives as hypotheses which can be assessed during field implementation to gain a greater sense of their applicability and efficacy under differing field conditions. This will likely lead to a more diverse array of protocols, policies, and practices and a reduced reliance upon single, rule-based solutions that attempt to address complex issues.
- Adaptive management is an inherently integrative undertaking; it joins science and society, multiple forms of knowing, and acknowledges the political nature of effective decision-making. However, these activities are often segmented and separated in organizational structures and processes. Moreover, the dominant "expert-driven" model can tend to marginalize knowledge from non-scientists. This not only results in a loss of important ways in which the environment is understood but likely also contributes to a diminution in trust among the parties.

Effective Adaptive Management Requires a Suite of Capacities – Skills, Processes, Policies and Laws, and Resources

- Adaptive management, as Walters (1997) reminded us, is a complex, time-consuming, and expensive business and organizational leaders, politicians, and citizens need to be reminded that it is unlikely this approach will be easy or cheap. However, it is also important to document the potential opportunity costs that can occur in the absence of appropriate, reasoned action; e.g., even a "no action" alternative constitutes an action, with associated costs and impacts. In extreme cases, no action can have adverse, even irreversible effects.
- The challenge of a lack of a sustainable organizational commitment to an issue, once the current crisis has subsided, remains a pervasive issue. The long-term commitment of personnel and financial resources often is lacking and, indeed, is difficult to maintain in contemporary short-term political and management environments. This suggests the need for continued efforts in working with the

political sector to foster understanding of the need for long-term support, particularly personnel and funding.

Collectively, a Host of Barriers Confront Efforts to Implement Adaptive Management and Overcoming Them Will Demand Leadership and Innovation

- The debate revealed sharp differences among scientists and experts as to the appropriate course of action and the studies needed to implement those actions. The conflict mirrors those that often reveal themselves between experts and citizens and has its roots in variety of factors, including disciplinary backgrounds, methodological orientation, but also in fundamental belief systems. Experts often disagree and although this might be seen as further confounding an already confusing situation, it can also be an essential component of fashioning alternative hypotheses for investigation under an adaptive model.
- In contrast to the above, there are stultifying effects on adaptive management from the reliance upon rule-based decisionmaking, particularly with a reliance on legislatively or statutorily imposed processes and solutions. Under a risk-averse management environment, there is often an inappropriate reliance upon such codified approaches, although it must be acknowledged that such risk aversion is often a rational approach, given the legal context within which resource management finds itself (especially an issue in the American context). Even the requirement of environmental impact statements, mandated by the National Environmental Policy Act, has become an issue. Originally intended to provide a broad policy-level endorsement and recognition of the importance of accounting for environmental effects of developments, the process has become a stultifying, limiting activity that often works to discourage innovation and change (Caldwell, 1998). Plans are framed primarily from a perspective to avoid litigation, rather than to be creative and original.
- Related to the above, there is a dis-connect between the complexity and uncertainty that tends to characterize many on-the-ground realities and the political pressures for simplicity and efficiency. This calls for improved efforts to involve a broad spectrum of interests and players in adaptive management, including contrary interests, differing technical experts, politicians, and regulators. For example, in the case of the Northwest Forest Plan, little outreach and early interaction to explain the rationale and approach to adaptive management was given to the regulatory agencies, even though these organizations effectively held "veto power" over subsequent decisions.

Federal agencies managing forests in the Northwest have demonstrated some adaptation based on learning historically. For example, research on nursery culture combined with legislation-mandated field monitoring led to increased success in reforestation of desired crop trees. The question is more about whether agencies are capable of asking the big questions and responding to them fast enough in advance of the next inevitable crisis or outside mandate, and here there is little evidence that agencies have such capability. In effect, they have lacked a functional change mechanism. An adaptive management framework has the potential to fulfill this need.

References

Administrative Procedures Act [APA]. (1947). United States Code - Chapter 5, sections 511–599.
Agee, J.K. (1993). *Fire ecology of Pacific Northwest forests*. Island Press, Washington, DC.
Bardwell, L. (1991). Problem-framing: a perspective on environment problem-solving. *Environmental Management* 15(5): 603–612.
Bormann, B.T., & Kiester. A.R. (2004). Options forestry: acting on uncertainty. *Journal of Forestry* 102(4): 22–27.
Bormann, B.T., Martin, J.R., Wagner, F.H., Wood, G., Alegria, J., Cunningham, P.G., Brookes, M.H., Friesema, P., Berg, J., & Henshaw, J. (1999). Adaptive management. Pages 505–533 in Johnson, N.C., A.J. Malk, W. Sexton, and R. Szaro (eds.) *Ecological stewardship: A common reference for ecosystem management*. Amsterdam: Elsevier.
Bormann, B.T., Lee, D.C., Kiester, A.R, Spies, T.A., Haynes, R.W., Reeves, G.H. & Raphael. M.G. (2006). Synthesis—interpreting the Northwest Forest Plan as more than the sum of its parts. Chapter 3 in: Haynes, R.W., B.T. Bormann, and J.R. Martin (eds.) *Northwest Forest Plan—the first ten years (1994–2003): Synthesis of monitoring and research results*. PNW GTR 651. USDA Forest Service, Pacific Northwest Research Station, Portland, Oregon.
Bormann, B.T., Haynes, R.W. & Martin, J.R. (2007). Adaptive management of forest ecosystems: some rubber hits the road? *BioScience* 57(2):187–192.
Bormann, B.T., Homann, P.S. Darbyshire, R. & Morrissette, B.A. (2008). Intense wildfire sharply reduces soil C and N: the first direct evidence. *Canadian Journal of Forestry Research*, 38: 2771–2783.
Caldwell, L.K. (1998). *The National Environmental Policy Act: an agenda for the future*. Bloomington, IN: Indiana University Press..
Campbell, J., Donato, D., Azuma, D., & Law. B. (2007). Pyrogenic carbon emission from a large wildfire in Oregon, United States. *Journal of Geophysical Research*, 112: G04014.
Clark, R. N. & Stankey, G.H. (2006). Integrated research in natural resources: the key role of problem framing. Gen. Tech. Rep. PNW-GTR-678. Portland, OR: U.S. Department of Agriculture, Forest Service, Pacific Northwest Research Station. 63 p.
Dombeck, M. (2001). A national fire plan for future land health. *Fire Management Today*, 61: 4–8.
Forest Ecosystem Management Assessment Team [FEMAT]. (1993). Forest ecosystem management: an ecological, economic, and social assessment. U.S. Department of Agriculture, Portland, OR.
Final Environmental Impact Statement [FEIS]. (2004). Biscuit Fire Recovery Project. Rogue River - Siskiyou National Forest, Medford, OR. [online: www.biscuitfire.com]
Gedalof, Z., Peterson, D.L., & Mantua, N.J. (2005). Atmospheric, climatic, and ecological controls on extreme wildfire years in the Northwestern United States. *Ecological Applications*, 15: 154–174.
Government Accounting Office [GAO]. (2004). Biscuit Fire: Analysis of fire response, resource availability, and personnel certification standards. GAO-04-426. Washington, DC.
Government Accounting Office [GAO]. (2006). Biscuit Fire recovery project. Analysis of project development, salvage, sales, and other activities. GAO-06-967. Washington, DC.

Gunderson, L. (1999). Resilience, flexibility and adaptive management—antidotes for spurious certitude? *Conservation Ecology* 3(1): 7 [Online at http://www.consecol.org/vol3/iss1/art7].
Harma, K.J. & Morrison, P.H. (2002). Analysis of vegetation mortality and prior landscape condition, 2002 Biscuit Fire Complex. Pacific Biodiversity Institute, Winthrop, WA. 23 p. [Online: http://www.siskiyou.org/issues/pbivegetative.pdf].
Haynes, R.W., Bormann, B.T., Lee, D.C.,& Martin, J.R., tech. eds. (2006). Northwest Forest Plan—the first 10 years (1994–2003): synthesis of monitoring and research results. Gen. Tech. Rep. PNW-GTR-651. Portland, OR: U.S. Department of Agriculture, Forest Service, Pacific Northwest Research Station..
Hogan, M. 2004. Siskiyou Regional Education Project et al. v. Linda Goodman. U.S. District Court Ruling, Case No. 04-3058-CO.
Holling, C.S. (1995). What barriers? What bridges? In: Gunderson, Lance H., Holling, C.S., Light, Stephen S. (eds.) *Barriers & bridges to the renewal of ecosystems and institutions*. New York: Columbia University Press:3–34.
Light, S.S., Gunderson, L.H., Holling, C.S. (1995). The Everglades: Evolution of management in a turbulent environment. In: Gunderson, Lance H.; Holling, C.S.; Light, Stephen S. (eds.) *Barriers & bridges to the renewal of ecosystems and institutions*. New York: Columbia University Press:103–168.
Lynch, D.L. (2004). What do forest fires really cost? *Journal of Forestry*, 102(6): 42–49.
Mckenzie, D., Gedenof, Z., Peterson, D.L. & Mote, P. (2004). Climate change, wildfire, and conservation. *Conservation Biology*, 18:890–902.
National Environmental Policy Act [NEPA]. 1969. Pub. L. 91–190, 42 U.S.C. 4321–4347, January 1, 1970, as amended by Pub. L. 94–52, July 3, 1975, Pub. L. 94–83, August 9, 1975, and Pub. L. 97–258, § 4(b), Sept. 13, 1982
Neuenschwander, L.F., Menakis, J.P., Miller, M., Sampson, R.N., Hardy, C., Averill, R., & Mask, R. (2000). In: Sampson, R.N., Atkinson, R.D., and Lewis, J.W. (eds.) *Indexing Colorado watersheds to risk of wildfire*. Mapping Wildfire Hazards and Risks.. pp. 35–56. The Haworth Press, Inc: NY.
Pipkin, J. (1998). *The Northwest Forest Plan revisited*. Washington, DC: U.S. Department of the Interior, Office of Policy Analysis.
Sessions, J., Bettinger, P., Buckman, R.. Newton, M. & Hamann, J. (2004). Complex forests following fire: The consequences of delay. *Journal of Forestry*, April/May, 102(3): 38–45.
Stankey, G.H. (2003). Adaptive management at the regional scale: Breakthrough innovation or mission impossible? A report on an American experience. In: Wilson, Benjamin J; Curtis, Allan (eds.) *Agriculture for the Australian environment*. Albury, New South Wales, Australia: Johnstone Centre, Charles Sturt University, 159–177.
Stankey, G.H. & Shindler, B. (1997). Adaptive management areas: achieving the promise, avoiding the peril. Gen. Tech. Rep. PNW-GTR-394. Portland, OR: U.S. Department of Agriculture, Forest Service, Pacific Northwest Research Station.
Stankey, G.H, Clark, R.N., Bormann, B.T., eds. (2006). Learning to manage a complex ecosystem: adaptive management and the Northwest Forest Plan. Res. Pap. PNW-RP-567. Portland, OR: U.S. Department of Agriculture, Forest Service, Pacific Northwest Research Station. 194 p.
Walters, C. (1997). Challenges in adaptive management of riparian and coastal ecosystems. *Conservation Ecology* 1:1. http://www.consecol.org/vol1/iss2/aet1
Westerling, A.L., Hidalgo, H.G., Cayan, D.R., & Swetnam, T.W. (2006). Warming and earlier spring increase western U.S. forest wildfire activity. *Science* 313: 940–943.
Westley, F. (1995). Governing Design: The Management of Social Systems and Ecosystem Management. In: L.H. Gunderson, C.S. Holling & S.S. Light (eds.), *Barriers and bridges to the renewal of ecosystems and institutions* (pp. 391–427). New York: Columbia University Press.
Yaffee, S. L. (1994). *The wisdom of the spotted owl: Policy lessons for a new century*. Washington, DC: Island Press.
Yankelovich, D. (1991). *Coming to public judgment: making democracy work in a complex world*. Syracuse, NY: Syracuse University Press. 290 p.

Loop Learning in Adaptive Management

Chris Jacobson

Adaptive management offers an approach to learning particularly suited to where conditions of high uncertainty exist. Learning occurs when we consider the implications of outcomes that result from actions we undertake. As a result, we make a conscious decision to either adapt or retain management practice. Learning theorists commonly refer to this process as single-loop learning (Argyris, 1999). It requires us to be explicit about the purpose for taking action by detailing what we hope to achieve, taking action and reflecting on it. The question remains whether there is better way to achieve the outcome. Consider the simple example of a dog who wants to stay warm. The dog moves into the sun to warm up. This is single loop learning, but is there a better way?

Double loop learning accelerates learning. It requires identification of assumptions about what will or will not lead to what we hope to achieve and reflection on them (Argyris, 1999). The dog in the example assumes a number of things, including that sitting outside is a good way to stay warm. While this might be true, a double-loop learning dog would consider a range of sources of warmth, their likely future availability and the importance of warmth in relation to other needs. Instead of choosing a sunny place, the dog might choose the companionship of a human who can light a fire. In adaptive management, the use of models helps to synthesise existing information and identify assumptions about the likely outcomes of decisions. Experimentation enables different options to be tested and to refine models.

Two additional types of learning are evident in the literature, although they are not often explicitly linked to adaptive management. Bateson introduces the notion of deuteron learning – the simultaneous learning about the outcomes of an action and the context within which it occurs (Visser, 2003). In adaptive management, this requires consideration of a broad range of factors that lead to project success, including social and institutional factors in addition to ecological ones. Triple loop learning is sometimes considered synonymous with deuteron learning (King & Jiggins, 2002). It involves learning about the predispositions to learning in particular ways (Ison et al., 2000). In adaptive management, it would involve reflecting on assumptions about epistemology; e.g., what counts as knowledge upon which to adapt? Why was it that the management team chose to learn as a collective at particular stages in their project? (Jacobson, 2007).

The key is not that one type of learning is more important than another. Adaptive management can involve a number of types of learning that can be applied either concurrently or in succession depending on the nature of uncertainties, the perceived benefit of reflection on different aspects of the learning process and the capacities of the individuals involved to do so.

References

Argyris, C. (1999). *On Organisational Learning – Second Edition.* Malden, MA: Blackwell.
Ison, R. L., High, C., Blackmore, C. P., & Cerf, M. (2000). Theoretical frameworks for learning-based approaches to change in industrialised-country agricultures. In Anon (Ed.), *Cow Up a Tree. Knowing and Learning for Change in Agriculture. Case Studies from Industrialised Countries* (pp. 31–55). Paris: Institut National de la Recherche Agronomique.
Jacobson, C. L. (2007). *Towards Improving the Practice of Adaptive Management in the New Zealand Conservation Sector.* Unpublished Ph.D. thesis, Lincoln University, Christchurch, New Zealand.
King, C. & Jiggins, J. (2002). A systemic model and theory for facilitating social learning. In C. Leeuwis & R. Pyburn (Eds.), *Wheelbarrows Full of Frogs: Social Learning in Rural Resource Management* (pp. 85–104). Assen, The Netherlands: Koninklijke Van Gorcum.
Visser, M. V. (2003). Gregory Bateson on deutero learning and double bind; a brief conceptual history. *Journal of History of Behavioural Science, 39*(3), 269–278.

Part III
Tools for Adaptive Management

Chapter 9
Modelling and Adaptive Environmental Management

Tony Jakeman, Serena Chen, Lachlan Newham, and Carmel A. Pollino

Abstract Models can be used to synthesise our understanding of a system and facilitate the exploration of possible impacts of changes in management, climate and other factors. Modelling can also be an effective process in helping to identify knowledge gaps and prioritising monitoring requirements and management options. Accordingly modelling can be a valuable tool in assisting adaptive management. Model development should follow a rigorous approach to enhance relevance and credibility, particularly when models are used to guide management decisions which require defensibility. Appropriate stakeholder involvement throughout the model development process can be an effective means of social learning and consensus building. Working in collaboration with all stakeholders helps to ensure the model is appropriately focussed, and is more likely to produce recommendations acceptable to the decision makers and community.

Integrated modelling is useful in informing decision making for systems involving complex, multi-sectoral issues. These models can also be applied for purposes such as prediction, forecasting, system understanding and social learning. The main integrated modelling approaches include Bayesian networks, coupled components models, expert systems, agent-based models and system dynamics. The selection of approach must depend on the purpose of the modelling exercise, the available knowledge and data on the system, the timeframe and the technical resources available. Integrated approaches promote stakeholder engagement, systems thinking and transparency, and can therefore be an effective tool in adaptive management.

Introduction

Observational data and existing knowledge are rarely sufficient in fully assessing the health of a complex natural system and its causal relationships. At best, data tend to be too limited in scope and knowledge too compartmentalized for this purpose.

T. Jakeman, S. Chen, L. Newham, and C.A. Pollino
Integrated Catchment Assessment and Management Centre, Fenner School of Environment and Society, Australian National University, Canberra Australia

Modelling is a scientific process of simplifying reality to enhance understanding through the structuring of data, knowledge and assumptions in a creative and disciplined way for a specific purpose. Models can systematically integrate and capture our understanding of how changes in management, climate, demographics and other factors affect selected indicators of system health so that the consequences of management options can be clarified. Models can be qualitative, quantitative, or a combination of the two. The more complex the system being considered, the greater the role for models to account for the interactions among drivers, processes and associated outcomes. On the other hand, models are imperfect representations and the nature of the systems they attempt to describe may change over time. But new knowledge and data, appropriately targeted, almost always benefits model usefulness, as does a rigorous approach to the model selection and development procedure.

The challenge is to view and implement modelling as an ongoing scientific and participatory process that serves adaptive management. In this connection the first aim of modelling should be to identify new knowledge and data needs that will lead to further understanding, if not direct clarification, of the impacts of various courses of action on system health. As far as possible another aim should be to assist obtaining consensus on, or defensibility of, the management decisions to be taken. As argued in section "A Rigorous Approach to Modelling Practice" of this chapter, such aims demand good modelling practice, especially in the selection of a model type and approach that recognizes context, a topic covered in section "Participatory Processes and Adaptive Management". It also calls for the use of participatory processes (section "Selecting the Appropriate Modelling Approach") and analytic tools (section "Sensitivity Assessment") that help identify the type of data, knowledge and associated experiments that gives leverage to achieving the first aim. For practical reasons, it also should take into account the cost-effectiveness of acquiring this new knowledge.

Few would disagree that adaptive management needs more emphasis and strategic research. To this end, modelling and its incorporation in information or decision support systems (section "Information and Decision Support Systems"), can aid the development of: (i) ways to gather, record and share conventional and unconventional environmental system information; (ii) improved tools to capture and express qualitative as well as quantitative knowledge; (iii) methods for testing knowledge, identifying gaps and designing experiments; (iv) monitoring techniques able to distinguish the effects of changed management practices from the large natural variations associated with most systems; and (v) approaches to screening and testing a broad range of alternative policies.

Two examples are given in this chapter of the value of integrated modelling for adaptive management. In section "An Illustrative Problem in Catchment Management" we report on the management of sediments and nutrients to protect catchment and estuarine water quality, and in section "Vegetation Management in a Conservation Reserve" we examine a study of vegetation management in a conservation reserve. Some of the barriers to enhancing the use of modelling for adaptive management are discussed in section "Barriers to Modelling for Adaptive Management". Section "Conclusions" constitutes the conclusions.

Modelling in the Adaptive Management Process

Modelling can be a useful tool at various stages of the adaptive management process outlined in Chapter 2 of this volume (Fig. 9.1). During the planning stage, models can be used to scope the problem and synthesize available knowledge and data on the system. Models are inherently subjective, and should incorporate uncertainties in the understanding of system processes and in the parameter estimates (Sutherland, 2006). Walters et al. (2000) suggested that the two key roles of modelling in adaptive management were: (1) to reveal gaps in understanding and data, and (2) to guide prioritization of experimental management options and monitoring. Identifying the knowledge gaps may subsequently help direct the experimental design to further system understanding.

Simulation models can be valuable tools in linking science and management (Rivers-Moore & Jewitt, 2007). McLain and Lee (1996) examined three case studies of adaptive management: (1) spruce budworm management in New Brunswick, Canada, (2) fisheries management in British Columbia, Canada and (3) fisheries management in Columbia River Basin, US. In each of the three cases, the most useful role of modelling was found to be in facilitating the exploration of possible impacts of different management scenarios. Although model predictions are never completely reliable, they can help to stimulate thought and further research. For example, if a model failed to predict an extreme event, understanding the reasons why can help in improving subsequent predictions and thus system understanding (Sutherland, 2006). Given that the outcomes are limited by model assumptions, McLain and Lee (1996)

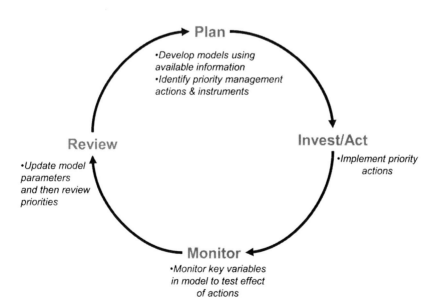

Fig. 9.1 Models as tools in the adaptive management process

also suggested that comparing multiple competing models may be useful in understanding how these underlying assumptions affect model results.

The Coastal Lake Assessment and Management (CLAM) tool is an integrative decision support tool for exploring tradeoffs associated with the management of coastal lakes in New South Wales, Australia (Ticehurst, 2008; Ticehurst et al., 2007). Using a Bayesian network framework (see sections "Selecting the Appropriate Modelling Approach" and "Vegetation Management in a Conservation Reserve"), the model can be used to determine which management actions are likely to produce the most desirable outcomes, and should thus be implemented. The CLAM tool can also help identify which system components are likely to be affected by these actions (given current knowledge), and thus identifies which key variables should be monitored. The resulting monitoring data can then be used to update the model parameters, to improve reliability of the tool for future runs. Models can also facilitate social learning and consensus building, as discussed further in section "Participatory Processes and Adaptive Management".

A Rigorous Approach to Modelling Practice

Good modelling practice can only increase the relevance, credibility and impact of the information and insight that modelling aims to generate. It is a necessity for long-term, systematic accrual of a good knowledge base for both scientific understanding and decision making. Jakeman et al. (2006) recommend a minimum set of standards for good modelling practice that includes:

- Clear statement of the objectives and clients and interest groups of the modelling exercise
- Documentation of the nature (identity, provenance, quantity and quality) of the data used to drive, identify and test the model
- A strong rationale for the choice of model families and features (encompassing alternatives)
- Justification of the methods and criteria employed in model calibration
- As thorough analysis and testing of model performance as resources allow and the application demands
- A resultant statement of model utility, assumptions, accuracy, limitations, and the need and potential for improvement and
- Fully adequate reporting of all of the above, sufficient to allow informed criticism

Adoption of these standards by modellers would benefit not only the model-building community but also those relying on model-based insight and model recommendations to make decisions. Models must also be applied according to their purpose and capabilities. Their misuse can lead to results of low confidence and invalid conclusions.

Participatory Processes and Adaptive Management

Managers and interest groups can also potentially benefit from use of a model to define the scope of a problem, to make assumptions explicit, to examine what is known and what is not, and to explore possible outcomes beyond the obvious ones. If models are accessible enough, they can act as a medium for wider participation in environmental management. Accessibility might be enhanced in different ways, such as through a reliable technical platform for the scientific users or in user friendly software such as decision support systems for stakeholders (see section "Sensitivity Assessment").

Aside from equity and justice principles, there are two main reasons for increased stakeholder participation in model development. The first is to improve the modellers' understanding, allowing a broader and more balanced view of the management issue to be incorporated in the model. The second is to improve adoption of results from the assessment, increasing the likelihood of better outcomes, as model development becomes an opportunity for stakeholders to learn about interactions in their system and the likely consequences of their decisions.

Stakeholder participation in the past has often been limited to researchers wishing to exploit the results of the modelling exercise. A better approach, increasingly employed, is to involve all stakeholders throughout model development in a partnership, actively seeking their feedback on assumptions and issues and exploiting the model results through feedback and agreed adoption. This approach is expensive in effort, time and resources, but the aim of modelling is often to achieve management change, and the learning process for modellers, managers and other stakeholders inherent in this approach is essential to achieving change. Examples of such participation in model development can be found in, Hare et al. (2003), Letcher and Jakeman (2003) and Newham et al. (2006).

Selecting the Appropriate Modelling Approach

Selecting a modelling approach depends firstly on our objectives for the problem of interest. Let us assume that broadly the purpose is prediction and/or decision making that involves complex, multi-sectoral issues and tradeoffs where impacts are a function of potential management interventions and uncontrollable drivers such as climate. In such a case we would also be interested in social learning of interest groups (see section "Participatory Processes and Adaptive Management") to be a necessary component to achieve our objectives. In this way the accrual and sharing of knowledge and the potential for adoption of the management are enhanced. Along with purpose, the choice of a modelling approach also depends on the type of knowledge and data at our disposal, the resources for the modelling exercise and timeframe for an outcome or decision to be achieved. Generally we are not talking about hard science or disciplinary models here.

Nor are we talking about the significance or power of a hypothesis test as to whether an intervention will have an individual effect of some sort or not (e.g., Field et al., 2007). We are using integrated models of systems to guide our understanding.

Several approaches suggest themselves for such objectives and five major types are discussed by Jakeman et al. (2007). Table 9.1 indicates their suitability in different contexts. Two very different approaches are the Bayesian Networks (BNs) and the Coupled Component models (CC) approaches.

BNs are well-suited to integrating multiple issues, interactions and outcomes. A network connects system variables (inputs, internal states and outputs), portraying the cascading influences and interactions, starting with key drivers affecting internal system states and their influence on outcomes that are the object of our management. They allow the variables in models to be characterised according to the level of knowledge we have, so they can be expressed either quantitatively, qualitatively (e.g. categorically such as high-medium-low) or both. Relationships between variables are represented probabilistically. The latter allows uncertainty to be characterised and propagated through the network. BNs are well-suited to static representations of systems that are lumped in space but some detail in time and space can also be accommodated. A real strength is their utility to map the system representation with stakeholders and 'parameterise' them iteratively. Because interactions between adjacent variables are related by probabilities, one can populate them with whatever information is at hand, and as it accrues. One can utilize outputs from component models to assign the probabilities, or indeed use best qualitative knowledge as might be elicited from experts. Sensitivity and other analyses (see section "Sensitivity Assessment") can be used to infer where the weak points in the network model of the system are and subsequently design the 'experiments' to improve this information. The example in section "Vegetation Management in a Conservation Reserve" illustrates some of the valuable utility of BNs for adaptive management.

CCs on the other hand are most suited to handle problems where a small number of issues are being investigated at a high level of spatial and/or temporal detail (see the example in section "An Illustrative Problem in Catchment Management"). This approach combines detailed component models from different disciplines, which are linked through the sharing of outputs and/or inputs. Uncertainties typically are handled by brute force such as undertaking multiple runs through sampling from the distribution of their parameters. Because of their complexity, they are less suited to social learning and knowledge sharing, but careful handling such as through meta-modelling can overcome the difficulties. A simple but powerful approach could be using information from CCs to populate a Bayesian network that captured the key interactions, and using this version of the system with interest groups.

Both approaches are very well-suited to adaptive management, as illustrated in the following case studies. Agent-based models are an approach somewhere between BNs and CCs. They are essentially a form of CC that is concerned with the interactions between 'agents' (individuals) in a system. The agents are software components containing code and data, and adapt to changes to their environment. Multi-agent systems comprise a network of interacting agents, where individual

Table 9.1 Appropriate use of integrated modelling approaches

Approach	Data types	Resources required (technical and data)	Suitability to handle uncertainty	Space treatment	Time treatment	Breadth versus depth of interactions	Application
Coupled components	Quantitative	Medium to high	Computationally demanding	High	High	Depth of processes	Prediction
Expert systems	Qualitative and quantitative	Low to high depending on the state of the knowledge base	Explicit	Medium	Variable	Compromise	Forecasting Decision making System understanding Social learning Prediction
Agent-based	Quantitative	Medium to high	Implicit	Medium	High	Both	Forecasting Decision making System understanding Social learning System understanding Social learning
Bayesian networks	Qualitative and quantitative	Low to high	Explicit	Medium	Low	Both	Prediction Decision making System understanding Social learning
System dynamics	Qualitative	Low to high	Implicit	Medium	High	Both	Decision making System understanding Social learning

agents share information, request services and negotiate with each other. Agent-based models are capable of modelling complex systems containing multiple interactions among dynamic and autonomous entities. They are useful for social learning, particularly at revealing large-scale outcomes resulting from local interactions between individuals. Uncertainty is handled by brute force. The models are generally hypothetical and tend not to be as suitable as BNs or CCs for prediction. Rather they are useful for representing complex systems involving several stakeholders and allow exploration of the possible effects of alternative management options.

System dynamics are a popular modelling approach used by applied scientists because they allow investigation of complex feedback systems, such as food webs. System dynamic models can represent complex links within the human-biophysical environment, including nonlinearities, feedback loops, and spatial and temporal lags (Costanza & Ruth, 1998). Such phenomena can be difficult to represent with most other modelling approaches. They can also include poorly understood processes, represented as 'plausible' connections. System dynamic models are most commonly applied to improve system understanding, to compare alternative system assumptions, or for social learning. Theoretically, system dynamics approaches can also be used for decision-making and policy development.

Another type of qualitative model is expert systems, which simulates the problem-solving behaviour of domain-specific experts. Prior knowledge is encoded into a knowledge base and the expert system then uses logic to infer conclusions to the given problem. The success of the expert system is thus determined by the knowledge base (Forsyth, 1984). This approach is therefore unsuitable for modelling systems containing complex interactions and poorly understood processes. Expert systems can be applied to problems where there is little interaction between variables and where experts can articulate decisions with confidence. Uncertainty can be incorporated into expert systems by assigning uncertainty values (e.g. probabilities, belief functions, membership values) to the facts and rules. A powerful attribute of expert systems is their ability to explain, by retracing the steps of reasoning used to arrive at the conclusion.

A range of model integration frameworks and analysis tools already exist as a basis to help address the needs of adaptive management and in particular the iterative and integrated assessment of the environmental, social and economic impacts of management decisions. Their potential to clarify options and generate partnerships has already been demonstrated in various case studies (e.g., Jakeman & Letcher, 2003) but there is still a need to develop them further by adapting them to new case studies tuned to the needs of adaptive environmental management.

Sensitivity Assessment

In classical experiment design, the approach to gain knowledge is to choose some forcing to optimise a measure of accuracy of estimated parameters (usually based on an estimated covariance matrix). For natural resource management this approach is problematic as it assumes some freedom in choice of the forcing (size, timing, location, form), and requires explicit modelling of error sources, structure and probabilistic properties. NRM modelling tends to involve heterogeneous errors and at best offers very limited control over the forcing that can be exercised. So we need an iterative experiment design procedure, focused largely on designing and/or revising the measurement regimes. In the example in section "An Illustrative Problem in Catchment Management" this involves integrating into the modelling the results of geochemical tracing and the collection and analysis of event-based water quality data in the waterways.

Sensitivity assessment (SA) of models is a key tool to inform modelling for adaptive management. It provides an objective means of assessing and improving modelling and is particularly useful in identifying high leverage data collection activities i.e. what new knowledge is most needed. Equipped with such information the necessary measurement and/or monitoring program can be designed within resource constraints. There is a wide body of techniques (see Saltelli et al., 2000) now available for SA but essentially they are all aimed at identifying the relations between changes in "factors" (values of forcing, parameters and/or boundary conditions) and resulting changes in model outputs, conclusions or internal variables. It is worth emphasizing that SA is a first step in formal uncertainty analysis which additionally requires the prescription of uncertainties in causes, usually as probability distributions, so that output distributions can be calculated.

So why perform SA?

- To see which factors are critical, which are uninfluential or redundant in determining modelled outcomes; check against prior knowledge
- To examine interactions, see what parts of the model are more or less independent
- To see how model structure could be improved: combine parameters, remove or simplify sections, add sections
- To identify data needs

Therefore SA helps to ascertain the areas of uncertainty that most affect management outcomes, making it essential for adaptive modelling.

Information and Decision Support Systems

Models can play a key role in decision making for environmental management and policy, by providing a means of formulating and exploring problems and bounding the range of uncertainty. Models can be used to discriminate and compare different options or strategies in a qualitative, if not, quantitative way, to assist in the selection of the 'optimal' alternative. The priorities and value judgments of decision makers can also be incorporated into the decision making process in a transparent manner.

There are some distinct advantages in performing integration methods within an appropriately-designed framework and tools where the audience and needs are well-identified for their prescription. They can constitute (Jakeman & Letcher, 2003):

a way of investigating tradeoffs and explaining them to interest groups; a readily accessible collection of models, methods and visualisation tools that can be updated; a focus for integration across researchers and stakeholders; a training and education function; an exploratory aid capable of adoption and further development by stakeholders; a permanent summary of the project methods; and a means of making the management analysis transparent.

An Illustrative Problem in Catchment Management

Here we provide an example of the development and application of a model to inform the process of adaptive management. The example is taken from a modelling-focused study of water quality in the catchments of the Eurobodalla region of southeastern NSW. The study was undertaken in a partnership between the Eurobodalla Shire Council, the NSW Department of Environment and Climate Change and the Australian National University. The objective was to identify high leverage management options for the control of sediment and nutrient inputs to streams, drinking water sources and estuaries. Such information is required to inform the adaptive management process whereby local managers seek knowledge of how their actions are expected to improve water quality.

The modelling is based around a Coupled Components model known as CatchMODS. CatchMODS combines hydrologic, sediment, nutrient and economic models in an integrated framework that is suitable for use by managers to test scenarios of alternative management. The model is constructed in accordance with the general principle of good modelling practice, evolving over an iterative cycle of application, evaluation and redevelopment within the study and as a legacy of previous investigations. In the Eurobodalla region, the model is constructed at a scale commensurate with the management objectives, available data and the level of existing process knowledge.

The structure and parameters of the model have been modified over the course of the study as process understanding and data become available. Driving this process are inputs from targeted scientific studies and monitoring programs. The initial steps in the modelling and assessment process for the Eurobodalla region were the identification of potential pollutant sources and the construction of coarse temporal and spatial scale hydrology and pollutant budgets. These early investigations enabled critical information and knowledge gaps to be identified and targeted. An important outcome was the establishment of an event-based water quality sampling program. The monitoring program enabled (i) identification of dominant pollutant generation process via interpretation of pollutant concentration and hydrograph data, and (ii) estimation of pollutant loadings. A lack of knowledge of pollutant inputs from intensive agriculture and from unsealed roads in the catchments led to the commissioning of a farm-scale nutrient budgeting study and to the application of measurement and modelling programs to estimate sediment yields from unsealed roads.

The information garnered from these studies was incorporated into the CatchMODS model. This was achieved in several ways including via improved estimates of parameter values, changes in the structure of the model where appropriate and the addition of new components of the model specifically for estimating sediment inputs from unsealed roads. Independent data to evaluate the effectiveness of the modelling was also sought ultimately to improve the modelling process and to inform the adaptive management cycle. In the study described here this was achieved via the comparison of model outputs against sediment tracing results at selected stream tributaries. SA techniques to investigate high leverage data gathering activities and more fully understand the behaviour of the model have been used in the study.

Through the iterative process described above the model produces relatively detailed spatial representations of pollutant sources. Importantly, estimates of the costs of implementing the various management scenarios that are tested are also available and provide valuable information to prioritise management effort.

The development and application of the model and its various supporting studies have all incorporated consultation to varying degrees with local experts, decision makers and the local community. The effect has been the development of a model that is appropriately focused, accepted by decision makers and is far more likely to produce management recommendations acceptable to the local community. The challenge remains to keep cycling through the adaptive management loop, using the model to inform continuing data collection and evaluation activities and to over time refine the underlying model. This has been at least partly achieved in the Eurobodalla case study but further progress is possible. It requires longer term research projects and a balance between data collection, evaluation, model development and consultative activities.

Coupling complex models are especially well-suited to problems where a small number of issues are being investigated (e.g., water allocation policy and environmental flows) and much detail is required spatiotemporally in terms of drivers, processes and impacts. Care needs to be taken in their direct use with a non-technical audience. But they can be hybridised with other approaches, such as Bayesian networks, to make their value more accessible.

Vegetation Management in a Conservation Reserve

Bayesian decision support tools are becoming increasingly popular as a modelling framework that can analyse complex problems, resolve controversies, and support future decision-making in an adaptive management framework. Traditionally, environmental management strategies have sought to avoid addressing uncertainty (Walters, 1997), leading to poor environmental outcomes and profound ecological and economic impacts (Halpern et al., 2006). Today, the importance of quantifying uncertainty in decision-making is being increasingly recognised as providing a mechanism for describing realistic outcomes, adding flexibility to the decision process, and dealing with variable systems, where our knowledge is poor or incomplete.

In representing and communicating our knowledge of complex systems, Bayesian networks are valuable aids. They are especially useful integration tools when there is strong connectivity among several issues and policies; for example, water allocation, supply, biodiversity, industrial production, social and cultural impacts. They also should be considered strongly in a planning context and/or when participation goes beyond technical interest groups and generation of trust and sharing of knowledge are quintessential. Because of their probabilistic nature they can represent uncertainties and can also be used in a risk assessment framework. Bayesian networks also can be updated when new information comes to light, or our understanding of a system changes.

A Bayesian network model was designed to assist in the management of an endangered Eucalypt species, the Swamp Gum (*Eucalyptus camphora*), found in the Yellingbo Nature Conservation Reserve (YNCR), an isolated patch of forest in the Yarra Valley (Victoria, Australia) (Pollino et al., 2007). The eucalypt community provides both habitat and food for a variety of threatened and endangered flora and fauna. Over the last 20 years the *E. camphora* has become increasingly threatened by dieback. In order to protect existing trees and encourage regeneration, management strategies and investments in restoration have focussed on restoring the hydrological regime, which has been altered due to agricultural activities within the catchment. However, there is still much debate on what the causal factors are resulting in dieback.

A Bayesian network model was constructed to integrate information on all the factors (Fig. 9.2) that were perceived by the scientists and reserve managers as affecting the eucalypt community (Pollino et al., 2007).

The model acted as an evidence base to better characterise the breadth of possible threatening processes affecting the eucalypt, and this was used to explore the strengths between the various hypotheses of the causal factors resulting in dieback. Using SA, we found that the key drivers in determining the condition of the eucalypt were strongly related to the group conducting the study in the reserve. Consequently, the drivers of eucalypt condition were biased by research group.

Unfortunately, given the poor quality of data and knowledge available, further research is required to definitively identify the causal factors of dieback. Instead the

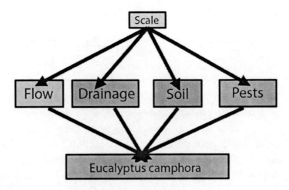

Fig. 9.2 Model schematic, where scale has two components: the locality in the conservation reserve and the group who conducted a study in the reserve

model acted as a demonstration of how important it was to undertake studies that were integrative, holistic and designed well. With future monitoring and research, the BN can be iteratively updated. This will inform and assist the development of future management strategies.

In the eucalypt study, assessment and management strategies had focussed on simplistic cause and effect hypotheses. Expensive interventions in the reserve had been undertaken based on this simple hypothesis, with little improvement in the condition of the eucalypt community. In focussing on a simple solution, it was evident that this had been to the detriment of the reserve. The BN approach offered a solution by addressing the need to better understand a complex system, while acknowledging the uncertainties that exist in our understanding of the functioning, and the variability that is inherent within systems.

As recognised by Walters (1997), although a wide range of alternative models can be equally 'valid' in ecology, they can also have wildly different predictions about the effects of various management policies. This inability to discriminate among alternative hypotheses based on historical data does not imply that modelling and analysis of historical data are useless exercises (Walters, 1997). Rather models should be used to direct more efficient field testing processes for further testing and evaluation.

At this stage, more efforts in generating alternative or additional models and hypotheses about dieback of *E. camphora* would be unproductive and a wasteful expenditure of limited resources and effort. What is needed is robust and systematic collection of further (new) data, routine analysis of data, and responsive management plans. BNs, and more generally Bayes' theorem, can provide a framework for the efficient accumulation and use of such evidence (Newman & Evans, 2002) to formulate improved and adaptive management strategies.

Barriers to Modelling for Adaptive Management

The high complexity of environmental problems, and the dynamic and poorly understood nature of many system processes, mean that no model can be perfect. Typically, models are specific to the dataset used to calibrate the input parameters, and thus are not necessarily transferable to other sites or time periods. There are also technical issues related to spatial and temporal cross-scale linkages between physical, chemical and biological processes (Walters, 1997). For example, change takes time to manifest itself especially in ecological systems, making it difficult to monitor let alone model the processes. Although these limitations must be appreciated, they should not discount the value of models in adaptive management, as the alternative is a less disciplined and accountable process.

In addition to adhering to standards, the education of modellers on further aspects is warranted; for instance, on how to engage with clients and stakeholders, on the need to develop more flexible models and on understanding the context in which the model will be used. The perceived credibility and accessibility of models by stakeholders and end-users such as managers can pose a barrier to their use. This challenge can be overcome through conforming to the good modelling practice protocols mentioned above and collaborative working between modellers, stakeholders and end-users, which can also help ensure that the model is built for the right purpose and used appropriately.

The development of long-term partnerships between modellers and model end-users is fundamental to promoting ongoing adaptive management. If trust is built between the two groups and the broad problem issues being addressed are ones of continuing concern (e.g. ecosystem health), then it is likely that funding sources for the collaborative efforts in improving management outcomes can be identified. Even as sources and topics for funding change, ways can often be found to reorient the work to capture such funding. Pursuing robustness in the relationships should be a major aim, making sure that it is not dependent on one individual champion.

Integrated assessment is becoming increasingly accepted and applied to inform environmental decision making. As an iterative and inclusive process focussed on better sustainability outcomes, it has much in common with adaptive management. IA exercises promote engagement by stakeholders, systems thinking and transparency, which can only assist people in appreciating one another's perspectives. Put crudely, adaptive management would benefit more from IA's principles, while IA would benefit from an adaptive management focus that uses models, analytic tools and processes that accelerate or increase our understanding and management of a system for the long term.

Conclusions

- Modelling is an important tool for adaptive management. It can assist communication, identifying knowledge gaps and prioritizing management options and monitoring requirements.
- Selection of a modelling approach must be purposeful and targeted to the issues, stakeholders and users.
- Approaches to modelling fall along a continuum that ranges from expert/reductionist/specialist to participatory/multidisciplinary/systems focused.
- Each modelling approach has capabilities and limitations, which must be appreciated if the tool is to be used effectively.
- Stakeholder participation, and preferably partnerships, throughout the modelling process can be an effective means of sharing and communicating knowledge and values, thereby enhancing the success of adaptive management.
- Models will always need adapting to the context of the given task/problem.

References

Costanza, R. and Ruth, M. (1998). Using dynamic modeling to scope environmental problems and build consensus. *Environmental Management* 22(2): 183–195.

Field, S.A., O'Connor, P.J., Tyre, A.J. and Possingham, H.P. (2007). Making monitoring meaningful. *Austral Ecology*, 32: 485–491.
Forsyth, R. (1984). The expert systems phenomenon. In: Forsyth, R. (Ed.), *Expert Systems: Principles and Case Studies*. Chapman & Hall, London, pp. 3–8.
Halpern, B.S., Regan, H.M., Possingham, H.P. and McCarthy, M.A. (2006). Accounting for uncertainty in marine reserve design. *Ecology Letters* 9: 2–11.
Hare, M., Letcher, R.A. and Jakeman, A.J. (2003). Participatory modelling in natural resource management: A comparison of four case studies. *Integrated Assessment*, 4: 62–72.
Jakeman, A.J. and Letcher, R.A. (2003). Integrated assessment and modelling: Features, principles and examples for catchment management. *Environmental Modelling and Software*, 18: 491–501.
Jakeman, A.J., Letcher, R.A. and Norton, J.P. (2006). Ten iterative steps in development and evaluation of environmental models. *Environmental Modelling and Software*, 21: 602–614.
Jakeman, A.J., Letcher, R.A. and Chen, S. (2007). Integrated assessment of impacts of policy and water allocation change across social, economic and environmental dimensions. In: Hussey, K. and Dovers, S. (Eds.), *Managing Water for Australia: The Social and Institutional Challenges*. Melbourne: CSIRO, pp. 97–112.
Letcher, R.A. and Jakeman, A.J. (2003). Application of an adaptive method for integrated assessment of water allocation issues in the Namoi river catchment, Australia. *Integrated Assessment*, 4: 73–89.
McLain, R.J. and Lee, R.G. (1996). Adaptive management: Promises and pitfalls. *Environmental Management*, 20(4): 437–448.
Newham, L.T.H., Jakeman, A.J. and Letcher, R.A. (2006). Stakeholder participation in modelling for integrated catchment assessment and management: An Australian case study. *International Journal of River Basin Management*, 4(3): 1–13.
Newman, M.C. and Evans, D.A. (2002). Enhancing Belief during Causality Assessments: Cognitive Idols or Bayes's Theorem. In: Roberts, J.M.H., Newman, M.C. and Hale, R.C. (Eds.) Overview of Ecological Risk Assessment in Coastal and Estuarine Environments. Lewis Publishers, New York, USA, pp. 73–96.
Pollino, C.A., White, A.K. and Hart, B.T. (2007). Examination of conflicts and improved strategies for the management of an endangered Eucalypt species using Bayesian networks. *Ecological Modelling*, 201: 37–59.
Rivers-Moore, N.A. and Jewitt, G.P.W. (2007). Adaptive management and water temperature variability within a South African river system: What are the management options? *Journal of Environmental Management*, 82: 39–50.
Saltelli, A. Chan, K. and Scott, E.M. (Eds.) (2000). *Sensitivity Analysis*. Wiley, Chichester, 475pp.
Sutherland, W.J. (2006). Predicting the ecological consequences of environmental change: A review of methods. *Journal of Applied Ecology*, 43: 599–616.
Ticehurst, J. (2008). Evolution of an approach to integrated adaptive management: The Coastal Lake assessment and management (CLAM) tool. *Ocean and Coastal Management*, 51: 645–658.
Ticehurst, J.L., Newham, L.T.H., Rissik, D., Letcher, R.A. and Jakeman, A.J. (2007). A Bayesian network approach to assess the sustainability of coastal lakes. *Environmental Modelling and Software*, 22: 1129–1139.
Walters, C. (1997). Challenges in adaptive management of riparian and coastal ecosystems. *Conservation Ecology*, 1(2): 1 [online] URL: http:// www.consecol.org/vol1/iss2/art1/.
Walters, C., Korman, J., Stevens, L.E. and Gold, B. (2000). Ecosystem modeling for evaluation of adaptive management policies in the Grand Canyon. *Conservation Ecology*, 4(2): 1 [online] URL: http://www.consecol.org/vol4/iss2/art1/.

Chapter 10
Lessons Learned from a Computer-Assisted Participatory Planning and Management Process in the Peak District National Park, England

Klaus Hubacek and Mark Reed

Abstract In order to support stakeholders in adapting to socio-economic, environmental and policy pressures a group of researchers and key stakeholders joined forces to develop an iterative social learning process supported by computer models designed in a participatory modeling process. We report on an ongoing research project in the peak district national park, UK. This chapter details the genesis, development and operation of this approach to enabling adaptive management in a complex socio-ecological landscape. Instead of experimenting with new management activities and learning from the results of these actions, we used formal computer models to tell the stakeholders what the implications of their actions might be in terms of their own economy and also environmental effects such as different growth patterns of plant species, biodiversity, as well as soil erosion, water quality and carbon fluxes. Such modeling of scenario modelling is assumed to enable decision making (and eventually activity) in 'risky' situations, or in a context of high risk aversion. Including stakeholders in all stages of the process increases acceptance of the work and allows the inclusions of relevant multiple views and can enhance shared understanding. A flexible approach that can react to participants' needs is a precondition. Participatory scenario modelling was found to be very useful as it enables surprises and changes in emphasis to be incorporated in the process thus providing flexibility to deal with social surprises such as linguistic ambiguity and physical surprises such as bird flu and foot and mouth disease, both of which reappeared on the agenda during this process. We also learned that the selection of stakeholders was important as well as developing a strong understanding of the context; and having a good facilitator. To have a chance for the learning and adaptive management process to survive beyond the project duration a certain set of attitudes and organisational cultures are required that can facilitate processes where goals are negotiated and outcomes are necessarily uncertain.

K. Hubacek
Sustainability Research Institute, School of Earth and Environment,
University of Leeds, Leeds, UK

M. Reed
Aberdeen Centre for Environmental Sustainability and Centre for Planning and Environmental Management, School of Geosciences, University of Aberdeen, St Mary's, Aberdeen, UK

Context and Problem Description

The Peak District National Park is typical of upland regions around the UK and Europe that are faced with challenges such as demographic change (here especially aging of the population), policy reforms (especially the single farm payment delinking production from subsidies) and environmental problems (especially climate change). It is also one of the UK's most visited National Parks due to the area's natural beauty and high local population (20 million people living within 1 h drive), and abundant opportunities for outdoor recreation.

Tourism plays an important role in the regional economy, comprising 15% of all businesses in the area (Derbyshire Chamber of Commerce, 2005). In addition to an economic structure similar to urban areas, the local economy depends on the Peak District's water provision to close-by large conurbations and the revenue that is received from farming and hunting (Hubacek et al., 2008).

The Park is home to 38,000 residents and a relatively high proportion of second homes. All these different ways of using the area cause some level of stress to the environment. Thus, recent environmental assessments of the Sites of Special Scientific Interest (SSSIs), home to a number of rare and fragile habitat types, have been characterized as being in unfavorable condition due to a combination of overgrazing and 'inappropriate' burning (English Nature, 2003).

Such problems are compounded by historic atmospheric pollution and deposition (accumulating since the industrial revolution) and increased climatic variability both of which have been blamed for increased erosion, declining water quality, and negative effects on carbon fluxes. Peat soils are important carbon stores (Worrall et al., 2003), however, Bellamy et al. (2005) found alarming evidence that 80% of UK soils carbon losses might be from upland peat soils.

A further layer of complexity is added through socio-economic, legal and other institutional changes such as the ongoing global economic crisis, an ongoing process of closing important rural services, and policy changes such as Europe's single farm payment and the European Union (EU) water framework directive.

In order to support stakeholders in adapting to this range of socio-economic, environmental and policy pressures a group of researchers and key stakeholders joined forces to develop an iterative social learning process supported by computer models designed in a participatory modeling process (Prell et al., 2007). Instead of experimenting with new management activities and learning from the results of these actions, we use formal computer models to tell the stakeholders what the implications of their actions might be in terms of their own economy and also environmental effects such as different growth patterns of plant species, biodiversity, as well as soil erosion, water quality and carbon fluxes. The resulting model should assist us in assessing important management options including grazing intensity and heather burning, as well as the threat to carbon stocks, biodiversity or water quality. Scenarios can then be developed to demonstrate the consequences of targeted management intervention or business-as-usual. This chapter details the genesis, development and operation of this approach to enabling adaptive management in a complex socio-ecological landscape.

The Approach

The Initial Idea and Goals

The Peak District National Park had seen a number of attempts to create stakeholder networks and public discussion forums dealing with fragile upland ecosystems and their human use and interferences. One of the most recent attempts was a network established by Moors for the Future (MFF) (for a summary see MFF, no date), a non-governmental environmental stakeholder with administrative ties to the National Park authority and good contacts to researchers interested in upland issues and to land managers and farmers alike. The main remit at the time was to restore heavily degraded moorland areas but the organization felt that they would want to be more proactive in helping to avoid future degradation and better prepare for such changes. This is where the Sustainable Uplands project entered the picture, based on an invitation of one of the stakeholders, and offered the expertise of a large project group consisting of about 20 social and natural scientists funded by the Rural Economy and Land Use program. The aim of this 4-year project, beginning in 2005, was to combine knowledge from local stakeholders, policy-makers and social and natural scientists to anticipate, monitor and sustainably manage rural change in UK uplands.

This process was to be supported by computer models which help to simulate potential responses to external threats and new land management activities and provide indicators that help assess progress. The computer models are seen as heuristic and starting points for discussions rather than providing exact predictions. These have been developed in an iterative process consisting of stakeholder meetings, expert interviews, and site visits where scientists and stakeholders explored and learned together (Prell et al., 2007).

Stakeholder participation is an integral part of the process that included involvement in the grant writing stages, the design of the project, as well as in the production of reports and policy recommendations (e.g. Reed et al., 2005). A number of national conservation agencies had independently embarked on the development of upland scenarios to inform the development of their future work. At the same time, a number of stakeholder groups felt that their voices were not heard by those taking high-level decisions about the future of upland landscapes. Thus, the Sustainable Uplands project aimed to bring these different groups together to investigate likely upland futures and identify strategies for policy and practice that could help different stakeholders prepare to better harness future change (Dougill et al., 2006). In the following, we describe key components of this process, including stakeholder selection, model building and the 'futures workshops.'

Stakeholder Selection

Problem definition and stakeholder identification are interacting processes. If the issues are defined without consulting stakeholders, then the issues may not be relevant to their needs and priorities. At the same time, the issue must be defined before it is possible to identify those who hold a stake. To complicate matters further, many stakeholders might not be interested in participating due to time constraints or lack of interest or understanding about how the issues under discussion relate to their daily lives (Hubacek et al., 2006; Prell et al., forthcoming 2009). We, i.e. the sustainable uplands team together with key stakeholders, tackled these problems through an iterative process of stakeholder analysis in focus groups, combined with semi-structured interviews, follow-up phone interviews with original focus group participants and social network analysis. We started by conducting a focus group with members of our core partner MFF we previously knew, and two key stakeholder organizations that they had identified. To avoid bias arising from initial group composition, focus group data were triangulated using semi-structured interviews with other stakeholders identified during the focus group to represent different land management perspectives. The aim of the focus group and interviews was to evaluate and adapt the proposed aims of the project in order to ensure it was focusing on issues relevant to the key stakeholders and identify and categorize stakeholders (Dougill et al., 2006; Hubacek et al., 2006). Initially we focused on a single issue of interest to the stakeholders – the review of national legislation dealing with burning on upland areas – which is also an important management tool for grouse managers. Later, we included a wider set of scenarios and issues. This process of adding scenarios and issues and developing responses is ongoing and will be supported by the computer tool.

The focus group and interviews also identified over 200 relevant stakeholder organizations, so it was necessary to develop a selection or sampling strategy. For different participatory activities we were using different group sizes. For example workshops at conferences were only limited by the size of the available facilities whereas site visits on farms and places of interests out in the field were limited to rather small groups of key stakeholders to allow for group discussion reflecting on what people saw and experienced despite, for example, adverse weather conditions.

Given this need for small sample sizes in some instances, achieving fair representation can be a major challenge. If sampling is deemed unrepresentative, then the legitimacy of the process can be undermined. We developed distinct categories to stratify our sample, so that sampling can be used within each stakeholder category, ensuring all the major groups are represented. To do this, stakeholders were initially categorized during the focus group, and information was elicited about the most effective way to gain the support and involvement of these stakeholders. Eight stakeholder group categories emerged from this process: water companies; recreational groups; agriculture; conservationists; grouse moor interests (consisting of owners/managers and game keepers); tourism-related enterprises; foresters; and statutory bodies. These categories were then used to guide our 'snowball' sample. We had to use a snowball sample as it was difficult to get addresses and phone numbers of potential interviewees. These interviews were used to deepen our knowledge of the current needs and aspirations of those who work, live and play in the Park (Dougill et al., 2006; Prell et al., forthcoming 2009). Stakeholders were asked about their relationship with other stakeholders in the park; and a social network analysis (SNA) was applied to identify which stakeholders were key in this area and the degree to which other stakeholders trusted these individuals and organizations (Prell, 2003; Prell et al., 2008). This identified both powerful players whose opinions may influence the wider community and stakeholders who are typically marginalized on a given issue; in our case this was initially heather burning for grouse management. These groups were then brought into the research to strengthen the legitimacy of the process and to add a variety of relevant knowledges (Prell et al., 2008). To achieve participation of all selected stakeholders we had to start with the 'key players' who would be able to pull other participants into the process.

Depending on the stage of the process and the main purpose at hand we created different group compositions based on discussions with our core stakeholder group. SNA provided the possibility of selecting individuals who are either different or similar to each other; where similar individuals are typically better able to communicate tacit, complex information, as there tends to be higher mutual understanding between them. On the other hand, if the purpose is to elicit a wider range of views, it may be appropriate to select individuals who are different from each other. In our research, stakeholders wanted focus groups to be composed of individuals who did not know each other well in order to enhance learning between participants, and it was possible to do this using outputs from the SNA (Prell et al., 2008).

Participatory Model Building

By taking a more bottom-up approach to model development involving stakeholders from the outset, it may be possible to identify and prioritise the problems that need to be solved first, and use this to determine the scope and choice of models to apply. This participatory approach should develop models that can help address issues pertinent to stakeholders, and provide outputs that will justify their time investment in the process. Effectively communicating model outputs, can increase the likelihood that stakeholders can help interpret model outputs and refine model development in collaboration with researchers (Giordano et al., 2007; Prell et al., 2007).

Two conceptual modelling workshops were held with researchers from the team, to map out their understanding of system structure and function in relation to key drivers. This was further enriched through a literature review (Holden et al., 2007). Additional insights from this work and the site visits were then integrated with the initial conceptual model (developed from semi-structured interviews). Finally, the conceptual model that emerged from the integration of these different knowledge bases was used to trace the likely effects of different drivers through the upland system, to develop preliminary scenarios (Dougill et al., 2006).

After identifying the relevant issues and stakeholders and codifying and integrating stakeholder's inputs in the conceptual model we started developing a quantitative computational model based on this qualitative information. The conceptual maps developed in collaboration with the stakeholders have served as a framework for the development of the integrated model. For this, we had to step back from the complexity of the conceptual model and identify the subsystems, drivers and potential models that can be used to link human behaviour to biophysical effects, which in turn influence land managers in their decisions. This required the integration of various models, model components and integration of social and biophysical systems, as well as existing and new data derived within the research project (Chapman et al., 2009; Termansen et al., 2009).

At its core, this integrated set of models has an agent-based model (ABM) that models the response of land managers to certain policy and environmental scenarios and the effects of their action. Results are discussed in multi-stakeholder focus groups, which are central to reconciling and elaborating shared understandings of the interactions between social, economic and environmental systems, and will be used to assess future management scenarios and policy recommendations (Dougill et al., 2006).

Stakeholders have been involved in all stages of this research, from problem formulation to model development through to discussion of outputs. The scenarios developed from the conceptual model form the basis of a detailed questionnaire developed to elicit the decision-rules for the ABM. Respondents who are land owners, agents for large land owners, tenant farmers and grouse moor managers took part in the questionnaire for the ABM. Each scenario is presented to the respondents and they are asked how they would change their management strategies in relation to grouse and/or sheep farming under each scenario using choice experiments (see e.g. Hanley et al., 1998; Termansen et al., 2009).

At the core of the biophysical model is an existing hydrological model (PESERA) which has been adapted to better fit the stakeholders' needs. This model has bee used to predict runoff and erosion across Europe and has been developed and improved over the last 15 years (Kirkby & Neale, 1987; Schofield & Kirkby, 2003). It provides a core bio-physical platform with additional elements attached to it tailored to the project needs. The use of PESERA as the model platform enables us to ensure that information is available in a spatially explicit way with topography and other relevant factors incorporated. This should enable us to show how an activity in one part of the catchment will have a different environmental impact than the same activity would have in another part of the catchment. This will allow the development of spatially explicit decision-making and a move away from simple blanket policies (Prell et al., 2007).

The agent-based model will provide data on human responses to socio-economic and political drivers, and this will provide inputs to the bio-physical model. At the same time, environmental change will influence decision-making and the physical models will therefore provide inputs to the agent-based model. The ABM mainly deals with farmers' decision making processes given data on economic conditions, natural conditions, and institutional constraints. The biophysical model provides data for the natural constraints but also for the biophysical responses to the farmers' land management choices (Termansen et al., 2009).

There is a trade-off between capturing all the data we would like and the patience of our stakeholders and interviewees, between scientific interest and building the ideal model and the pragmatic interests of e.g. farmers and their other time commitments. Thus we prioritized and selected scenarios based on the conceptual model, which in turn was derived from interviews with stakeholders. As a consequence, the results of the choice experiments should, in theory, reflect stakeholder priorities rather than researcher's biases (Termansen et al., 2009).

Description of Workshops Developing Scenarios and Indicators to Monitor Progress

Scenarios can communicate complex information about socio-ecological change in ways that can be easily understood by stakeholders from a variety of backgrounds, giving people the opportunity to use this information to shape their future or adapt to changing conditions. The rationale for involving stakeholders in scenario development follows broader participation discourses that focus on normative and pragmatic reasons (Reed, in press). Normative arguments suggest that people have a democratic right to participate in analysis about their own futures; pragmatic arguments focus on participation as a means to an end, which can deliver higher quality scenarios, higher buy-in and identification with outcomes (Stringer et al., 2006).

On the other hand, a number of drawbacks and limitations of stakeholder participation in scenario development have been identified: For example, local knowledge is not always sufficiently robust or detailed enough to provide information about relationships between system components, necessary for scenario quantification (Walz et al., 2007). A number of studies noted the significant time necessary to engage meaningfully with stakeholders (for a summary see MS Reed et al., under review). However, many of the limitations identified in the literature simply reflect poorly practiced participatory methods (MS Reed et al., under review). For example, the choice of stakeholders who are involved has the potential to significantly affect the outcome of scenario studies. This is particularly relevant when stakeholders are involved in both initial scenario development and the evaluation/selection of scenarios. Hence, without systematic and representative stakeholder selection, there is a danger that participation can bias results (see our earlier discussion).

In order to minimize bias and facilitate the process we designed the process around a number of key elements such as agreed upon objectives and agreed upon social learning activities that would most fit their needs:

Objectives were developed with the stakeholders from the beginning of the project. Stakeholders proposed their own sustainability goals for the upland system and suggested indicators (e.g. water quality indicators such as dissolved organic carbon was suggested by water companies) that could monitor progress towards these goals.

Together with our core group of stakeholders we developed a number of participatory events that would be most conducive for learning given the variety of backgrounds of participants. For example stakeholder-led site visits were suggested and became a core element of engaging stakeholders. The outdoor context and facilitation style significantly reduced the discrepancies in power that were witnessed in the initial workshop, with all participants feeling comfortable engaging in discussion. The site visit programme was designed by a steering group of stakeholder representatives who selected the issues to be covered and the most appropriate sites to stimulate discussion. The steering group also suggested the development of information sheets about each issue, to ensure all participants had similar levels of information about each issue and could engage in debate at a similar level with one another. The scope of each information sheet was decided through discussion with stakeholders, and drafts were peer-reviewed by stakeholders prior to distribution (Sustainable Uplands Project, no date).

Discussion focussed around future drivers of change in the different landscapes that were visited, how these might play out in the upland system, and how stakeholders might be able to adapt to these changes. Two scribes took notes to capture the discussion and summarised key points at the end to provide participants with an opportunity to correct misinterpretations and/or add important missing points. Thus information about drivers of change and their potential effects on system dynamics was obtained in collaboration with stakeholders though individual semi-structured interviews, group site visits between stakeholders and researchers, and a conceptual modelling workshop to check our assumptions and findings from an extensive literature review. This information was then used to develop preliminary scenarios (Reed et al., under review).

Preliminary scenarios were evaluated by a cross-section of stakeholders in the Peak District National Park, in October 2007. Scenarios were then prioritised and ranked, and alternative scenarios that had not been evaluated were elicited and discussed. In addition, there was a desire to see more "surprise" scenarios that were unlikely to happen, but that would have a major impact if they did occur. The scenarios were considered general in nature, and participants requested more site-specific, spatially explicit components. This will be provided through the model outputs in the next stage of the research. Participants also suggested that there should be more socio-economic scenario components. A number of such components were added by participants (Reed et al., under review).

More detailed implications of each scenario are currently being explored using integrated computational models (see above). Likely feedback will be investigated including potential interactions between scenarios that could occur concurrently. Outputs from this process relating to the scenarios short-listed in each study area will then be communicated to stakeholders supported by visual aids. This will then form the basis for discussion to identify innovative adaptation options that could help maintain livelihoods and the ecosystem services upon which they depend under each scenario. In this way the ultimate goal is to inform future decision-making that could enable effective adaptation to upland change.

Monitoring of adaptive options in this project takes place in two ways. First, indicators were developed to monitor progress towards sustainability goals. These are designed for use by stakeholders, given the relative inaccessibility of computer models to them for this purpose. Second, the project uses an integrated biophysical and socio-economic model of the upland system to evaluate the likelihood that adaptive options will achieve sustainability goals. This replaces the sort of monitoring that traditionally takes place in adaptive management. Because the results are based on models rather than empirical data, they can only be used in a heuristic capacity to guide decision making but also inform further research needs. On the other hand, the experimental approach usually adopted in adaptive management also has limitations. Notably, it is only possible to monitor and evaluate a limited number of adaptive options at limited spatial and temporal scales. In contrast, the modelling approach adopted by this project facilitates the evaluation of far more options across far greater spatial and temporal scales than is normally possible in adaptive management. Proposed adaptive options can then be changed in response to model outputs, in collaboration with stakeholders, before they are ever implemented on the ground. In theory, this should significantly enhance the likelihood that proposed options contribute towards sustainability goals, and reduces the length of the adaptive/learning cycles in the adaptive management process.

Reflections

Stakeholder participation and the iterative nature of the adaptive management cycle means that outcomes are necessarily uncertain and dynamic; given the involvement of diverse group of people the responses of the participants and the direction of the process is difficult to predict. This means that decision-makers may feel uncomfortable committing themselves to implement and resource the as-yet unknown outcome of an adaptive management process. In many cases, to do so would represent a radical shift in the organisational culture of government agencies and other relevant institutions, especially funding agencies. Although this means adaptive management may be perceived as a high risk strategy by those with power, there is growing evidence that if the process is well designed, these perceived risks may be well worth taking. Building on experience from this case studies and other evidence, Reed et al. (under review) (2008) concluded that (participatory) adaptive management needs to be underpinned by a philosophy that emphasises empowerment, equity, trust and learning. Stakeholder participation should be considered as early as possible and throughout the adaptive management process, representing relevant stakeholders. The adaptive management process needs to have clear objectives from the outset, and should not overlook the need for highly skilled facilitation. Local and scientific knowledge can be integrated to provide a more comprehensive understanding of complex and dynamic socio-ecological systems and processes. Such knowledge can also be used to evaluate the appropriateness of potential technical and local solutions to environmental problems. Finally, Reed et al. (under review) argue that to overcome many of its limitations, adaptive management must be institutionalised, creating organisational cultures that can facilitate processes where goals are negotiated and outcomes are necessarily uncertain.

Most of the organisations and participants in the case study supported the adaptive management approach adopted by the project (although for different reasons). For example, conservationists might see such processes as an opportunity for influencing land owners after everything else has failed; or land managers would like to like to get involved in such a process to know what is going on and also to have the possibility to influence the research teams and see the project as a way to get research done that supports their own interest. Despite the fact that there are these special interests involved in the process we found that there is an understanding and a common vision of a larger common good and an interest in working towards it, even though the means to achieve it might differ.

There is a growing recognition that the complex, uncertain and multi-scale nature of environmental problems demands transparent decision-making that is flexible to changing circumstances, and embraces a diversity of knowledges and values. To achieve this, adaptive management approaches based on stakeholder participation are increasingly being sought and embedded into environmental decision-making processes, from local to international scales (Stringer et al., 2006). Following these larger trends, there has been an increase in partnerships (e.g. the Moors for the Future partnership and similar bodies being established in other upland areas in the UK), and a growing interest in the sorts of outputs that adaptive management can offer. For example, two large national-level stakeholder organisations independently initiated their own "upland futures" programmes after the start of this project, informed by project outputs, in an attempt to better anticipate and adapt to future drivers of change.

Despite the project's focus on a wide variety of ecosystems services and their trade-offs, there has been substantial stakeholder interest especially in climate change related issues. Many degraded upland peat soils currently lose more carbon than they absorb through gaseous and fluvial pathways. Peatlands represent one of the few long-term stores of carbon that can accumulate on the land surface through good management, so the identification and restoration of damaged peatlands to functioning ecosystems could have significant beneficial impacts (Worrall et al., 2003). This focus on carbon was not there at the beginning of the project but we found that our future scenarios were strongly influenced by what was in the minds of people at the moment such as a specific regulation, like the ongoing discussion of the grass burning code in the earlier stages of the project or the bird flu and reoccurrence of the foot and mouth (MS Reed et al., under review).

Operationalising this adaptive management project to date has been far from straightforward, and a number of problems have been encountered and led to an adaptation of the process. For example, participatory model building was attempted unsuccessfully in an initial multi-stakeholder workshop. This was due to the highly heterogeneous composition of the group in terms of their views/interests and formal education level, coupled with inadequate facilitation. First, although experienced in facilitating workshops with high-level stakeholders in other countries, the professional facilitator was not sufficiently familiar with the local issues and stakeholders to be able to adequately follow and hence facilitate discussion. In addition, the wide range of educational backgrounds, ranging from those who were illiterate to those with PhDs, presented significant facilitation challenges and methods based on reading and writing had to be abandoned. Before this limitation had emerged a discussion had been facilitated about system structure and function on the basis of a number of pre-prepared, highly simplified and linked system components drawn from the initial system model. However, this discussion was dominated by more formally educated stakeholders who attempted to add complexities and feedbacks back in, with significant debate over a very small number of system components and links. The lack of alternative, more appropriate facilitation tools that could be used by illiterate participants, led to a power dynamic where more educated participants felt more comfortable and authoritative, and less formally educated participants felt marginalised and disempowered. As a result, little constructive progress was made during this workshop. All participants, even those who remained quiet during the workshop, indicated their willingness to participate in further activities, which is another indicator that the format was not appropriate.

Learning from this experience and building on suggestions from stakeholders, a series of site visits was developed to initially replace workshop activities, using the landscape as 'classroom'. Investment was made in professional facilitation training for two project members, who then shadowed a UK-based professional facilitator (with extensive knowledge about regional issues) on site visits, and then led site visits under observation before conducting facilitation unaided. The site visit programme was designed by a steering group of stakeholder representatives who selected the issues to be covered and the most appropriate sites to stimulate discussion. The steering group suggested the development of information sheets about each issue, to ensure all participants had similar levels of information about each issue and could engage in debate at a similar level with one another.

Site visits were designed to bring stakeholders with different interests and backgrounds together with researchers as equal partners to discuss the upland management issues that were perceived to be most important. The outdoor context and facilitation style were seen as reducing the discrepancies in power that were witnessed in the initial workshop, with all participants feeling comfortable engaging in discussion. Again, facilitation was an important element in the success of each of these participatory tools to overcome the divide and facilitate an open exchange of ideas. For example, we found that initially in the workshops and site visits discussions that the environmentalists held back because they need to work with people whereas the land owners would be much more vocal in voicing their views or frustrations.

Often in meetings we also found that people started citing various studies and the process was in the danger of degrading to an argument of 'my evidence versus your evidence.' To overcome this process of pulling evidence out of the bag we early on decided to develop the integrated model to have a commonly agreed evidence base. This acceptance of one source was to be achieved through the participatory modelling approach (see above). Science this is an ongoing process we cannot report her if it actually works in this particular case and what new problems might emerge because of this approach. Despite the advantages that participatory model building potentially offers, it is far more time-consuming than more traditional approaches. It has required modellers from very different disciplinary backgrounds to learn how to work together, and work out how their models can be meaningfully integrated. For biophysical modellers used to describing and modelling the environmental system through deductive science, the participatory approach was both refreshing and frustrating. Partly, this was because the biophysical modelling could not progress until the issues emerged from the participatory process; for natural scientists to begin a project without knowing what would be modelled was an unfamiliar experience (Prell et al., 2007).

Outputs from the integrated computation model still need to be communicated to stakeholders in a transparent way, so that the model is not seen as a "black box" and stakeholders have the capacity to interact with and modify the model if necessary and see and accept it as tool and support. At the moment the two processes, participation and model building have had little interaction. The site visits and scenario workshops have engaged the stakeholders and policy briefings have been developed and discussed but 'the model' has not entered the stage yet despite the fact that outputs of the participatory activities have been used to build the computer models. In the next step the outputs of the model will be used as a basis for discussion in workshops. The model should help to learn about the structure and dynamics of the system, the implication of land manager's responses, and as a basis to develop new responses to policy or environmental drivers.

Overall, we have experienced an eagerness of people to get involved and talk about issues relating to land management issues of the future. We have also been invited to submit our outputs to planning and futures exercises of statutory bodies. If the project develops into a truly adaptive management project with a continuation of experimental learning after the research project itself has ended remains to be seen. We are hopeful that we contributed towards such a process.

Acknowledgements This work was carried out as part of project RES-224-25-0088, funded through the joint Rural Economy and Land Use programme (RELU) of the UK Economic and Social Research Council, Biotechnology and Biological Sciences Research Council and Natural Environment Research Council, with additional funding from the UK Department for Environment, Food and Rural Affairs and the Scottish Executive Environment and Rural Affairs Department. The research was conducted in close collaboration with staff from the Moors for the Future Partnership, which is supported by the Heritage Lottery Fund.

References

Bellamy, P. H., Loveland, P. J., Bradley, R. I., Lark, M. R., Guy, J. D., & Kirk, G. J. D. (2005). Carbon losses from all soils across England and Wales 1978?2003. *Nature, 437*, 245–248.

Chapman, D. S., Termansen, M., Jin, N., Quinn, C. H., Cornell, S. J., Fraser, E. D. G., Hubacek, K., Kunin, W. E., & Reed, M. S. (2009). Modelling the coupled dynamics of moorland management and vegetation in the UK uplands. *Journal of Applied Ecology. 46*, 278–288. doi: 10.1111lj.1365-2664.2009.01618.x.

Derbyshire Chamber of Commerce. (2005). *Survey of Businesses in the Peak District and Rural Action Zone. Report for the Peak District National Park*. Chesterfield.

Dougill, A., Reed, M., Fraser, E., Hubacek, K., Prell, C., Stagl, S., Stringer, L., & Holden, J. (2006). Learning from doing participatory rural research: Lessons from the Peak District National Park. *Journal of Agricultural Economics, 57*, 259–275.

English Nature. (2003). *England's best wildlife and geological sites; the condition of SSSIs in England in 2003*. Peterborough: English Nature.

Giordano, R., Passarella, G., Uricchio, V. F., & Vurro, M. (2007). Integrating conflict analysis and consensus reaching in a decision support system for water resource management. *Journal of Environmental Management, 84*(2), 213–228.

Hanley, N., Wrigth, R. E., & Adamowicz, V. (1998). Using Choice Experiments to Value the Environment: Design Issues, Current Experience and Future Prospects. *Environmental and Resource Economics, 11*(3–4), 413–428.

Holden, J., Shotbolt, L., Bonn, A., Burt, T. P., Chapman, P. J., Dougill, A. J., Fraser, E. D. G., Hubacek, K., Irvine, B., Kirkby, M. J., Reed, M. S., Prell, C., Stagl, S., Stringer, L. C., Turner, A., & Worrall, F. (2007). Environmental change in moorland landscapes. *Earth-Science Reviews, 82*(1–2), 75–100.

Hubacek, K., Prell, C., Reed, M., Bonn, A., & Boys, D. (2006). Using Stakeholder and Social Network Analysis to support participatory processes. *International Journal of Biodiversity Science and Management, 2*(3), 249–252.

Hubacek, K., Dehnen-Schmutz, K., Qasim, M., & Termansen, M. (2008). Description of the upland economy: areas of outstanding beauty and marginal economic performance. In A. Bonn, T. Allott, K. Hubacek, & J. Stewart (Eds.), *Drivers of environmental change in uplands*, London: Routledge.

Kirkby, M. J., & Neale, R. H. (1987). A soil erosion model incorporating seasonal factors. *International Geomorphology 1986: Proceedings of the First International Conference on Geomorphology, II*, 289–310.

MFF. (no date). Moors for the Future Website. *www.moorsforthefuture.org.uk*. Moors for the Future. Retrieved 31.10.2008, from the World Wide Web:

Prell, C. (2003). Community networking and social capital: early investigations.. *Journal of Computer-Mediated-Communication, 8*(3), http://jcmc.indiana.edu/vol8/issue3/prell.html

Prell, C., Hubacek, K., Reed, M., Burt, T., Holden, J., Jin, N., Kirby, M., Quinn, C., & Sendzimir, J. (2007). If you have a hammer everything looks like a nail: 'traditional' versus participatory model building. *Interdisciplinary Science Review, 32*(3), 263–282.

Prell, C., Hubacek, K., Quinn, C., & Reed, M. (2008). 'Who's in the social network?' When stakeholders influence data analysis. Special Issue *Systemic Practice And Action Research. 21*, 443–458. DOI: 10.1007/s11213-008-9105-9.

Prell, C., Hubacek, K., & Reed, M. (forthcoming 2009). Stakeholder analysis and social network analysis in natural resource management *Society and Natural Resources*.

Reed, M., Arblaster, K., Bullock, C., Burton, R., Fraser, E., Hubacek, K., Mitchley, J., Morris, J., Potter, C., Quinn, C., & Swales, V. (under review). Using scenarios to explore UK upland futures. *Futures*.

Reed, M., Bonn, A., Broad, K., Burgess, P., Burt, T., Fazey, I., Hubacek, K., Nainggolan, D., Quinn, C., Roberts, P., Stringer, L., Thorpe, S., Walton, D., Ravera, F., & Redpath, S. (under review). Participatory scenario development for environmental management: a methodological framework. *Journal of Environmental Management*.

Reed, M., Prell, C., & Hubacek, K. (2005). *Sustainable Upland Management for Multiple Benefits: a multi-stakeholder response to the Heather & Grass Burning Code consultation*. Project report submitted to DEFRA's consultation on the review of the Heather and Grass Etc. (Burning) Regulations 1986 and the Heather and Grass Burning Code 1994.

Schofield, R. V., & Kirkby, M. J. (2003). Application of salinization indicators and initial development of potential global soil salinization scenario under climatic change. *Global Biogeochemical Cycles, 17*(3), 1078.

Stringer, L. C., Dougill, A. J., Fraser, E., Hubacek, K., Prell, C., & Reed, M. S. (2006). Unpacking 'participation' in the adaptive management of socio-ecological systems: a critical review. *Ecology and Society, 11*(39), [online].

Sustainable Uplands Project. (no date). Website of the Sustainable Uplands Project. *URL: www.see.leeds.ac.uk/sustainableuplands*.

Termansen, M. (2009). Combining choice models and habitat succession models for land use change modelling. International Choice Modelling Conference. Harrogate, 2009.

Walz, A., Lardelli, C., Behrendt, H., Gret-Regamey, A., Lundstrom, C., Kytzia, S., & Bebi, P. (2007). Participatory scenario analysis for integrated regional modelling. *Landscape & Urban Planning 81*, 114–131.

Worrall, F., Reed, M. S., Warburton, J., & Burt, T. (2003). Carbon budget for a British peat catchment. *Science of the Total Environment, 312*(1–3), 133–146.

Chapter 11
Signposts for Australian Agriculture

Jean Chesson, Karen Cody, and Gertraud Norton

Abstract Signposts for Australian Agriculture was initiated by the federal government to help provide a better sense of how the agricultural industry contributes to sustainable development. It offers profiles of six agricultural sectors and how those sectors are performing with regard to specific objectives, such as the conservation of natural biodiversity. As such, it helps identify priorities for investments in natural resource management. From an adaptive management perspective, Signposts is an example of a more passive form of the continuum of learning-based approaches. A major finding in an evaluation of the project has been the underutilisation of data in some situations and the lack of data in others.

Introduction

Agricultural industries are significant environmental managers. In Australia, agriculture occupies just over 61% of the land area and uses approximately 65% of extracted water (AGDAFF, 2008). Environmental management, however, is not the only role of agriculture. Agricultural industries produce food and fibre to feed and clothe humans. In doing so, they have significant economic and social impacts on local, regional, national and international communities.

Signposts for Australian Agriculture (Signposts) is a project initiated by the Australian Government Department of Agriculture, Fisheries and Forestry in 2004 to demonstrate and communicate industry performance. Signposts represents an essential part of the adaptive management cycle described in Chapter 2 of this volume, in this instance with a focus on the 'learn' and 'describe' components. If performance is not meeting expectations, what else should be done? Should performance, or expectations, or both, be altered?

J. Chesson and G. Norton
Bureau of Rural Sciences, Canberra, Australia

K. Cody
National Land & Water Resources Audit, Canberra, Australia

The Signposts Partnership

Signposts is a partnership between industry and government. It asks the question 'How does an agricultural industry contribute to sustainable development?' Sustainable development is interpreted as an increase in the value of our assets over time where assets are interpreted in the broadest sense to include, for example, natural capital, produced capital, human capital and social capital (Hamilton & Atkinson, 2006). The Signposts framework distinguishes assets 'held' by the industry such as land and industry institutions from assets held by others such as the atmosphere and social capital associated with regional communities. For assets held by the industry, industry performance is measured by the change in value of those assets. Change in value can be measured in a variety of ways and does not necessary involve monetary valuation. For assets held by others, industry performance is measured by the industry's impact on those assets. The components of the framework are progressively sub-divided until it is possible to specify a desired outcome against which performance can be measured. The resulting 'component tree' (Fig. 11.1) is tailored to the needs of each industry.

The Signposts framework was developed through a series of workshops and formal and informal review as documented in a series of reports published by the National Land & Water Resources Audit (LWA, n.d.) The framework was designed to evolve, and the first 2 years saw progressive refinement of the asset-based approach and considerable development of the social components. More recent changes have been primarily in the detail of the lower-level components.

In collaboration with industry representatives, web-based profiles of six agricultural industries (grains, beef, horticulture, dairy, wine and cotton) have been created and are being prepared for public release at www.signposts4ag.com.au. The profiles are intended to provide readily accessible information for government and industry policy makers. Each profile is based on an industry-specific component tree. When a user clicks on a component they are provided with the desired outcome for that component, an indicator to measure performance and a summary measure on a scale between 0 and 1 where 0 represents unsatisfactory performance and 1 represents ideal performance. The desired outcome is based on stated industry and government policies wherever they can be identified. Indicators are derived from available data and may be replaced when better alternatives become available.

For example, one of the components of the value of land managed by an industry is its capacity to conserve native biodiversity. In the web-based profile, there is a desired outcome based on government and industry policy. The indicator might be the area of native vegetation in good condition and the performance measure might be this area as a proportion of an agreed target. Different indicators may be used for different industries depending on the specific desired outcome and data availability. The primary objective is to show progress over time.

Each component also includes a description of the responses and management practices that have been or could be applied to improve performance and cross references to other components that might interact with the component of interest.

11 Signposts for Australian Agriculture

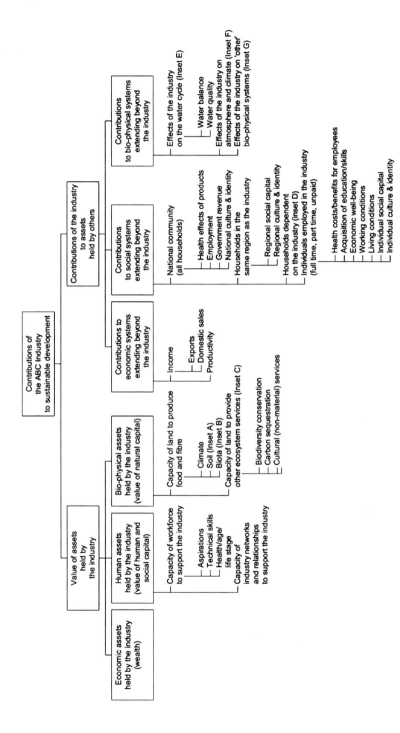

Fig. 11.1 Generic components of the Signposts framework. The component tree is tailored to the needs If each agricultural industry.

Agriculture exhibits all the properties of a complex and complicated management problem listed in Chapter 2 of this volume. A key achievement of Signposts has been to articulate the multiple objectives for agriculture from society and industry perspectives. This encourages policy decisions regarding agriculture to be made more holistically rather than in the context of a single issue such as water use or regional employment. An important use of Signposts is in prioritising investment in natural resource management. Signposts identifies:

- The extent and condition of natural resources managed by an industry
- Current management practices and
- The associated social, economic and environmental contributions of the industry

From an adaptive management perspective, Signposts is currently operating at the passive end of the continuum. This is not surprising, as Signposts looks at the net aggregate effect of all policies, strategies and actions carried out by government, industry and individuals. Coordinated experimentation at this scale will rarely be practicable. Signposts does however provide for 'natural experimentation' in the same sense as ongoing monitoring of bio-physical systems can provide valuable comparisons before and after a volcanic eruption or with or without the presence of an invasive species. The Signposts framework can be used to ask whether performance increased or decreased following some policy intervention. This type of question may lead to more specific investigation and possibly to more active forms of adaptive management focussing on a particular issue but within the overall context of a complex, interacting system. Signposts helps identify questions that need to be answered either through data collection, formal or informal modelling or a combination of all three.

Lessons

A major finding of the Signposts project has been the underutilisation of data in some situations and the lack of data in others. There are opportunities to make better use of industry-specific social data collected by the Australian Bureau of Statistics and through programs such as the former Agriculture Advancing Australia program. Signposts has been able to extract industry-specific bio-physical information by superimposing information on industry location on existing spatial bio-physical data. However, much of the bio-physical information is for one time period only and is not being collected over time. This issue has been raised repeatedly, most recently by the Wentworth Group of Concerned Scientists (Boully et al., 2008) and through the 2020 Summit (DPMC, 2008), resulting in discussions about the creation of a national environmental information system. Without systematic collection of information over time, the opportunity for effective adaptive management is lost.

The development of the Signposts framework required time but otherwise was not particularly resource hungry. The framework has multiple and continuing uses.

For example, the Australian grains industry used it to help develop their environmental plan (GRDC, 2008). The creation of the web-based profiles required more resources, but the focussed nature of the framework allowed them to be established within a modest budget. The goal was to make better use of existing information and provide the basis for strategic data collection in the future. The extent to which the existing profiles are updated and expanded will depend on having industry and government support for future funding.

References

AGDAFF (2008) Australia's agricultural industries 2008: At a glance. Australian Government Department of Agriculture, Fisheries and Forestry, Canberra.

Boully, L., Cosier, P., Flannery, T., Harding, R., Karoly, D., Lindenmayer, D., Possingham, H., Purves, R., Saunders, Thom, B., Williams, J. and Young, M. (2008) Accounting for nature – A model for building the national environmental accounts of Australia. Wentworth Group of Concerned Scientists, Sydney.

DPMC (2008) Australia 2020 – Australia 2020 Summit final report. Department of Prime Minister and Cabinet, Canberra.

GRDC (2008) A responsible lead: an environmental plan for the Australian grains industry. Grains Research and Development Corporation, Canberra.

Hamilton, K. and Atkinson, G. (2006) Wealth, welfare and sustainability: Advances in measuring sustainable development. Cheltenham, UK

LWA (n.d.) Australian Government National Land & Water Resources Audit, website www.nlwra.gov.au. Land & Water Australia, Canberra.

Chapter 12
Environmental Management Systems as Adaptive Natural Resource Management: Case Studies from Agriculture

George Wilson, Melanie Edwards, and Genevieve Carruthers

Abstract There are strong parallels between Environmental Management Systems (EMS) and Adaptive Management (AM); both focus on a cycle of continuous improvement through planning, doing, checking and acting and they both enable the modification of management practices based on monitoring. AM is a science-based structure for natural resource management. The strength of AM is that it brings a scientific approach to the management of complex biological, ecological, economical and social processes and that is what agriculture is. EMS can be based on an international standard. A manager using EMS identifies likely environmental impacts and legal responsibilities and implements and reviews changes and improvements in a structured way. EMS was developed so it could be used in all business sectors. The complexity of issues facing agricultural managers can provide a challenge to the application of EMS within that sector, however at the same time the process involved in developing an EMS can assist greatly in reducing and clarifying the complexity. An understanding and application of AM can also assist the application of EMS in agriculture. Importantly, in both AM and EMS the modifications are continual and can be determined mid-course. This chapter draws on an analysis of a group of 17 agricultural EMS case studies as examples of adaptive management in an industry that uses natural resources.

Adaptive Management and Environmental Management Systems

Why Introduce EMS in a Book on Adaptive Management?

Chapter 2 describes the components of adaptive management (AM), and provides a framework and a set of operational methods for application to complex natural

---

G. Wilson and M. Edwards
Australian Wildlife Services, Canberra, ACT

G. Carruthers
NSW Department of Primary Industry, Wollongbar, NSW

resources management problems. Thus AM is used when impacting factors are so complex there is no clear management path to 'solve the problem'. The application of AM to agriculture is a particular application in which the 'problem' is how to reduce the environmental impact of agricultural production and enhance sustainability of the natural resource. An Environmental Management System (EMS) is a tool with many similarities to AM and there is a growing interest in its implementation in the agricultural sector.

The Origins of EMS

In 1946, the International Standards Organisation (ISO) was founded in Geneva to facilitate international trade and increase the reliability and accuracy of the descriptions of goods and services (see www.iso.org/iso/about/the_iso_story.htm). ISO now operates in over 150 countries around the world. Initially involved in the development of technical standards, the ISO committees changed their focus to management practices in 1979 (Tibor and Feldman, 1996). ISO went beyond its initial focus on product specification standards and prepared ISO 9000 – the international management quality standard or in other words, the 'how' of doing business (Gilpin, 2000). Since then, questions have been increasingly asked about the adverse impact humans have on the environment and the sustainability of the use of resources and production practices. At the same time many companies implementing environmental management and pollution control found that savings came from the critical evaluation of these processes and their businesses benefited. As a result a range of management standards specifically relating to EMS proliferated worldwide.

In 1996, ISO issued ISO 14000, a set of environmental management guidelines for activities that have an effect on the environment and provide a cost effective, means of complying with environmental regulations and integrating economic and environmental performance (Lamprecht, 1996). Within these documents, ISO 14001, the Standard for EMS development, has become the most widely used EMS standard worldwide, and 112 countries have begun to issue certifications pursuant to its guidelines (Bellesi et al., 2005). It includes implications for food safety, international trade, consumer purchasing preferences and financial and legal risk management (Heinze, 2000).

Various EMS Schemes

In addition to ISO 14001, other environmental management approaches such as best management practices, codes of practice, catchment management targets and regional and local scale plans, have been developed. These differ in how prescriptive they are in terms of performance outcomes and practices. ISO 14001 does not specify particular environmental targets; instead these are set by the person/company/business setting up the EMS. ISO 14001 however, in contrast to a range of other documents does require that the EMS developed must as a minimum, meet standard legislated requirements, and/or (if available) industry codes of practice or best management practice.

While quality control and assurance programs focus mainly on the consistent production of goods and services, EMSs have a broader focus. That is, not only will goods be produced according to the same set of standard procedures, but the effects of producing that product, be it impact of pollution, rate of use of resources, or impacts of transport of the completed object or delivery of the service, are also taken into account and addressed in order to minimise, or where possible, eliminate these impacts.

The EMS Cycle

EMS is described as the application of a systematic management approach used by an enterprise or business to manage its impacts on the environment. It seeks continual environmental improvement (Gleeson & Carruthers, 2006). It can be applied in many different types of industries but each follows much the same path. A manager uses EMS to identify likely environmental impacts and legal responsibilities, then implements and reviews changes and improvements in a structured way. Figure 12.1 shows the 'predict, do, learn, describe' cycle of AM and Fig. 12.2 the similar continuous improvement cycle of EMSs' – plan, do, check, and act in a systematic process.

Under the International Standard (ISO 14001) there are five major components for developing and using an EMS. These steps are:

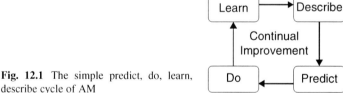

Fig. 12.1 The simple predict, do, learn, describe cycle of AM

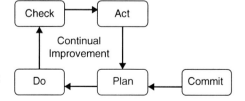

Fig. 12.2 The simple commit, plan, do, check cycle of an EMS

- **Commitment and policy development** – In agriculture this is where a producer or farm manager commits to undertaking an EMS, and develops a policy statement which makes explicit the areas that are to be addressed under the EMS.
- **Planning** – this is where the policy is translated into things to be done. A number of steps are usually involved, from a review of the environmental aspects and impacts of the business, the identification of legal requirements for compliance, the setting of objectives and targets, through to establishing the environmental management program.
- **Implementation** – this is the 'doing' of the plan. This phase requires the provision of resources and support mechanisms to ensure that the environmental management plan is achieved, and may include staff training programs to ensure that the objectives of the policy and plan can be met.
- **Measuring and evaluation** – this phase checks to see if the objectives and targets previously established are being met. Such methods as environmental performance evaluation, laboratory analyses of emissions, financial records examination and staff understanding of training programs may be used to assess whether the environmental plan is being met.
- **Review and improvement** – here the data gathered in the previous phase are put to use. Were targets met? If not, why not? What can be improved? What worked well and why? ISO 14001 specifies that continuous improvement of the management system (note: not the environmental performance) is required.

EMS Can Extend to Agriculture

Farmers are intuitively adaptive. They are always trying out new ways of doing things; if their methods don't work they modify them slightly and try again. This is AM in its simplest form.

In the case of the application of EMS in agriculture, the growing of beef cattle is a suitable example to illustrate the point. Not only does (or should) the farmer want to be able to produce a consistently high quality product, but s/he should also consider the way in which the cattle are treated during the production period, the effect they have on the soil, water and air quality on and off the farm, ways in which the cattle are transported, the potential impact other farm activities might have on the cattle (e.g. use of pesticides), the packing of beef produced and so on. This consideration of the entire cycle of production is the essence of an EMS, and thus farmers are well placed to adopt the adaptive management cycle of an EMS, in order to capture not only business management but also potential market differentiation benefits.

Importance of EMS in Agriculture

For a long time Australian Agriculture has endured criticism about practices that adversely impact on the environment. The impacts include a decrease in biodiversity, water quality and availability and increased soil erosion and salinity. During 1999–2000 the debate on the impact of natural resource degradation in Australia began in earnest. Amongst other things, this resulted in the establishment of a new Natural Resources Management Ministerial Council (NRMMC).

In 2002 the NRMMC published the Australian National Framework for EMS in Agriculture (NRMMC, 2002). It nominated three key drivers as objectives for implementing EMS in Agriculture:

- Natural resource management and environmental improvement
  - Such as conservation of soil, water, vegetation, and biodiversity
- Competitiveness objectives
  - Such as input–output efficiencies, better prices, lower costs, more efficient production
- Social objectives
  - Landholder and community values such as cultural heritage and occupational health and safety matters

While these objectives are relevant to most producers, agriculture has a very variable nature, as a result of the many biological production processes. The application of EMS to agriculture is thus complex with the diversity of agricultural enterprises complicating the process further. However, the process, rather than prescription approach, means that the complexity can be addressed and built upon in any EMS developed, reflecting the true diversity of enterprises using the EMS tool.

The remainder of this chapter examines reports of case studies of EMS in agriculture by farmers who responded to the funding opportunities and new programs which were funded by governments. The discussion considers what works, what doesn't, why and what can be done to improve the EMS process in agriculture.

Review of Case Studies of the Adoption of EMS in Agriculture

In order to explore the concept of EMS as an AM process we have drawn on a review of EMS in agriculture by one of us – Genevieve Carruthers. She conducted interviews with over 40 farmers and farm managers adopting either a recognised EMS process or less formal environmental management programs (EMP) in Australia and New Zealand (Carruthers, 2003a, 2005). Here we focus on the results from the EMS group which included 17 farms in Australia and New Zealand. The EMS farmers were all full time farmers or farm managers, with all of them but one having agricultural careers between 10 and 40 years. As the interviews were conducted quite early after the ratification of ISO 14001 worldwide, the EMS farmers were at the cutting edge of adoption, with most having been involved in the use of EMS for less than 3 years. The study included the first farmer in the world to gain ISO 14001 certification, an Australian cotton farmer. The aim of the report was to assess the usefulness of applying an EMS within an agricultural context rather than the more usual context within secondary industry (Carruthers, 2003a, 2005). We also examine the EMS National Pilot Study, which examines the risks that impact agriculture in general and those that are relevant to EMS.

The EMS National Pilot Program 2003–2006 launched by the Australian Government was followed by the Pathways to Industry Environmental Management Systems Program from 2004–2007 which assisted 19 industry bodies to develop and implement EMS and other environmental assurance approaches. It was believed that this would position them to achieve the adoption of profitable and sustainable farming practices, improved natural resource management and environmental outcomes, and provide them with ability to demonstrate environmental stewardship to domestic and international markets.

Table 12.1 shows the range of producers across sectors which have carried out EMS activity. It also lists a number of industry organisations, large and small, with EMS projects. Although it is not exhaustive it shows the scope of industry sectors that can and do carry out EMS activity. DAFF (2007) and Gillespie et al. (2008) list more detailed examples and case studies.

Table 12.1 Industry, sectors and company/organisation examples carrying out EMS

Industry	Sector	Company/organisation
Plant-based broad-acre industries	Cotton	Cotton Research and Development Corporation
	Grains	GRDC
	Rice	Ricegrowers' Association of Australia
	Sugar	Canegrowers
Horticultural industries	Vegetables	Horticulture Australia Limited
	Fruit	AUSVEG Limited
		Mount Lofty Ranges Watershed
		Apple and Pear Growers of South Australia
	Nursery	Nursery & Garden Industry Australia
	Cut flowers	
	Extractive crops	
	Turf	Turf Producers Australia Ltd.
	Nuts	Victorian Wine Industry Association
	Wine and grapes	Grape and Wine Research and Development Corporation
	Bananas	Pacific Coast Eco-bananas Pty. Ltd.
Extensive animal industries	Sheep wool	Meat and Livestock Australia
	Lamb	Gippsland Beef and Lamb
	Beef	Northern Australian Pastoral Company
		Western Downs group
		YNot Beef Group
Intensive animal industries	Dairy	Australian Egg Corporation Ltd.
	Poultry eggs	Australian Chicken Growers' Council
	Poultry meat	Poultry Cooperative Research Corporation
		Baiada Poultry Pty. Ltd.
	Pigs	Australian Pork Ltd.
Forestry, Fisheries and Aquaculture	Forestry	Forest and Wood Products Research and Development Corporation
	Fisheries and Aquaculture	Fisheries Research and Development Corporation
		Seafood Services Australia
Mixed industries		NSW Murray Catchment/NECMA

Case Studies of Farmers Using EMS in Agriculture – Scope of the Case Studies

The review by Carruthers (2003a) targeted participants who were leaders in environmental management. They were identified by recommendations from farmer and industry groups, state agricultural, resource conservation and regulatory agency staff, certification companies, and by publicity about some of the farmers who had won environmental awards. Other areas of interest in selecting potential candidates were farm businesses making use of environmental labelling or marketing, and those who were selling into known 'environmentally sensitive' markets. The enterprises included in the study covered the full range of sectors and are shown in Table 12.2.

In the examination of case studies conducted by Carruthers, personal desire to improve the sustainability of the farm, and ensure the health of both personnel and resources were the specific drivers for improved environmental management. Some farmers were motivated by public pressure, market/consumer demands, desire for management improvement, erosion concerns, new development/license regulations or fear of sprays. Self choice, resource conditions such as drainage, salinity and water allocations, and regulatory agencies were also nominated by farmers as factors that had stimulated their change from previous environmental management.

Getting Started on EMS

Documentation

An EMS manual provides direction for an EMS and provides links to important information. Information generated by the EMS allows landholders to have their systems independently audited and certified to ISO 14001, if they desire. They can also provide information needed for Quality Assurance (QA) systems and provide evidence for 'clean and green' marketing claims (NRMMC, 2002). Fear of paperwork is considered a major barrier for farmers contemplating an EMS (Tinning & Carruthers, 2002). However, in Caruthers' study, fears of being document

Table 12.2 Sectors covered by the case studies

Sector	Examples
Intensive livestock	Poultry, pigs and feedlots,
Extensive livestock	Dairying, beef and sheep
Horticulture	Fruit, vegetables, essential oils and nursery
Broadacre farming	Rice, cotton and grain
Mixed enterprises	Usually a mix of livestock and cropping
Wine	Vineyards and wineries
Aquaculture	Prawns and salmon

controlled were not an issue. Instead, record-keeping was found to be useful as it enabled farmers to provide proof of stewardship efforts. In addition, by following the directed 'plan, do, check, act' approach of an EMS, farmers often found that their record keeping was rationalized, and that they put the data collected to better use. Records collected also allowed more informed decision making, based on fact, rather than recollection. Carruthers suggests that perhaps fears regarding paperwork and EMS may come from those who do not have firsthand experience in EMS and are commenting from a theoretical perspective. It is unknown, but would be interesting to know, how the farmers from these studies felt about documentation before they implemented EMS. This is an area currently being pursued by Carruthers in on-going research.

Identifying Indicators

Measuring the impact of management is central to EMS and the most useful indicators are those that a manager can easily comprehend and respond to. Table 12.3 shows indicators the farmers undertaking EMS from the case studies found useful.

Training, Advice and Group Membership

Seven farmers undertook EMS training and used ISO 14001 as a guide to change, half of them relied on individual observation and the rest relied on group membership and shared learning when developing changes to practices. They also made use of consultants who helped with the development and design of the EMS and advice on technical issues.

Other forms of support included best management practices, Codes of Practice (e.g. Farmcare Code of Practice) and a quality assurance (QA) approach. Two

Table 12.3 Key indicators used for EMSs

Environmental indicators	Business indicators	Output indicators
Water quality	Community relations	Waste water
Water table	Complaints	Nutrient run-off
Water use efficiency	Financial performance	Effluent nutrients
Salinity	Market appraisal	Soil erosion
Soil nutrients	Production statistics	Odour
Soil health	Training	Noise
Vegetation		Vehicle/machinery use
Ground cover		Fuel use
Chemical fertiliser use		Solid waste
Integrated pest management		Leaf tissue analysis
Bird counts		
Biodiversity		
Climatic conditions		
Environmental performance		

farmers drew on the Cotton Best Management Practices for pesticides, even though they did not farm cotton. An important finding of the study was that farmers using an EMS were more likely to adopt best management practices in general, and environmentally focused best management practices specifically, than their non-EMS using counterparts. This suggests that EMS use may lead to enhanced adoption of innovation for management throughout agriculture.

Community and Departmental Attitudes

EMSs demonstrate greater responsibility for the environment and they help develop community goodwill by increasing the Australian communities' confidence in agriculture. Many of the case study farmers thought the community was seeing their industry in a more positive light and reported that EMS improved relationships with neighbours and community. Around 30% of farmers stated that they had received some form of community recognition for their environmental work.

However, support from government agencies and decision makers was inconsistent. The natural resources departments were mentioned by less than 40% of farmers as being supportive, with slightly higher levels of support perceived from industry groups. Carruthers suggests that this may be because agencies were often seen to be on the 'back foot' in regard to provision of EMS advice specifically, with many of the farmers indicating that they felt that they were better informed about EMS than agency staff, and were 'teaching them'. In the case of industry groups, farmers were sometimes seeking a market benefit as an incentive to adopt EMS, and the industry groups were seen as being able to assist with achieving this goal. However, these groups again were seen as supplying little practical EMS information. The top five supporters were considered to be staff, family, environmental protection agencies, researchers, and agricultural departments of government, with different information sought from each.

Independent and Longer Term Advisers and Facilitators

Farmers often found it difficult to source information on EMS uncoupled from provision of EMS services. Early EMS information often came from consultants that had assisted the businesses with the establishment of QA programs, and EMS was seen to be an 'added service'. Farmers appreciated innovation of (the then) NSW Agriculture in providing a full-time officer to assist with EMS development. However, in general the EMS projects undertaken across Australia as part of the EMS Pilots Program were generally staffed by short-term appointed officers, who frequently had a specific enterprise focus to their work. Carruthers suggested that the establishment of a national network of EMS facilitators (similar to the Landcare network) would be a great step forward (Carruthers, 2005), and advocated that such a network be established as part of the development of the National EMS training program (Final report, Carruthers, 2003b).

Expectations and Learning Experiences

Many farmers saw improvements in environmental performance as positioning them to take advantage of the market for 'clean and green' as they become available. At the time of the study, however, market benefits are not what farmers might have hoped for. Consumers were more focused on food safety and quality than environmental attributes. There is now emerging evidence of market benefit, in particular market access. In addition, a range of groups (Pacific Coast Eco-bananas, The Gippsland Enviromeat Groups and the Merino Group) are now gaining price premiums for products from farms with an EMS in place. The European markets in particular are looking for third-party certified systems to be in place as a key market entry requirement (Carruthers, 2007b).

Outside Support

Farmers in the case studies obtained support from a variety of funding programs for environmental improvement activities. Sources included State governments, the Australian Government (through the Natural Heritage Trust), and a New Zealand community/EPA/local council conglomerate also contributing. Individual farms also obtained funding from natural resource management bodies, Greening Australia, CSIRO and a native vegetation covenant agreement. Such funding was usually used for fencing, revegetation and in a few cases it covered auditing and assistance with documentation and monitoring. Funding very rarely was provided for EMS development and implementation activities directly.

What Were the Benefits of EMS in the Case Studies?

Ecosystem Benefits

Many farmers reported an improvement in their resource base. Changes in flora and fauna on-farm became apparent and biodiversity was the most commonly nominated area of ecosystem benefit, followed by improved soil structure and condition. Farmers reported that they needed to spend less time fixing the consequences of adverse environmental impacts, which led to financial benefits of the EMS process.

Social Benefits

The three major categories arising from changed farm practices in terms of social benefits were greater peace of mind, more confidence in management and improved human health and safety. Confidence was the most frequently observed benefit and 'more planning' was the predominant reason for this. Confidence also grew from industry adoption of environmental management, supply of good-quality produce, gaining stakeholder support, and using a range of technology, research and infrastructure.

Carruthers and Vanclay (2007) summarise the social benefits of EMS. They report that EMS can affect neighbours and neighbouring communities, increase social interactions, create personal and family time, improve succession planning, help negotiate family partnerships and roles on the farm, help farmers gain a voice, increase social standing, legitimacy and proof of claim and last but not least create personal satisfaction.

Financial Benefits

The most commonly reported financial benefit was savings in input costs. Improved stock/crop health, better profits and yields were frequently commented on and a reduced workers' compensation insurance premium was also considered a saving. Some farmers obtained approved supplier status on the basis of their environmental performance. One farmer reported gaining a price premium because of their 'environmental credentials'.

Another farm experienced a 40% expansion in sales, at a time when other farmers in their industry reported a significant downturn. Approximately 60% of farmers were differentiating their products in some way in the market place, using either quality or environmental certification logos as a means of product differentiation.

Contrary to a commonly held belief in Australia that EMS adoption always involves great additional expenditure, the case studies showed that there was often little difference in reporting of expenditure on infrastructure, development, monitoring, and auditing costs between adoption of EMSs, EMP and other management systems (including QA).

What Didn't Work in the EMS Case Studies?

Auditors

The auditing process is carried out by an external third party and consists of five stages (Carruthers, 2003b). The first is to complete an application form, the second is a preliminary document review which includes a review of the key EMS documents and the third is a preliminary audit which includes reviewing environmental assessments, environmental policies and management system documents. The preliminary audit serves to highlight any deficiencies before the certification audit is carried out. The fourth stage is the certification audit which is a detailed audit of an EMS to determine if it is performing as required and to assess that it meets the ISO 14001 Standard. The final stage is ongoing surveillance audits which aim to ensure that the EMS is functioning effectively over time. Amongst the EMS farmers who had undertaken external auditing, many thought that few EMS auditors had a thorough understanding of the agricultural sector, with most auditors coming from a quality and industrially focused background. There is a small, but growing, number of EMS auditors in Australia who do have agricultural backgrounds available to work with farmers (Carruthers, 2005). EMS consultants are also developing more specific agricultural expertise.

Sourcing Support

Farmers from the case studies found sourcing support and information specific to EMS difficult. At the time of the study, there were virtually no agricultural EMS facilitators available and other community groups were unable to assist because they were also unfamiliar with the EMS approach. However, while the development of a national training course specific to agriculture on EMS (Carruthers, 2003a) has been an important step forward, there is still a need for a national network of skilled trainers and facilitators to assist farmers develop and implement agricultural EMSs. Such people not only provide a focal point, but should also be able to gather and share EMS experience across a range of industries, and provide encouragement and support to those newly approaching use of EMS. The formation of an EMS Association for Australia was promoted by the need to develop a community of practice amongst EMS practitioners in Australia (see www.ems.asn.au).

Consumer Demand

Around 40% of the farmers from the case study were hoping for improved market access and found that their EMS positioned them to take advantage of consumer demand for 'green' produce. However, most reported that current consumer demand was still focused on food safety, and QA audits were sufficient to meet the demands. Many farmers felt that more formal approaches to environmental stewardship would be 'required' in the future and thus saw their adoption of increased environmental management as 'getting ahead of the pack'. Premiums are now being achieved by groups such as Eco-bananas, Enviromeat and the Merino Group, as discussed above.

Measuring Benefits

Some farmers felt that some of the benefits were difficult to quantify, particularly outcomes such as confidence in management, improved community relationship/perceptions, and the like. They also found that it was difficult to estimate the financial benefits they were gaining from the information arising from improved monitoring of indicators. The case study questionnaire was not designed to determine economic benefits but instead to determine environmental and business management benefits. The EMSs developed by farmers were typically *not* designed or implemented in ways that focused specifically on gathering economic data. Rather the achievement of environmental outcomes was seen as the higher priority. There is a need to research the economic benefits that can accrue from EMS, but this requires specific targeted research that identifies this aspect of EMS at the outset, with the use of a multi-disciplinary team to conduct the research.

EMS National Pilot Program

In 2003, the Australian Government launched an EMS National Pilot Program. It was funded by the Natural Heritage Fund and included 1050 farmers and fishers in 16 projects from both extensive and intensive agriculture. Projects were backed by their industry associations and research and development corporations. The objectives of the program were to:

- Develop and assess the value of EMS as a management tool to improve natural resource management from the enterprise level to the catchment scale
- Assist industry competitiveness and production efficiency
- Help primary producers meet emerging market demands for quality and environment assurance

Findings

The EMS Pilots found that the EMS AM process provided farmers with a better understanding of their business and its effect on natural resource base, both at the farm and catchment level (DAFF, 2007). EMS also provided a framework which landholders could use to better manage their environmental risks, and become more aware of their legal requirements towards the environment and how to comply with regulation.

Carruthers (2006) summarised the outcomes of EMS implementation on Australian farms and found additional benefits. Improved relationships with neighbours, service providers, extension staff and regulators were reported. Many farmers actively engaged in providing advice and information to other farmers, regulatory agencies and industry groups and EMS contributed to an improved image for the whole of the industry. Further benefits such as enhanced information exchange, clear articulation of targets and ease of access to information are discussed by Carruthers elsewhere (Carruthers, 2007a). Financial performance was improved as a result of improved stock and crop health, and production efficiencies – increasing yields and profits. EMS also provided information regarding the true cost of production. A reduction of inputs has been associated with cost savings from more efficient evaluations and judicious application of inputs such as fertiliser and pesticides. An enhanced ability to attract funding and support for natural resource management initiatives and to demonstrate successes was also a benefit from the use of EMS.

A Summary of Risks and Issues

In another review of EMS, Quinn (2009) noted that EMS improved links between sections of an enterprise; this reduces the risk of counterproductive approaches as the integration between an enterprise and the natural systems that support it is better understood. There is also an increased knowledge base and in particular increased knowledge of the environment in which the enterprise operates and the natural systems supporting the environment. This encourages the adoption of the most progressive approaches, thus keeping ahead of regulatory pressures and at least some consumer preference changes.

Quinn reviewed reports produced in the course of the EMS Pilots and a subsequent EMS Pathways Program and literature about other experience in Australia and overseas. His report highlights risks and issues confronting agriculture and practical issues for environmental management. Monitoring their impacts and where possible managing them is a component of effective AM. Below we summarise Quinn's findings.

Risk Factors That Adversely Impact Agriculture in General

- Climatic variations are unpredictable beyond very short periods.
- Agriculture and associated land clearing are major contributors to greenhouse gas in Australia, accounting for about one third of total emissions.
- Most agricultural emissions are from livestock production, but all the other agricultural sectors contribute.
- Loss and damage due to pests and diseases is increasing in Australia due to imports of agricultural goods.
- Failure to understand the relationships among an enterprise, the natural environment, community expectations, regulatory systems and market needs.
- Biodiversity is fundamental to human activities and is constantly at risk from agricultural practice.
- Loss of biodiversity follows habitat destruction, growth by pest plants and animals, pollution and hunting.
- Costs of production tend to continue to rise while prices for the raw products fall.
- Need to increase the profitability of the enterprise, either by including all the costs in product prices or by otherwise rewarding the enterprise owners.

Risk Factors That Reduce Support for EMS

- Perceptions of a lack of relevance of EMS, as being unnecessary, time consuming and unrelated to income, or even conflict with business plans.
- Some landowners are wary of schemes that may lead to publicising information about threatened species locations. This is linked with the failure of the community to pay landowners for eco-system services or to compensate them for restricting their commercial activities in support of an environmental objective
- Concern about documentation and costs of paperwork, particularly for fully certified and audited systems.
- Consumers in Australia and overseas place a higher priority on food safety and product quality.
- Many consumers continue to expect that the products they buy and use have minimal or no environmental impact, and that producers respect the environment.
- Persistence with traditional ways of doing business can also hamper attempts to make beneficial changes.
- A continuing search for systems to capture value for the enterprise because of its use of EMS.
- EMS process being too rushed, reinforcing the need to tailor introduction of processes to the needs, style, interests and level of knowledge of the prospective audience.
- Enterprise isolation or an absence of networks can also be a handicap, reinforcing the desirability of working through compatible existing structures such as landcare networks and industry associations.
- Difficulties in gaining access to the wide array of public and private AM, EMS provides a structure for collating and using information about the management of natural resources, in order to achieve improved outcomes. Achieving this brings with it several other benefits but some costs.

Conclusions

Like AM, EMS provides a structure for collating information about the management of natural resources. Achieving this brings with it several other benefits but some costs.

- AM is useful in natural resource management when a problem has no clear or obvious causative factors, and there is no clear management path to 'solve the problem'.
- EMS follows a similar, systematic process: plan, do, check and act to AM.
- Drivers of EMS in agriculture include personal desire to improve sustainability, ensuring health of personnel and resources, public pressure, market/consumer demands, management improvement, erosion concerns, new development, license regulations and fear of sprays.

- EMS adoption does not involve great additional expenditure and there is often little difference reported of expenditure on infrastructure, development, monitoring, and auditing costs between adoption of EMSs, EMP and other management systems (including QA).

The following is a summary of benefits and difficulties which have become apparent from case studies and pilot projects applying EMS in agriculture.

Benefits

- The documentation that is required for EMS is not as intimidating as once reported and farmers consider better documentation to be useful in the more profitable operation of their enterprise. This arises from both a change in the type and amount of information gathered and used, and how it is used by farmers.
- EMSs can demonstrate environmental responsibility by agricultural producers to the community.
- Like AM, EMSs allows new information to be uncovered which can then be used to improve the process.
- Market benefits for recognised or certified EMSs are not always there, but when they are EMS farmers are a step ahead of those not carrying out EMS.
- EMSs provide not only environmental benefits but social and financial benefits as well.

Difficulties

- The application of EMS to agriculture is complex because agriculture involves many biological production processes, many of which are not well understood.
- There are unresolved goals and roles for all parties involved including industry organisations, governments, landholders, agri-businesses, researchers and catchment management bodies.
- EMS facilitators and auditors need more experience in the agricultural sector and agricultural advisors need more experience with EMS.
- Support for EMS can be difficult to source.
- Benefits can be difficult to quantify.
- Increased funding for assistance across industries.
- Risks that reduce support for EMS include:
  - Declining terms of trade
  - Local, national and international regulatory pressures
  - Consumer expectations
  - Change processes and capturing value
  - Isolation

The Way Forward

EMSs are a form of AM that can facilitate greater predictability about key elements affecting the enterprise, generate better understanding of all the issues affecting the enterprise and reduce costs, including the costs and impacts of local and externally generated emergencies and crises. They are one means of integrating social, environmental and financial elements of a business when used as part of a whole business planning system. While there is still much to learn there is sufficient evidence to support a redirection of effort towards greater consistency to achieve EMS. EMS parallels the AM cycle while providing a means to gain an internationally credible outcome based on ISO 14001. This allows users to not only capture business management benefits, but also to enter global market places with a well supported 'green' story on which to market their products.

References

Bellesi, F., Lehrer, D., & Tal, A. (2005). Comparative advantage: the impact of ISO 14001 environmental certification on exports. *Environmental Science and Technology, 39*, 1943–1953.

Carruthers, G. (2003a). *Adoption of Environmental Management Systems in Agriculture Part 1: Case Studies from Australian and New Zealand Farms* (03/121). Canberra: New South Wales Agriculture.

Carruthers, G. (2003b). *Introduction to Environmental Management Systems in Agriculture: National Course Manual*. Canberra: Department of Agriculture, Fisheries and Forestry and Department of the Environment and Heritage.

Carruthers, G. (2005). *Adoption of Environmental Management Systems in Agriculture Part 2: An Analysis of 40 Case Studies* (05/032). Canberra: NSW Department of Primary Industries.

Carruthers, G. (2006). Outcomes of EMS implementation on Australian farms. *Farm Policy Journal, 3*(4), 33–45.

Carruthers, G. (2007a). Using the EMS process as an integrative farm management tool. *Australian Journal of Experimental Agriculture, 47*, 312–324.

Carruthers, G. (2007b). Overseas Travel Report, NSW Department of Primary Industries.

Carruthers, G., & Vanclay, F. (2007). Enhancing the social content of Environmental Management Systems in Australian agriculture. *International Journal of Agricultural Resources, Governance and Ecology, 6*(3), 326–340.

DAFF. (2007). *Environmental Management Systems National Pilot Programme*. Canberra: Department of Agriculture, Fisheries and Forestry.

Gillespie, R., Dumsday, R., & Bennett, J. (2008). *Estimating the Value of Environmental Services Provided by Australian farmers*. Surry Hills: Australian Farm Institute.

Gilpin, A. (2000). *Dictionary of Environmental Law*. Cheltenham: Edward Elgar Publishing.

Gleeson, T., & Carruthers, G. (2006). What could EMSs offer land management in rural Australia? *Farm Policy Journal, 3*(4), 1–13.

Heinze, K. E. (2000). Credible "Clean and Green", An Investigation of the International Framework and Critical Design Features of a Credible EMS for Australian Agriculture. Canberra: CSIRO Land and Water.

Lamprecht, J.L. (1996). ISO 14000: *Issues and implementation guidelines for responsible environmental management*. American Management Association, New York.

NRMMC (2002). *Australia's National Framework for Environmental Management Systems in Agriculture*. Canberra: Department of Agriculture, Fisheries and Forestry.

Tibor, T. and Feldman, I. (1996). *ISO 14000: A guide to the new environmental management standards*. Irwin Professional Publishing, Chicago, IL.

Tinning, G., & Carruthers, G. (2002). *Develop Your Own EMS: a grain farming example*. Tocal: NSW Agriculture, CB Alexander College.

Chapter 13
The Adaptive Management System for the Tasmanian Wilderness World Heritage Area – Linking Management Planning with Effectiveness Evaluation

Glenys Jones[*]

Abstract This paper provides a 30 year retrospective on the development of the adaptive management system for the Tasmanian Wilderness World Heritage Area (Australia). It describes the historical background, key influences and stages that paved the way to establishment of adaptive management. It outlines how effectiveness monitoring, evaluation and reporting are integrated with the management plan for the Area to establish an ongoing adaptive management cycle. The chapter presents figures and tools for adaptive management, including 5 useful questions for guiding the integration of effectiveness monitoring, evaluation and reporting into management plans and programs. Strengths and weaknesses of the adaptive management system are discussed. Key lessons and insights distilled from this experience are offered, including the importance of planned monitoring of management effectiveness; the role of stakeholder assessments; and the factors that can assist in sustaining longterm strategic programs despite ongoing institutional change. The chapter concludes with suggestions for fostering an enabling environment for adaptive management.

Introduction

About This Case Study

This chapter describes how effectiveness evaluation and reporting has been linked to management planning for the Tasmanian Wilderness World Heritage Area (Australia) to establish an ongoing adaptive management cycle. The adaptive management cycle is supported by two key documents – the statutory management

---

[*] The views expressed in this paper are those of the author and do not necessarily reflect the views of the Parks and Wildlife Service.

G. Jones
Parks and Wildlife Service, Hobart, Tasmania, Australia

plan for the area; and a linked 'State of the Tasmanian Wilderness World Heritage Area Report' which evaluates the effectiveness of management under the plan, and identifies opportunities and proposed actions for improving management.

The first comprehensive evaluation of management effectiveness for the area has been published (Parks and Wildlife Service, 2004) and the findings and recommendations are being used by the managing agency and others to guide adjustments to ongoing management. The evaluation report is also being used by a variety of stakeholders as a consolidated reference source for detailed accurate information about management of the area.

Defining Terms

The meanings of the following terms as used in this chapter are defined below.

Adaptive management: An approach that ensures management not only plans and carries out actions to achieve objectives, but also measures the results so that everyone can see what's working and what's not, and consequently make informed decisions and adjustments to enhance the achievement of objectives and the delivery of desired outcomes. (This definition is similar to that provided in Chapter 2 of this volume.)

The adaptive management process is well suited to evidence-based management approaches. The process can accelerate organisational learning and improvement, and can also provide a mechanism for providing public transparency and accountability in management.

Evidence-based management: An approach that deliberately grounds management decisions and practices on the latest and best available facts, especially as established by scientific method, rather than on untested suppositions, negotiated positions or long-standing practice. Evidence-based management had its origins in medical practice and is now applied in a range of fields including education, public management, business and increasingly environmental management.

Management effectiveness: The extent to which management objectives are achieved.

Evaluation: The structured gathering, documentation and critical review of evidence against criteria, such as management objectives, statements of management intent, and targets or limits for performance indicators to determine the quality and/or effectiveness of management. The process of evaluation usually involves monitoring and documenting data and other evidence against the criteria, and identifying opportunities and recommendations for improving ongoing management.

Assessment: A judgment or opinion expressed by a person or group of people, especially in relation to management performance. The process of assessment usually involves the consideration of various aspects of management performance by relevant experts, staff, stakeholders, and users, and can involve the use of questionnaires or workshops.

Management Area and Context

The Tasmanian Wilderness World Heritage Area is a vast and globally significant area of protected temperate wilderness located in southwestern Tasmania, Australia. The area comprises approximately 1.38 million hectares (about 3.42 million acres) of contiguous National Parks and reserves, and covers approximately 20% of the island state of Tasmania.

The conservation significance of the area has been formally recognised through listing as a World Heritage Site. The core area was inscribed on the World Heritage list in 1982 on the basis of all four natural criteria and three cultural (Aboriginal) criteria, and an expanded area was accepted for listing in 1989.

The area is managed under joint federal-state government arrangements. The principal managing agency is the Tasmanian Parks and Wildlife Service which, amongst other things, prepares and implements the statutory management plan for the area.

Development of the Adaptive Management System

In Tasmania, as in most parts of the world, the concepts of effectiveness evaluation and adaptive management are relatively new to protected area management.

Commencing in a time and context where there was no management awareness of, or resources for evaluation or adaptive management, the following section sets out the timeframe and the key influences and milestones that were vital to the development of the adaptive management system for the Tasmanian Wilderness World Heritage Area. A relatively detailed account is provided because adaptive management is still rare in protected area management, and the demonstration of how adaptive management can progressively be built into management processes will assist the broader uptake of the approach.

With hindsight, it can be seen there were five main stages in the development of the adaptive management system:

1. Establishment of enabling management arrangements
2. Capacity building
3. Management planning
4. Effectiveness evaluation and reporting
5. Establishment and consolidation of adaptive management

The story begins almost 3 decades ago, when Tasmania's southwest wilderness was the centre of the nation's biggest-ever conservation battle.

Establishment of Enabling Management Arrangements

Legislation and High Court Decision

In the early 1980s, public and political controversy concerning the planned construction of dams for hydro-electric power generation in Tasmania's southwestern wilderness National Parks escalated dramatically and culminated in the proposed Franklin River dam becoming a state and federal election issue.

Following a change in federal government, legislation was passed in 1983 which gave federal powers to prevent damage or destruction of World Heritage properties (*World Heritage Properties Conservation Act*, 1983). These powers applied to Tasmania's recently listed wilderness World Heritage Site, and a State challenge through the High Court of Australia resulted in a legal decision that halted construction of the Franklin River dam.

Establishment of Joint Federal-State Arrangements for Management

This stormy history paved the way for the establishment of joint federal-state arrangements for management of the Tasmanian Wilderness World Heritage Area. These arrangements, which continue today, include an inter-government Council of Ministers called the Tasmanian Wilderness World Heritage Area Ministerial Council (TWWHAMC) (recently replaced by the Environment Protection and Heritage Council), which includes Ministers from each of the Federal and State governments; a Standing Committee of government officials to advise the Ministerial Council and oversee policy, programs and administrative arrangements; and a 16-member external management advisory committee called the Tasmanian Wilderness World Heritage Area Consultative Committee (TWWHACC) which reflects a broad range of community interests (including scientific, conservation, industry, Aboriginal and local government) to provide advice to the Ministerial Council and Standing Committee. Half the members of the Committee are appointed by the State Government and half by the Federal Government, with the Chair and conservation advocate being appointed jointly by both governments.

Significant Increase in Funding for Management

Following the establishment of joint federal-state arrangements, funding for management of Tasmania's World Heritage Area significantly increased. For example, total funding for management of the area increased from less than $AU1 million per annual in the early 1980s to around $AU3 million per annual by the mid-1980s, and subsequently rose to around $AU9 million per annual in the mid-1990s. Last financial year (2007–2008), the budget was over $AU11 million.

Capacity Building

Appointment of Additional Staff

Significantly increased funding for management enabled extra staff to be appointed to assist management of the World Heritage Area. The new staff included natural resource scientists, cultural heritage specialists and management planners. This team increased the capacity of the managing agency to develop management systems and approaches which were informed by professional expertise and sound scientific inputs. By 1991, 18 staff were employed in the professionally-based Resources, Wildlife and Heritage Division of the managing agency, while a further 65 staff were employed in operationally-focused roles and field centres for the World Heritage Area.

Management Planning

Preparation of the First Statutory Management Plan

Preparation of the first statutory management plan for the Tasmanian Wilderness World Heritage Area commenced in 1989. The purpose of the management plan was to provide an explicit framework of objectives, policies and prescribed actions to guide long-term management of the entire World Heritage property, which covered a variety of reserves proclaimed under Tasmanian legislation, viz. ten state reserves (including five National Parks), nine conservation areas, and two protected archaeological sites.

Advocacy by Key Management Advisory Committee for an Evaluative Approach to Management

During development of the first management plan for the World Heritage Area, the key management advisory committee for the area (TWWHACC) saw the potential value of evaluation to ongoing management, and advocated for the adoption of an evaluative approach to management.

Uptake of Evaluation by the Managing Agency

Planning staff within the managing agency recognised and embraced the need for evaluation, and began to develop a way of evaluating management effectiveness. Our focus was on evaluating management effectiveness rather than other aspects of management performance (such as process, inputs, activities). The rationale was simply that if the fundamental purpose of management is to achieve objectives, then the principal measure of management performance should be the extent to which the management objectives are achieved (i.e. management effectiveness).

Working in a context where there was low agency priority for evaluation, and no designated staff positions, we focused effort on effectiveness evaluation as the best use of limited resources to guide adaptive management for better on-ground outcomes.

An important first step in furthering development of effectiveness evaluation was to include prescribed actions in the management plan for designing and implementing an evaluation system. This was done, and the first statutory management plan for the World Heritage Area was approved in 1992 (Department of Parks, Wildlife and Heritage, 1992).

Clarification of Management Intent

The next step in developing the evaluation system was to clarify the outcomes that management was seeking to achieve.

A consultant with expertise in evaluation was engaged to work with staff of the managing agency to develop a framework for monitoring and evaluation which identified a range of outcomes and indicators of management effectiveness for the management plan. The consultant's report was submitted to the managing agency in 1994 (Hocking, 1994).

Integration of Clear Statements of Management Intent into the Management Plan

Drawing on the above report, formal statements of management intent ('Key Desired Outcomes') were included against each management objective in the next edition of the statutory management plan (Parks and Wildlife Service, 1999). This inclusion served several important purposes. It ensured that:

- The management intent of the plan was clear.
- The statements of Key Desired Outcomes were subject to public review as part of the public consultation process for development of the management plan.
- The statements of Key Desired Outcomes were formally endorsed and approved as part of the statutory management plan.
- The statements of Key Desired Outcomes established a stable framework of criteria against which management effectiveness under the plan would be evaluated, thereby avoiding potential issues associated with 'shifting goalposts' (i.e. changing objectives).

The contents of the management plan included:

- Management objectives and statements of Key Desired Outcomes
- Prescribed management strategies and actions for achieving the objectives

- Identified high priority areas for implementation ('Key Focus Areas')
- A formal process for considering new proposals which were not covered by the plan
- Requirements for monitoring, evaluating and reporting on management performance
- Requirements for review of the management plan

Effectiveness Evaluation and Reporting

Preparation of the First Evaluation of Management Effectiveness

Preparation of the first evaluation of management effectiveness for the Tasmanian Wilderness World Heritage Area commenced in 1999. Tools and templates for evaluation were developed and measured data and other evidence which addressed identified needs for evaluation were collated and reported e.g. through evaluated case study reports; geographical information system (GIS) mapping, and preparation of text, tables, figures, and photographs. Where feasible, identified gaps in data for evaluation were addressed through new projects (e.g. market research polls of public opinion). Key stakeholders closely associated with management were also invited to provide assessments and critical comment on management performance through targeted questionnaires.

Linkages with International Initiatives in Evaluation

As planning staff undertook development of methodologies and tools for evaluating management effectiveness, linkages were established with national and international initiatives aimed at advancing evaluation for protected area management (e.g. the World Commission on Protected Areas (WCPA) Management Effectiveness Task Force; and the Australian and New Zealand Environment and Conservation Council (ANZECC) Benchmarking and Best Practice Program). These linkages resulted in the experience and lessons learnt from Tasmania's work in evaluation being reflected in the best practice guidelines developed by these bodies (Best Practice in Parks Management Planning, 2000; Hockings et al., 2000; Best Practice in Park Management, 2002).

International conferences and expert workshops in evaluation of protected areas provided important opportunities for presenting and sharing Tasmania's evaluation approaches and progress, and for gaining an understanding of the global context of evaluation in protected area management. These forums also provided a valuable catalyst for establishing ongoing professional networks in evaluation, and for preparing papers which captured progress and lessons to date e.g. Jones and Dunn (Hocking) (2000), Jones (2000, 2003, 2005a, b), Day et al. (2002), and The Nature Conservancy (2005).

Ongoing Support for Evaluation by Management Advisory Committees

Firm ongoing support for the agency's evaluation initiative by key external management advisory committees played a critically important role in maintaining the continuity of the evaluation program through potentially destabilising departmental restructures and other changes. For example, advisory committees championed budget allocations for evaluation, wrote letters of support for the program to senior government Ministers, and gave public recognition to the agency's leadership in evaluation through speeches and media releases.

Publication of the First 'State of the Tasmanian Wilderness World Heritage Area Report'

Publication of the full findings of the evaluation of management effectiveness was approved by the Director for National Parks and Wildlife, and the report was launched in 2004 under the title *'State of the Tasmanian Wilderness World Heritage Area – an evaluation of management effectiveness.* Report No. 1, 2004 (Parks and Wildlife Service, 2004).

The report presents over 300 pages of detailed information, data, maps and photos which document the extent to which management under the first statutory management plan achieved its objectives.

A separate summary report was also published to provide a brief overview of the key findings and recommendations of the evaluation. This booklet proved popular with the public and also provided a convenient reference source for senior managers and government.

The contents of the full report include:

- Scientific data and other evidence of management effectiveness against the objectives and Key Desired Outcomes
- Information and professional advice from specialists in natural and cultural heritage
- The views of the general public and on-site visitors to the area
- Assessments and critical comment on management performance by internal and external stakeholders closely associated with management of the area
- An overall indication of management effectiveness against each objective of the management plan by the managing agency and
- Identified opportunities and proposed actions for enhancing management performance

The report consolidated a vast amount of information and data about management matters. Publication of the full findings and data provided the evidence which underpinned the conclusions and recommendations of the report. This transparency established a high level of credibility for the evaluation, and contributed to general community acceptance of the findings.

The evaluation report has established a sound reference platform of data against which future management progress can be compared. This has laid the foundation for transparent evidence-based adaptive management for the Tasmanian Wilderness World Heritage Area into the future.

External Recognition for the Evaluation Report

'Feedback is a gift'; and the right feedback from the right people at the right time can play a pivotal role in nurturing management initiatives and programs. External recognition for the 'State of the Tasmanian Wilderness World Heritage Area Report' contributed significantly to local acceptance of evaluation as an integral component of protected area management. External recognition included:

- Acclaim for the significance and value of the evaluation report by a wide range of local, national and international authorities
- Recognition through a range of prestigious awards for innovation and excellence in management planning and evaluation for the Tasmanian Wilderness World Heritage Area (including the Australian Planning Minister's Award for overall winner across all categories of the 2003 PIA National Awards for Planning Excellence for the *Tasmanian Wilderness World Heritage Area Management Plan 1999*; and the 2005 Australasian Evaluation Society's Caulley Tulloch Award for best publication in evaluation for the *State of the Tasmanain Wilderness World Heritage Area Report*)
- General community acceptance of the evaluation's findings, which was reflected in neutral to positive media coverage for the report and a media focus on the issues identified in the report
- Local and international interest in, and uptake of, Tasmania's adaptive management approaches, figures and tools

Establishment and Consolidation of Adaptive Management

Global Recognition of Evaluation as a Key Strategic Direction for Protected Area Management

Explicit recognition – both internationally and nationally – of effectiveness evaluation as a key strategic direction for protected area management fostered increasing agency awareness of the role of evaluation in protected area management. (For example, the importance of evaluation for protected area management was expressed by the Vth IUCN World Parks Congress (2003); the Convention on Biological Diversity's Programme of Works on Protected Areas COP7 Decision VII/28; and in 'Directions for the National Reserves System' (Natural Resource Management Ministerial Council, 2005); 'The National Reserves System Programme 2006 Evaluation' (Gilligan, 2006); and was recommended by the Australian Senate Report on National Parks, Conservation Reserves and Marine Protected Areas (Standing Committee on Environment, Communications, Information Technology and the Arts, 2007).

Establishment of a Designated Staff Position for Evaluation

In 2005, the managing agency formally recognised the role of evaluation in protected area management through the establishment of a staff position with designated duties and responsibilities for performance monitoring, evaluation and reporting. This significantly increased security for the continuity and ongoing development of effectiveness evaluation and adaptive management for Tasmania's National Parks and Reserves.

Uptake and Application of the Findings and Recommendations of Evaluation

The 'State of the Tasmanian Wilderness World Heritage Area Report' is providing an informed basis for guiding improvements in management. In particular, the managing agency is using the findings and recommendations of evaluation to guide adjustments in the content of the next edition of the statutory management plan (currently in preparation). Prescribed actions in the new draft plan are specifically addressing the opportunities for improvement identified by the evaluation e.g. through establishment of a cross-agency program to address illegal activities; and through establishment of a new Key Focus Area for key threatening processes (including fire, disease, introduced animals & plants, and climate change). These responses mark the practical implementation of adaptive management. Feedback from evaluation is being used to adjust ongoing management to better achieve objectives.

Consolidation and Ongoing Development of Adaptive Management

The managing agency is continuing to consolidate the adaptive management system, and to broaden application of the adaptive management approach to all Tasmanian National Parks and Reserves. Ongoing work is focusing on:
- Developing a Tasmanian reserves monitoring and reporting system
- Fostering staff engagement in the adaptive management process through evaluated case studies of selected reserve management projects
- Developing web-based tools and templates to support adaptive management and continuous reporting on management effectiveness
- Integrating adaptive management into 'enduring' agency systems and processes e.g. through Information Management Systems

Figures and Tools for Adaptive Management

The following figures and tools have been developed to communicate the concepts of adaptive management to staff and stakeholders, and to foster the practical application of adaptive management to protected area management.

13 The Adaptive Management System

The figures and tools are:

1. The adaptive management cycle
2. Five useful questions for planners and managers
3. Reporting template for evaluated case studies
4. Performance snapshot
5. Framework for performance management

The Adaptive Management Cycle

The adaptive management cycle (Fig. 13.1) illustrates how effectiveness monitoring, evaluation and reporting can be integrated into an overall cycle of management to support evidence-based adaptive management.

The adaptive management process begins with determining the management objectives and articulating clear statements of management intent against each objective e.g. through formal statements of Key Desired Outcomes. With the objectives and key desired outcomes clearly articulated, management can focus on developing and implementing appropriate strategies and actions to achieve the objectives and deliver the desired outcomes.

Monitoring and evaluation is undertaken to reveal how management is progressing in relation to the objectives and Key Desired Outcomes. The findings

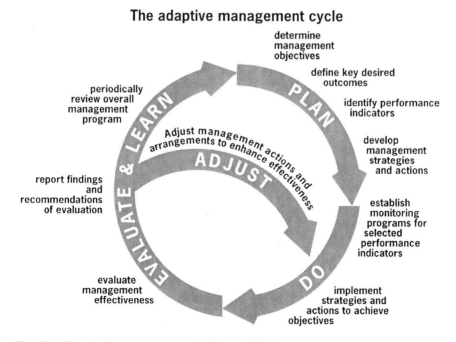

Fig. 13.1 The adaptive management cycle (Jones, 2005b)

and recommendations of evaluation are reported at appropriate point/s in the management cycle to inform and guide review of the management plan and/or other decisions affecting management directions, priorities and budgets. Summary Reports and/or special briefings can ensure decision-makers have timely, ready access to the key findings and recommendations of evaluation.

The period of the adaptive management cycle should suit the management context and purpose. In the case of the Tasmanian Wilderness World Heritage Area, the period of the management cycle is about 10 years (the term of the management plan) with a mid-term limited review at the 5-year point for minor adjustments. Long-term management and monitoring programs will span multiple iterations of the adaptive management cycle.

The adaptive management cycle for the Tasmanian Wilderness World Heritage Area operates through two key documents – the Management Plan for the area; and a linked State of the Tasmanian Wilderness World Heritage Area Report which evaluates the effectiveness of management under the plan and identifies opportunities and proposed actions for improving management. The Tasmanian Wilderness World Heritage Area Management Plan (Parks and Wildlife Service, 1999) and the first comprehensive evaluation of management effectiveness (State of the Tasmanian Wilderness World Heritage Area – an evaluation of management effectiveness Report No. 1 2004) are available online at http://www.parks.tas.gov.au/index.aspx?base=6174. The findings and recommendations of the evaluation are used to guide adjustments in management actions and arrangements to better achieve objectives. For example, adaptive responses can include:

- Immediate adjustments to operational activities
- Adjustments to strategic management directions and priorities
- Budget processes and resource allocations which take account of the findings of evaluation, e.g. by providing ongoing support for management programs that have been demonstrated to be effective; and by considering the relative merits of increasing or redirecting effort to, or from, areas of weak performance
- Adjustments to management arrangements related to identified positive and negative factors affecting management performance
- Targeting identified critical gaps in information required for sound management and
- Adjustments to the content of the next edition of the management plan

In addition, the findings of evaluation contribute to ongoing organisational learning which informs continuous improvement in management practice.

Five Useful Questions for Planners and Managers

The following five questions assist program planners and managers to integrate effectiveness monitoring and evaluation into the design of management plans or programs. Ideally, these questions should be considered during the early phases of program planning and approval.

1. **'What would we expect to see if management was working well?'** And the converse question: **'What would we expect to see if management was NOT working well?'**

The answers to these questions assist in developing clear statements of management intent (Key Desired Outcomes) and also assist in identifying appropriate performance indicators. Note that the main purpose of the converse question is to identify any additional issues that need to be reflected in the statements of management intent.

Statements of management intent help to focus management effort on achieving identified desired outcomes, and establish an explicit framework for evaluating the effectiveness of management.

Formal objectives for protected area management are often worded in general terms (as in the mandates of legislation or in management plans), and it is important for managers to provide an operational interpretation of these objectives. By carefully considering what management success and failure would be likely to look like, managers can articulate clearer and more specific statements of management intent.

As an example, Objective 6 of the management plan for the Tasmanian Wilderness World Heritage Area is 'To assist people to appreciate and enjoy the World Heritage Area in ways that are compatible with the conservation of its natural and cultural values, and that enrich visitor experience'. Key Desired Outcomes for this objective are:

- KDO 6.1 Ecologically sustainable management of human use of the World Heritage Area to within acceptable, and where necessary defined limits.
- KDO 6.2 High levels of community and visitor satisfaction with:
  - The range and quality of recreational opportunities and facilities available
  - The operations and services of the Parks & Wildlife Service, licensed tour operators, and concessionaires and
  - The quality of their experience in the World Heritage Area
- KDO 6.3 Cooperation of visitors and other users with the Parks & Wildlife Service, especially in caring for the World Heritage Area, its values, and assets.

Note that in poorly understood ecosystems (such as marine protected areas) it may be difficult to know where to begin in developing meaningful statements of Key Desired Outcomes. In such cases, it can be helpful to refocus the question to consider 'What is needed in order to progress the identification of Key Desired Outcomes?' For example, the acquisition of basic knowledge about natural resources, impacts and/or ecosystem processes may be a necessary intermediate step towards developing clear statements of the ultimate desired outcomes.

2. **'What could we monitor or measure** (or photograph, or map, or survey) **to reveal the outcomes that are being delivered?'**

The answers to this question help to identify a range of potential performance indicators that could be used to monitor management effectiveness.

If it is important to detect and/or demonstrate change over time (e.g. as a result of the management strategy or program), it is necessary to document the baseline or reference situation prior to commencing the management intervention. This may simply mean ensuring that 'before' photos are taken as well as 'after' photos so that changes can be documented and demonstrated. In some cases, fixed point photographic sequences and long-term aerial photographic monitoring can provide relatively low cost and easily demonstrable changes in on-ground outcomes. Management interventions that result in significant changes over short periods of time are relatively straightforward to demonstrate; however, where changes are slow and subtle, it can be challenging to know when sufficient time has elapsed to properly evaluate the effects of the management intervention.

3. **'Where would we realistically expect to see improvements or changes if management was working well?'** And the converse question: **'Where would we realistically expect to see things getting worse or changing if management was not working well?'**

The answers to these questions assist in identifying 'Indicators of Change', i.e. indicators that are sensitive to positive or negative change in management performance. These indicators suggest high priorities for monitoring programs.

As funding levels are rarely, if ever, sufficient to support a full and comprehensive monitoring program, careful choices need to be made to prioritise resources towards those indicators that are most likely to provide valuable feedback for guiding adaptive management.

Monitoring programs that detect hypothesised changes in management outcomes can provide important feedback about whether management strategies or programs are working as intended. For example, monitoring may reveal that the management strategy is delivering the anticipated changes and so provide endorsement for continuing the strategy. Alternatively, monitoring may reveal that the management strategy is not delivering the anticipated outcomes and so needs to be reviewed and changed. Monitoring may also be used to differentiate between two different hypothesised trajectories of outcomes from a particular management strategy, and so contribute to organisation learning about how the managed system works. Nevertheless, there are times when the interpretation of observed results can be difficult and in some cases it may not be possible to conclude whether the observed change is due to the management intervention or an independent event e.g. it may be a 'chance event' or a transitional occurrence that will change again over time.

4. **How will the findings of monitoring and evaluation be reported and/or used?**

The answers to this question help to ensure that the findings of monitoring and evaluation are useful and used.

The process of adaptive management relies on the findings and recommendations of evaluation feeding back into and influencing ongoing management. Consideration needs to be given to how this feedback and learning can best be achieved in the particular management context. For example, the findings and recommendations of evaluation may be used to:

– Inform and guide reviews of the management plan/strategic plan/program to improve its effectiveness

- Inform budget processes and resource allocations, e.g. through considering the relative merits of various programs in terms of their effectiveness and/or relevance to management objectives
- Provide transparency and accountability in management through public reporting of the findings of evaluations (e.g. through Annual Reports; State of Parks Reports; Periodic Reports on World Heritage Properties)
- Contribute to broader staff and community understanding and engagement in the adaptive management process

5. **Who will be responsible for doing the monitoring, evaluation and reporting** (including design of the monitoring program, data collection, data analysis and management, overall coordination and quality control)?

The answers to this question assist in identifying the roles, responsibilities and resources required for monitoring, evaluation and reporting.

For most protected area managers – perhaps most managers generally – effectiveness monitoring, evaluation and reporting is the 'missing link', the part of the iterative management process that just doesn't get done. As effectiveness evaluation for protected areas is still in its infancy worldwide, there is often no established practice of allocating funding and resources for evaluation. This needs to change. Realistic levels of resources and priority for effectiveness monitoring, evaluation and reporting are an essential ingredient of sound adaptive management.

Reporting Template for Evaluated Case Studies

The selection of a relatively small number of case studies of reserve management for detailed monitoring, evaluation and reporting on management effectiveness can be a practical and useful way of investing limited resources for monitoring and evaluation to gain significant returns in organisational learning for adaptive management.

The process of prioritising and selecting case studies for monitoring and evaluation will be affected by many factors. However, key considerations include the extent to which the management project or program is considered likely to contribute to organisational learning for adaptive management. For example, a 'good' case study is likely to contribute to organisational learning by providing:

- An example of effective management which can serve as a model for others to emulate
- Feedback about the effectiveness of a major or significant reserve management project or initiative or
- Monitoring data that will increase understanding about an unresolved or emerging management issue

In addition, the selection of a suite of case studies for evaluation should aim to provide a balance of projects across the range of management objectives and responsibilities.

The use of a standardised template for reporting the monitored results of evaluated case studies serves several important purposes. For example, it:

- Assists managers to plan and integrate effectiveness monitoring and evaluation into the design of conservation management programs and projects so that their effectiveness can be determined and documented
- Establishes a consistent format for reporting the monitored results of a diverse range of management programs and projects (e.g. dealing with different issues in different reserves) and
- Provides an easily accessible summary of the effectiveness of conservation management programs for policy-makers, funders, and the broader community

A standard template for reporting the monitored results of evaluated case studies of reserve management is provided in Fig. 13.2. This template can be used to plan and report on the effectiveness of management programs or strategies in achieving management objectives. To see worked examples of evaluated case studies, refer to sections 4.10.2, 4.10.3 and 5.7 of the 'State of the Tasmanian Wilderness World Heritage Area Report' (Parks and Wildlife Service, 2004), available online at http://www.parks.tas.gov.au/index.aspx?base=6174.

Note that a well-coordinated program of evaluated case studies requires appropriate coordination, quality control, and professional advice and support for the project managers of evaluated case studies.

Performance Snapshot

The Performance Snapshot tool (illustrated below) provides a simple 'traffic light' system for summarising and presenting the findings of monitoring in relation to identified performance standards e.g. as established by the Framework for Performance Management (see following section). Green, red and amber traffic light symbols are used to provide readers with an easily accessible overview of how management is performing in relation to identified targets and limits, and to identify any issues that need to be addressed by management in order to achieve the desired results.

The basis for assigning traffic light symbols to the monitored performance data, and the implications for management, are outlined in Table 13.1 below.

In the event that monitoring data are inadequate to inform a sound assessment of performance (e.g. the data are old, out-of-date, incomplete), a hatched circle overlay can be used to denote uncertainty about the traffic light symbol. Where no data are available to inform an assessment of performance, a white traffic light symbol can be used.

This tool was developed from a staff concept for a simple traffic-light system to guide sustainable management of the Overland Track, Tasmania's most popular long-distance walking track. While the performance management system for the Overland Track is currently under development, performance indicators are likely to include the monitored condition of walking tracks, water quality parameters, and social indicators.

REPORTING TEMPLATE FOR EVALUATED CASE STUDIES

PROJECT NAME:
About the threat or issue: *Brief description of the issue, what causes it, and what reserve values are affected by it. Photos and maps as appropriate (including captions & credits).*
Background to management: *Brief history of the issue and its management prior to the current period.*
Overall management goal: *Brief statement (e.g. one sentence) of what management is aiming to achieve.*
Management actions and significant events over the management period: *Bullet points of key management actions and any significant external events that may have affected the outcomes. Photos of management activities as appropriate.*

MONITORED RESULTS FOR PERFORMANCE INDICATORS

Performance Indicators	Targets or Limits	Detected Changes over the Management Period
1. CONDITION INDICATORS		
Indicator 1: *Name the performance indicator and describe how the indicator is monitored.*	**Target or limit for indicator:** *State any target or limit that has been established for the indicator. If a meaningful target has not been established, simply state 'no target established', 'target under development', or 'tracking only'.* **Assessment of performance**: *How will performance against the standard be determined?*	**Results:** *Describe any changes or trends detected for the monitored indicator i.e. any increase or decrease in the measured data; or evidence of stability.*
Indicator 2: *As for above. Insert additional rows as required*		
2. PRESSURE INDICATORS *As for above*		
3. OTHER INDICATORS (e.g. social or economic indicators) *As for above*		

OUTCOMES: *Bullet points of the key outcomes that have been delivered. Identify any further anticipated outcomes.*
Investment in this project: *Brief summary of the nature, level and sources of investment in this project, e.g. funding, staffing, volunteers.*
Commentary on management performance: *Bullet points of key factors that have contributed positively to, and/or have limited or threatened, management performance over the period. Suggestions for improving management effectiveness; lessons learnt and/or other comments relevant to ongoing management.*
Sources of information and/or comment: *Contact details of all sources. Photo of key personnel.*
Last updated: *Date this report was prepared and/or most recently checked and updated.*
For more information: *Links to more detailed sources of information, e.g. websites, references.*
Photo gallery, appendices: *Additional photos, maps and tables as appropriate.*

Fig. 13.2 Simplified reporting template

Table 13.1 Performance snapshot

Assessed performance	Implications for management and monitoring
● GREEN – *Management on target*: Indicator (or all indicators) within the green target zone	Management: Management is on target and is achieving the desired results. Current management approach is endorsed, although minor adjustments may be warranted if there is a declining trend
	Monitoring: Requires only low level periodic monitoring to affirm the stability of the performance status
● AMBER – *Caution/ some cause for concern*: One or more indicators outside the target zone but still within an acceptable range. May also include indicators in the unsustainable/ unacceptable zone provided they demonstrate measured evidence of an improving trend	Management: Requires ongoing management review and corrective action as appropriate to bring performance into target zone
	Monitoring: Requires regular ongoing monitoring
● RED – *Unsustainable/unacceptable*: One or more indicators in the unsustainable/ unacceptable zone	Management: Requires focused management attention and significant change to bring performance into acceptable range
	Monitoring: Requires regular ongoing monitoring
○ HATCHED RING: *Data deficient/ uncertain result*: Assessment based on old/out-of-date data or incomplete data to support a sound assessment of performance (Can be used as an overlay over coloured symbol)	Management: Requires a precautionary approach until more data become available
WHITE – *No data to support an assessment*	Monitoring: Priorities for baseline monitoring and repeat surveys need to be kept under review as part of the ongoing monitoring program

Framework for Performance Management

The Framework for Performance Management (Fig. 13.3) provides a structured approach to monitoring and adaptively managing performance to specified standards. This tool is suitable for application where the management system is sufficiently advanced to be able to identify appropriate targets and/or limits for performance indicators; and where the aim – and commitment – of management is to achieve the identified standards. Of course, this approach is only suitable where management has the capacity to influence the factors affecting management performance.

By identifying targets and limits for monitored performance indicators, three zones of management performance are established:

1. The target zone, where management performance is within the desired range of measurements for the performance indicator

Framework for performance management

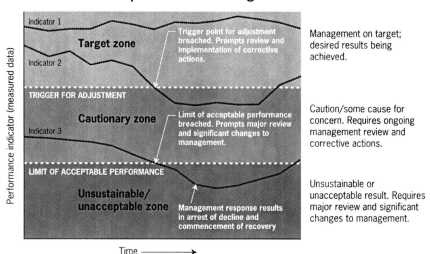

Fig. 13.3 The framework for performance management

2. A cautionary zone, where performance is less than desirable but still within an acceptable range of measurements for the performance indicator and
3. An unacceptable/unsustainable zone, where performance is outside the defined limit of acceptable performance for the performance indicator, and is deemed to be unacceptable

The boundaries between these zones are trigger points which prompt corrective actions and/or major review and significant change in order to restore performance to the required standard. For example, in the figure below, Indicator 1 is stable within the green Target Zone and demonstrates that management is achieving the desired results. Indicator 2 has declined from the Target Zone into the amber Cautionary Zone and this prompts management to implement corrective actions to halt the decline and return performance to the desired range. Indicator 3 has continued to decline from the Cautionary Zone into the red Unsustainable/Unacceptable Zone and this is the trigger for major review and significant change to management to address unacceptable performance.

To be successful, performance management systems must be underpinned by adequate and secure management and funding arrangements that provide the capacity to respond promptly and effectively to any breaches of the trigger points. Prior identification of agreed management responses and responsibilities in the event that a trigger point is breached is highly desirable and will significantly increase the likelihood that management will successfully deliver the desired results. In the absence of these enabling arrangements, a common but undesirable response to breaches of performance standards is for the management standards to be lowered and/or the performance management system to be allowed to lapse.

The Framework for Performance Management was developed to support sustainable management of walking tracks to specified standards, but is suitable for a wide range of applications.

Discussion

Practical Benefits of the Adaptive Management Approach

Overall, establishment of the adaptive management system for the Tasmanian Wilderness World Heritage Area has achieved a more integrated and strategic approach to achieving long-term desired outcomes. Tangible benefits and changes associated with adopting an adaptive management approach include:

- Managers and other decision-makers have a better and more readily accessible information resource to support good decision-making. In particular, the findings and recommendations of evaluation are being used to guide decisions and adjustments which are expected to improve the achievement of management objectives.
- All stakeholders now have ready access to detailed accurate information about management matters. This major shift in information sharing has benefited stakeholders directly by increasing transparency and public accountability of management. This is contributing to more informed public involvement in management planning processes for the area. Increased transparency may also be indirectly benefiting the managing agency as the broader community becomes more familiar with the nature and relevance of its work.
- Application of an evaluative approach to preparation of the management plan for the Tasmanian Wilderness World Heritage Area resulted in a more systematic and transparent linkage between management objectives and prescribed actions in the plan. In doing so, it brought to light several gaps that had previously existed between management responsibilities and management actions, which were consequently rectified.
- The findings of evaluated case studies of reserve management have in some cases strongly influenced management decisions and the allocation of management resources, which has resulted in demonstrably improved management outcomes, e.g. adaptive management of serious riverbank erosion on the lower Gordon River achieved the cessation of erosion in some areas and dramatically decreased rates of erosion in others. (Nonetheless in some other cases, the findings of evaluation have not as yet been acted upon, e.g. in relation to measured evidence of increasing degradation from walker impacts in some areas.)
- The process and findings of monitoring and evaluation have in some cases 'taken the heat' out of management decision-making through the systematic collection and use of information for decision-making, and the transparency of that process. In other cases, while not achieving resolution of controversial issues, the process has served to highlight social and/or political barriers to proposed management actions.

- Establishment of the adaptive management system is gradually bringing about positive change in the way the managing agency is approaching its roles and responsibilities for reserve management. For example:
  - The managing agency is increasingly taking responsibility for articulating and focusing on the outcomes it is seeking to achieve, and for evaluating the quality of its strategies and actions in relation to these goals.
  - There is a declining reliance on the paradigm of 'trust us, we're the experts' and a growing focus on being able to document and demonstrate the results of management, e.g. through evaluation reports and evaluated case studies of reserve management.
  - The simple knowledge that implementation and effectiveness of the management plan is being monitored and evaluated has in some cases acted as a prompt to invigorate and maintain staff focus on implementing the management plan's prescriptions.
  - The process of reporting on management effectiveness is driving a more systematic approach to information collection, collation and presentation, e.g. through the use of standardised reporting templates, data mapping, and the integration of effectiveness monitoring and evaluation into the design of management projects.
- The Australian government was able to cite Tasmania's adaptive management system and related documents as evidence of the high standard of management tools used to conserve and manage the Tasmanian Wilderness World Heritage Area in its State Party Report to the UNESCO World Heritage Committee Decision WHC-06/30.COM/7B (Australian Government, 2007).
- Nationally and internationally, the demonstration of a practical system of adaptive management for the Tasmanian Wilderness World Heritage Area is being used as a model for the broader application of performance-based management approaches to protected area management.

Key Factors Affecting Management Performance

In addition to the feedback provided by measured evidence of management effectiveness, understanding the factors that affect management performance can also provide important feedback to the adaptive management process by suggesting how management performance can be maintained and/or enhanced. The identification of positive influences on management performance suggests factors that should continue to be supported in order to maintain performance, while the identification of negative influences suggests factors that should be addressed in order to improve performance.

As part of the evaluation of management effectiveness for the Tasmanian Wilderness World Heritage Area, key stakeholders closely associated with management of the area were invited to identify the key factors that had contributed positively and negatively to management performance over the term of the management plan.

The stakeholders for this assessment included the external management advisory committee for the World Heritage Area (TWWHACC); the federal agency with responsibilities for World Heritage management; the representative organisation of the Tasmanian Aboriginal community; and staff of the then managing agency (including natural and cultural heritage specialists, planning and operational staff, and senior managers).

The following findings of that assessment are broadly relevant to protected area management elsewhere, and are therefore of interest to those with interests or responsibilities for protected area management.

Positive Factors

The key factors that stakeholders identified as having contributed positively to management performance for the Tasmanian Wilderness World Heritage Area over the management period were (in descending order of frequency of mention by stakeholders):

- The level of Federal-State funding for management
- Public support and cooperation in management
- Good staff
- An effective World Heritage Area Consultative Committee
- A good management plan and key management strategies
- Good science and
- No major wildfires over the period

Negative Factors

The key factors stakeholders identified as having limited or threatened management performance were:

- Inadequate resources and uncertainty of future funding
- Inadequate community engagement and support
- Political decisions were not always consistent with World Heritage management objectives
- Slow response/low priority to management of impacts and threats to values
- Inadequacy of fire management and
- Delays with site plans

In some cases, the same or similar factors were identified as having both supported and limited or threatened management performance e.g. funding and public support. While this initially appears a paradox, these findings simply reflect the strong correlation that exists between these factors and performance across the full range. Identification of the same factor as a key positive and negative influence also prompts a closer examination of how and why these factors are exerting their influences. In the case of funding, stakeholder assessments drew attention not only to the importance of the level of funding to management performance, but also to the importance of the ongoing security of funding.

Stakeholders identified the absence of major fires over the management period as a key factor which contributed positively to management performance, yet also identified the inadequacy of fire management as a key threatening factor. Again, the apparent paradox can be understood by a closer examination of the issues. The managing agency (along with other stakeholders) recognised that fire is arguably the greatest realistic threat that could cause rapid, large-scale major ecological impacts to the area. However, they also recognised that luck plays a significant role in the nature and success of fire management operations in the Tasmanian Wilderness World Heritage Area. For example, on a single day – Christmas Eve of 1998 – lightning started four separate fires in the World Heritage Area. However these, and every other potentially major fire over the management period, were doused by following rains, which avoided significant impacts. So while there were no major wildfires over the management period, staff and stakeholders recognised that there is a continuing significant risk of major impacts from wildfires, and considered that the existing level of knowledge and preparedness for fire management was inadequate.

Stakeholders' assessments of the key factors affecting management performance led to the development of proposed actions for enhancing management performance, some of which are currently being addressed.

Strengths and Weaknesses of the Adaptive Management System

Strengths of the Adaptive Management System

Key strengths of Tasmania's adaptive management system are:
- It delivers a transparent and credible evaluation of management effectiveness which provides an informed basis for guiding improvements in management.
- The transparency of the adaptive management process – including publication of detailed evidence of management effectiveness – provides for transparency and accountability in management and this helps foster community understanding and trust in management.
- The evaluation methodology takes account of both measured evidence of effectiveness and stakeholders' assessments of management performance, and this combination of approaches provides a balanced view of management performance.
- The adaptive management cycle 'gets monitoring and evaluation to happen'. Despite widespread acknowledgement of the importance of evaluation to sound adaptive management, evaluations of protected area management are still rare. Where they are undertaken, planning for the evaluation often only commences after-the-event of management. Such evaluations are usually severely limited by the lack of relevant time-series measured data to inform effectiveness evaluation. The adaptive management cycle ensures that performance indicators are identified

early in the management cycle so that appropriate monitoring programs can be established to generate the data required for informed evaluations.
- Evaluation of management effectiveness provides a basis for recognising successful management programs and strategies, and importantly for recognising the people behind those programs.
- The adaptive management process is 'owned' by the managing agency, and this confers a range of practical advantages for growing an organisational culture of adaptive management and continuous improvement. For example, the managing agency is often more readily able to:
    - Develop and/or modify management processes, systems and tools to integrate and embed an ongoing adaptive management cycle
    - Understand and take account of changes in the management context, issues and operational constraints and so tailor and adjust the adaptive management program to ensure it remains relevant and viable over time
    - Develop in-depth detailed knowledge of the conservation values and management issues in the area
    - Establish long-term monitoring programs for performance indicators and so provide the necessary data for informed evaluations
    - Ensure that data sets and information management systems are developed and maintained over the long-term
    - Access the data and the professional and technical support usually available within government agencies e.g. for GIS manipulations and map generation and
    - Facilitate the uptake of the findings and recommendations of evaluation into ongoing management decisions, processes and policy

Weaknesses of the Adaptive Management System

Weaknesses or disadvantages of Tasmania's adaptive management system include:

- Internal ('first party') evaluations can sometimes be less objective and credible than independent ('third party') assessments. In particular, governments and/or managing agencies can find it particularly challenging to publish the full results of an evaluation especially in the event that the evaluation reveals performance has been very poor in one or more areas, or generally poor across several areas of management responsibility. In these circumstances there can be a temptation for some degree of 'bureaucratic cleansing' to make the results – and the managers – look better. An explicit commitment by management to 'truth and honesty in reporting' reduces this risk; however it is important to recognise that this risk exists and may in some cases significantly degrade the value and integrity of an evaluation. Full disclosure of the findings and evidence underpinning an evaluation significantly increases the credibility of the evaluation and reflects well on the integrity of the managers. Nonetheless, a separate and complementary role exists for periodic independent audit and review of management to examine overall standards of practice, including the extent to which the organisation may be distorting information to avoid disclosing unfavourable matters (Bella, 1987, 1992, 2000).

- Evidence-based adaptive management usually involves long timeframes and a relatively high level of investment of resources. For example, the process of adaptive management involves sound strategic planning, long-term monitoring programs, and well-targeted scientific and other specialist inputs. These activities require significant professional capacity and secure ongoing funding arrangements. The benefits of adaptive management are significant and enduring – including better on-ground outcomes – however, adequate and secure ongoing resources are needed to support the programs that deliver adaptive management.
- Managers everywhere tend to give priority to urgent short-term needs over important long-term needs, and this bias disadvantages the long-term strategic programs that support adaptive management. Management planning, monitoring and evaluation are key components of sound adaptive management, yet these programs are often under-resourced, and also seem to be at greater risk of having their allocated resources diverted to meet immediate ad hoc needs. Long-term funding arrangements that quarantine staff and funding resources for adaptive management and its associated programs help to provide ongoing security and consistency of focus for these programs.
- The long-term strategic nature of monitoring and evaluation programs for adaptive management makes them particularly vulnerable to disruption associated with institutional change, as discussed later.

Lessons

Monitoring for Adaptive Management

Good quality scientific inputs and monitored evidence of management effectiveness are cornerstones of sound adaptive management for protected areas.

Monitoring for adaptive management sometimes aligns well with monitoring programs for other purposes, such as science-driven research. However this is not always the case, so it is important for protected area managers to clearly identify management's needs for monitoring. This helps to ensure that well-targeted monitoring programs for adaptive management are established and implemented.

Monitoring programs for the purposes of adaptive management for protected area management are likely to be focused on measuring evidence about:

- The effectiveness of key management projects, strategies and programs
- The condition of reserves and reserve values (including restoration of degraded values)
- The nature, extent and severity of threats, risks and impacts on reserves and their values (including new and emerging threats)
- The level and nature of human use in reserves, its environmental sensitivity and sustainability
- The views of the general public and/or on-site visitors
- Trends and changes in all of the above

Increasingly, as the significance of climate change and its consequences are recognised, monitoring for predictive modelling and adaptive management under climate change will become a key strategic direction for protected area and regional management. Having an adaptive management system in place will facilitate proactive management and evidence-based adjustments and learning.

Stakeholder Assessments for Adaptive Management

While the main focus of effectiveness evaluation is on measured evidence of achievement, stakeholders' qualitative assessments and critical comment on management performance can provide an additional and complementary dimension to the evaluation. For example, stakeholders' feedback can provide a social perspective of performance and allow important lessons and insights to be distilled from human observations and experience. Stakeholder assessments can be especially valuable for identifying the causal factors that explain the results delivered (Vedung, 1997).

The inclusion of external (as well as internal) stakeholders in assessments of management performance significantly enhances the credibility of the findings, and can sometimes result in the capture of views and insights which might not readily be sourced from within a managing agency.

Importantly for the practice of evaluation, the use of stakeholder assessments and critical comment on management performance provides a low-cost way of considering the impact of an almost infinite array of factors that could potentially have affected management performance without incurring the high costs of formally monitoring large sets of input and process indicators. This approach allows limited resources for monitoring and evaluation to be focused on measuring management effectiveness, threats and outcomes, with only a relatively small proportion being directed to monitoring key input and process indicators, such as funding levels and implementation of management actions. If management of protected areas is primarily concerned with achieving on-ground conservation outcomes, then monitoring for adaptive management of protected areas needs to focus on indicators of management effectiveness.

Maintaining Long-Term Strategic Programs Through Institutional Change

Change happens… and it keeps on happening. Governments change, Ministers change; departments get restructured; budgets change; key staff leave or arrive; and internal processes and systems bring about more change. Some level of institutional change is inevitable, and some changes will be warranted, but excessive levels of institutional change can compromise an organisation's ability to perform well.

Institutional change is usually accompanied by significant shifts in management direction, priorities and/or focus – a phenomenon known as 'shifting goal posts'. Typically, institutional change comes with new imperatives to be achieved (=high priority for funding and resources) while 'old' management programs may be seen as yesterday's news (=lower priority for funding and resources). These latter programs are at risk of being scaled down, terminated, or more often simply allowed to lapse through dwindling allocations of resources.

Management programs that are working to long-term objectives and timeframes are especially vulnerable to institutional change. At worst, institutional change can render years of careful planning and investment in long-term strategic programs irrelevant or discarded.

The Tasmanian Parks & Wildlife Service is no stranger to change. Over the past 25 years, the managing agency for National Parks has undergone numerous institutional changes, including being restructured to a different government department on average every 4 years (see Table 13.2).

So how can long-term strategic programs be buffered against the impacts of short-term institutional change?

There are no simple answers; however, the Parks and Wildlife Service's experience of institutional change may offer some insights and lessons.

Reflecting on those long-term strategic management programs which stayed on track through multiple institutional changes, the following factors seem to have contributed to their continuity and success:

- Certain activities and programs were explicitly required by legislation, international conventions, and/or formal agreements (including long-term funding

Table 13.2 Parks and Wildlife Service's history of institutional change

Department	Duration
1971: National Parks and Wildlife Service department established following enactment of the *National Parks and Wildlife Act* 1970	16 years
1987: Department of Lands, Parks and Wildlife	2 years
1989: Department of Parks, Wildlife and Heritage	4 years
1993: Department of Environment and Land Management	5 years, plus internal restructure at 2 years
1998: Department of Primary Industries, Water and Environment	4 years, plus internal restructure at 2 years
2002: Department of Tourism, Parks, Heritage and the Arts	4 years, plus new legislation (replacing the former Act) which separates the responsibilities for National Parks and reserves management from nature conservation. The latter responsibilities remain with the former department and Minister
2006: Department of Tourism, Arts and the Environment	2 years
2008: Department of Environment, Parks, Heritage and the Arts. The Parks and Wildlife Service is one of nine divisions in the current department	1+ years (current)

arrangements), and this usually ensured the continuity of these programs. This highlights the importance of core management objectives and responsibilities for protected area management being enshrined in legal instruments, specified in formal management mandates and/or in long-term funding agreements.
- Some staff positions (e.g. World Heritage specialist positions) were provided through long-term funding arrangements, and this provided increased security for the roles provided by these positions. This observation highlights the importance of funding arrangements for protected area management providing ongoing security for valued staff roles, positions, and/or individuals.
- The objectives of enduring programs were often closely aligned to relatively stable mandates for management such as legislation, the World Heritage Convention, and/or the provisions of a statutory management plan. The stability of these mandates provided some protection for these programs from institutional change and shifting goal posts. This observation highlights the importance of aligning long-term strategic programs to stable long-term management mandates.
- The long-term continuity and personal commitment of key staff to their programs often made the difference between a program surviving a period of destabilising change and lapsing. Staff with a strong personal sense of purpose and commitment to their program were often able to sustain the program – at least in a basic form – through periods of significant setbacks such as staff losses, budget cuts, or simply not being 'flavour of the month'. When circumstances became more conducive to the program, these staff were able to revitalise the program and move forward again. This observation highlights the importance of appointing capable and dedicated staff to key strategic positions with long-term tenure.
- Key external stakeholders sometimes played a critically important role in maintaining particular programs during times of major institutional change or other perturbation by voicing their firm support for continuation of particular programs. This observation highlights the importance of fostering external as well as internal support networks for long-term strategic management programs.
- Community and/or stakeholder expectations for maintaining or enhancing standards of management in some cases assisted the continuity of particular programs. This observation highlights the importance of authoritative standards and best practice principles for protected area management being produced and made widely available to those with responsibilities and/or interests in protected area management.

Creating an Enabling Environment for Adaptive Management

Protected area managers, governments, stakeholders and individuals all have a role to play in bringing about changes to foster sound adaptive management (Table 13.3).

Reflecting on the experiences of establishing an adaptive management system for the Tasmanian Wilderness World Heritage Area, it seems there are many things that could be done to create an enabling and supportive environment for adaptive management. Table 13.3 presents some suggestions.

Table 13.3 Enabling factors for adaptive management of protected areas

Enabling factors for adaptive management of protected areas	
1. A government and management culture that	• Focuses on the achievement of important long-term outcomes • Encourages evidence-based management approaches and informed decision-making • Values strategic management planning that establishes the adaptive management cycle • Is committed to transparency and accountability in management e.g. through effectiveness evaluation and outcomes reporting • Strives for and values excellence, achievement and continuous improvement • Welcomes feedback that assists in improving management effectiveness, e.g. from management advisory committees, evaluations, expert panels, etc. • Places high priority on addressing identified opportunities for improving management effectiveness • Celebrates significant achievements and the people behind them
2. Funding and investment that	• Provides secure ongoing funding realistic to the task of achieving the management mandates for protected area management • Gives appropriate priority and resources to adaptive management and its associated programs • Builds enabling management arrangements for adaptive management • Builds capacity for effectiveness monitoring, evaluation and reporting • Encourages the integration of effectiveness monitoring and evaluation into key management plans, programs and projects • Establishes requirements for reporting on the effectiveness of key management plans, programs and projects • Takes account of the findings and recommendations of evaluations to inform budget processes and decisions • Provides long-term secure allocated resources for programs that support adaptive management, including management planning and effectiveness monitoring evaluation and reporting • Supports well-targeted scientific and monitoring programs that support adaptive management • Encourages the development of information management systems and tools that support adaptive management, including predictive modelling • Encourages the demonstration of successful applications of adaptive management, and fosters the broader uptake of adaptive management
3. Management agencies that	• Integrate the adaptive management cycle into 'enduring' agency systems, processes and tools, e.g. planning and review processes, budget allocation processes, project approval processes, Information Management Systems, etc.

(continued)

Table 13.3 (continued)

Enabling factors for adaptive management of protected areas	
	• Establish designated staff positions and programs for effectiveness monitoring, evaluation and reporting
	• Foster staff and stakeholder engagement in the adaptive management process, e.g. through evaluating the effectiveness of selected management programs and projects
	• Identify needs for, and establish, monitoring programs to support adaptive management, e.g. through in-house programs or collaborative programs with research organisations, Universities, other agencies etc.
	• Are committed to achieving high to exemplary standards of protected area management
4. External stakeholders who	• Provide firm advocacy and support for evidence-based adaptive management
	• Expect transparency and accountability in management
	• Provide positive feedback, recognition and/or awards for outstanding achievements and/or exemplary practice in protected area management
	• Where appropriate, provide inputs and/or constructive feedback that helps improve management effectiveness and/or the standards of management practice
5. Government and intergovernmental legislation, agreements and policy that	• Provide clear and stable core mandates for protected area management, e.g. objectives of management specified in legislation
	• Establish formal requirements for evaluating and reporting on management effectiveness, e.g. in legislation and/or through long-term funding arrangements, etc.
	• Provide institutional stability and a consistent focus and priority on achieving core mandates
	• Create enabling management arrangements for adaptive management within and across government and non-government sectors
	• Establish mutually compatible and reinforcing objectives and arrangements for management across all tiers of government e.g. clear linkages between international conventions, national and state legislation, and regional policy frameworks
6. Individuals and groups who	• Are dedicated, capable and committed to establishing better management systems for better outcomes

Acknowledgments I gratefully acknowledge the following key people who contributed to the establishment of the adaptive management system for the Tasmanian Wilderness World Heritage Area, *sine qua non*: Helen Hocking/Dunn (consultant in natural area evaluation and honorary research associate, University of Tasmania); Anni McCuaig and Tim O'Loughlin (my colleagues in management planning, Parks and Wildlife Service); Bryce McNair (former Chairman of the Tasmanian Wilderness World Heritage Area Consultative Committee); and Keith Sainsbury (my husband and unexpected mentor in adaptive management).

References

Australian Government. (2007). The Tasmanian Wilderness World Heritage Area World Heritage Committee Decision WHA-06/30.COM/7B State Party Report. Department of the Environment and Heritage, Canberra.

Bella, D.A. (1987). Organizations and Systemic Distortion of Information, *Journal of Professional Issues in Engineering*, 113: 360–370.

Bella, D.A. (1992). Ethics and the Credibility of Applied Science (pp. 19–31) in *Ethical Questions for Resource Managers*, compiled by Reeves, G.H., Bottom, D. & Brookes, M.H., General Technical Report PNW-GTR-288, Pacific Northwest Research Station, Forest Service, US Department of Agriculture.

Bella, D.A. (2000). Faith, Responsibility and Knowledge. Paper prepared for a seminar series at Oregon State University, Fall 2000, Oregon State University.

Best Practice in Protected Area Management Planning. (2000). ANZECC Working Group on National Parks and Protected Areas Management Benchmarking and Best Practice Program.

Best Practice in Park Management. (2002). Report to the National Parks and Protected Area Management Committee. A review of current approaches to performance measurement in protected area management.

Day, J., Hockings, M. & Jones, G. (2002). Measuring Effectiveness in Marine Protected Areas – Principles and Practice. Keynote address, World Congress on Aquatic Protected Areas, August 2002, Cairns Australia.

Department of Parks, Wildlife and Heritage. (1992). Tasmanian Wilderness World Heritage Area Management Plan 1992, Hobart Tasmania.

Gilligan, B. (2006). The National Reserve System Programme: 2006 Evaluation. Department of the Environment and Heritage, Canberra.

Hocking, H. (1994). Tasmanian Wilderness World Heritage Area Management Plan. A Framework for Monitoring and Evaluation. June 1994, unpublished report to the Parks and Wildlife Service, Landmark Consulting, Hobart.

Hockings, M., Stolton, S. & Dudley, N. (2000). *Evaluating Effectiveness: A Framework for Assessing the Management of Protected Areas*. World Commission on Protected Areas (WCPA) Best Practice Protected Area Guidelines Series No. 6. IUCN in collaboration with Cardiff University. Gland, Switzerland/Cambridge.

Jones, G. (2000). Outcomes-based evaluation of management for protected areas – a methodology for incorporating evaluation into management plans. In *The Design and Management of Forest Protected Areas*, Papers presented at the Beyond the Trees Conference, May 8–11, 2000, Bangkok, Thailand/WWF, Switzerland, pp. 349–358.

Jones, G. (2003). *A Dummy's Guide to Evaluating Management of Protected Areas*. Unpublished paper for the Vth World Parks Congress, Durban, South Africa.

Jones, G. (2005a). Evaluating effectiveness in the Tasmanian Wilderness World Heritage Area. In *Assessing the Conservation Management Status of Biodiversity*, Workshop Proceedings, The Nature Conservancy, Dallas, TX, October 25–27, 2005.

Jones, G. (2005b). Is the management plan achieving its objectives? In Worboys, G., De Lacy, T., & Lockwood, M. (Eds.), *Protected Area Management. Principles and Practices*. Oxford University Press.

Jones, G. & Dunn (Hocking), H. (2000). Experience in outcomes-based evaluation of management for the Tasmanian Wilderness World Heritage Area, Australia. Case study 1, in *Evaluating Effectiveness: A Framework for Assessing the Management of Protected Areas*. World Commission on Protected Areas (WCPA) Best Practice Protected Area Guidelines Series No. 6. IUCN in collaboration with Cardiff University. Hockings, M., Stolton, S., & Dudley, N., Gland, Switzerland/Cambridge.

Natural Resource Management Ministerial Council. (2005). Directions for the National Reserve System. A Partnership Approach, Australian Government, Department of the Environment and Heritage, Canberra ACT.

Parks and Wildlife Service. (1999). Tasmanian Wilderness World Heritage Area Management Plan 1999, Hobart Tasmania. Available on the Parks and Wildlife Service website at: http://www.parks.tas.gov.au/index.aspx?base = 6158.

Parks and Wildlife Service. (2004). State of the Tasmanian Wilderness World Heritage Area – An Evaluation of Management Effectiveness. Report No. 1. Department of Tourism Parks Heritage and the Arts, Hobart Tasmania. Available on the Parks and Wildlife Service website at: http://www.parks.tas.gov.au/index.aspx?base = 6174.

Standing Committee on Environment, Communications, Information Technology and the Arts. (2007). Conserving Australia: Australia's National Parks, Conservation Reserves and Marine Protected Areas, April 12, 2007, Australia Parliament Senate, Canberra ACT.

The Nature Conservancy. (2005). Assessing the Conservation Management Status of Biodiversity. Draft Workshop Proceedings, October 25–27, 2005, Dallas, TX.

Vedung, E. (1997) *Public Policy and Program Evaluation*. New Brunswick, NJ/London: Transaction.

Part IV
The Importance of People

Chapter 14
Adaptive Management of Environmental Flows – 10 Years On

Tony Ladson

Abstract Many rivers in south eastern Australia have been degraded because of changes to flow caused by development of water resources. Environmental flows – water managed specifically to meet environmental objectives – have been proposed as a response to these concerns. In this chapter I look at the history of environmental flow policy and action and then consider the possible role of adaptive management in achieving better environmental flow outcomes. There are clear policies by state and federal governments to decrease over-allocation and provide water for the environment but the practical implementation of these policies has been fraught. Consumptive users have been given priority which, in a drying climate, means there is no water 'left over' to satisfy environmental needs. There is also a great deal of uncertainty around deciding on appropriate objectives for environmental condition and designing water releases to meet these objectives. Adaptive management is an appropriate response to reducing the uncertainty that plagues the relationships between management actions and outcomes, but first the legitimacy of environmental flows must be established. Unless we agree as a society that some water should remain in rivers for environmental purposes processes such as adaptive management of environmental flows, are unlikely to produce useful outcomes. They will be just a short term distraction until we work out how to exploit any remaining water resource.

Introduction

About 10 years ago I aimed find out about adaptive management of environmental flows by travelling to North America and South Africa, learning about the successes of the approach and bringing this knowledge back to Victoria, Australia.

T. Ladson
Institute for Sustainable Water Resources, Monash University

At the time, environmental flows seemed a solvable problem. We were only a few years into what at the time people took to be a temporary drought, but what now appears to be a new drier climate. I thought that a rational approach to the problem of water allocation, guided by the experiences of others could allow a balance between consumptive and environmental values. From the perspective of 2009, that doesn't look so possible.

In southern Australia, there is recognition that many rivers have been degraded because of the changes in flow caused by development of water resources. In particular, there are concerns about environmental damage to the Snowy, Goulburn and Murray Rivers that has been caused by the harvesting of water for irrigation (Ladson & Finlayson, 2002; Erskine et al., 1999). In a recent review of river health in the Murray-Darling Basin, the Goulburn was one of two rivers rated as 'very poor', the other being the Murrumbidgee which also has a high level of water diversions (Murray-Darling Basin Commission, 2008). Environmental flows – water managed specifically to meet environmental objectives – have been proposed as a solution to the problem of these degraded rivers. In this chapter I look at the history of environmental flow policy and action and then consider the possible role of adaptive management in achieving better environmental flow outcomes.

A Brief History of Environmental Flow Research and Policy

Environmental flows are a relatively new idea in Australia. In 1963 the Australian Academy of Sciences held a conference on water resource use and management. The discussion that followed the formal presentations is recorded in the proceedings and clearly there were some strong views and differences of opinion. There was concern about the environment but it was mainly restricted to salinity. By 1974 at a similar conference, there were proposals to use releases from weirs to mimic environmental events (Firth & Sawer, 1974). By the 1980s environmental flows were mainstream science (e.g. Walker, 1985) and the federal government sponsored a specialist workshop on instream needs and water users (Department of Primary Industries and Energy, 1986). In 1991 there was an influential conference on water allocation for the environment, but the 1990s are most notable for environmental flows becoming part of government policy.

In 1994, the Council of Australia Governments (representing all the states, territories and the federal government) signed up to a water reform framework that included recognition that water should be provided to the environment:

> *In cases where river systems have been over allocated, or are deemed to be stressed, arrangements will be instituted and substantial progress made by 1998 to provide a better balance in water resource use including appropriate allocations to the environment in order to enhance/restore the health of river systems.* (Working Group on Water Resources Policy, 1994)

In 1996 another important policy document was produced by the Standing Committee on Agriculture and Resource Management which included representatives of all Australian Governments. They agreed that:

> Where environmental water requirements cannot be met due to existing uses, action (including reallocation) should be taken to meet environmental needs.

In 2004 the National Water Initiative was agreed to by all jurisdictions and territories, and includes the imperative that governments should:

> Complete the return of all currently over allocated or overused systems to environmentally-sustainable levels of extractions.

The new Murray-Darling Basin Authority, set up in the NWI is now working on a Basin Plan, due by 2011. Central to this plan will be "sustainable diversion limits on water use in the basin to ensure the long-term future health and prosperity of the Murray-Darling Basin" (Overton & Saintilan, 2008). So, we have the right words when it comes to environmental flows, what about the corresponding deeds. Have these policies been translated in action?

A Brief History of Environmental Flow Deeds

Once an environmental flow has been determined it must be legally specified in some way. There are four approaches that may allow the translation of environmental flow policy into protection of water in rivers (Ladson & Finlayson, 2004). These approaches are reviewed along with examples of their application in Victoria and the Murray-Darling Basin.

1. 'Prior right' approach

Under this approach the environment has a prior right which needs to be satisfied before trade of water by consumptive users is allowed. This means that only the volume of water that exceeds that needed by the environment is made available for use through entitlements, and these also have conditions on them to ensure that environmental needs are met first. This is consistent with a 'cap-and-trade' approach to water markets where the use of the resource is capped; permits to take and use water are allocated, and can then be traded. This approach has the policy challenges of determining how much of the resource can be used, over what time frame and requires specification of exceptions for unusual conditions such as droughts (Colby, 2000).

This approach seems to be what was intended in the policy statements of the 1990s but it has not been achieved in practice. A parliamentary review of rural water usage comments that:

> the idea of taking the environmental requirements out of the market and specifying them as environmental needs is also a very clear statement from 1994, but it has been remarkably difficult for the state jurisdictions to do it (Ridgeway, 2004)

Consider the situation in Victoria. In 2005, the Victorian Government passed legislation to establish an 'Environmental Water Reserve' (Water Act, 1989 as amended).

This establishes a legally protected share of water in both rivers and groundwater systems for the environment with the objective to "preserve the environmental values and health of water ecosystems, including their biodiversity, ecological functioning and quality of water and the other uses that depend on environmental condition".

This sounds like a prior right for the environment, and the legislation reads that way but the accompanying policy document comments "In establishing the initial Environmental Water Reserve, the rights of existing entitlement holders will be recognised". It is also noted that water entitlements for consumptive uses will have secure tenure (Victorian Government, 2004). Consumptive entitlements can be reviewed but only every 15 years and next review is not planned until 2019 (Victorian Government, 2008). In most cases, the environmental water reserve is water 'left over' rather than legally protected in advance of satisfying consumptive needs (see below).

2. 'Equivalent right' approach

Environmental flows can be specified in a form that is equivalent to water for consumptive use. This would imply that the environment would have the same security of entitlement as other users and the same system of title would be used to specify and record its entitlement. This approach would allow environmental allocations to be traded on the water market along with consumptive entitlements. The Wentworth Group, for example, has proposed the creation of 'environmental water trusts' that could buy and sell water in pursuit of environmental objectives (Wentworth Group of Concerned Scientists, 2003).

The Victorian Government has granted a small number of high reliability bulk entitlements that provide water for the environment. The largest, is an annual entitlement of 27.6 GL that has been granted for wetlands in the Kerang Lakes area and along the Murray River (Victorian Government, 2008). In practice though, much of this water has not been made available to the wetlands. The water has to be delivered to these wetlands by using the channels that are operated by a water authority and which also supplies water to irrigators. Irrigators pay a fee for water delivery and a similar fee is charged to deliver the environmental flow. For example in 1997–1998, 13,700 ML of environmental water was delivered at a cost of $155,802 (or $11.37/ML) (Tan, 2001). The way that funds have been raised to pay these delivery charges is to sell some of the environmental water, which means less is available to meet environmental objectives. The water, once it is sold, can be used by irrigators. The end result is that a volume of water, specifically allocated for environmental use, ends up being used for irrigation.

A similar situation could arise with the environmental water reserve where stated government policy is that where water is delivered via a distribution system an 'Environmental Water Reserve Manager' would be expected to pay costs to an irrigation authority (Department of Sustainability and Environment, 2006).

3. 'Rules' approach

Environmental water allocations can be stipulated as operating rules that specify minimum passing flows, or conditions that would trigger an environmental release. This approach is likely to be appropriate for unregulated rivers where peak consumptive demands coincide with periods of low streamflow. Flow triggers may be used to curtail pumping or diversions to ensure minimum flows are maintained for the environment.

Often a rostering system is used to ensure that all the licensed diverters get some water although the volume available may be much less than the licensed amount. In these systems, specifying environmental flows as an equivalent right is not appropriate. Protecting environmental values requires a more secure entitlement than is available to consumptive users. It is in the driest time of a dry year that an environmental allocation is likely to be required and it is then that consumptive users have the least chance of getting their water.

Instead, a more secure way of specifying environmental flows is through operating rules. Most commonly, this will be a flow rate below which rostering will be implemented and a minimum flow where further diversions are banned.

As an example, consider the case in Victoria, where the process to specify the rules around water use in unregulated rivers is through the development of a Streamflow Management Plan. Consideration of environmental requirements does take place through an environmental flow study that aims to determine the minimum flow in a river that will protect environmental values. This environmental recommendation is then considered by a consultative committee. The stated aim of the current streamflow management plans is to provide a: 'balanced and sustainable sharing of stream flows between all water users in unregulated catchments. Particularly how water is shared between consumptive use and what is left in the stream' (Goulburn-Murray Water, 2003). However, these environmental recommendations haven't been given much weight and the outcomes from many streamflow management plans have not protected or enhanced the environmental qualities of streams. Three examples that reveal the deficiencies of this process are the SFMPs for the Yea, Kiewa and Ovens Rivers (Ladson & Finlayson, 2004).

On the Yea River, which contains the endangered Macquarie Perch, an environmental flows study recommended a minimum flow of 40 ML/day and commented that below this level there would be high risk of environmental degradation. The consultative committee adopted a minimum flow of only 10 ML/day increasing to 20 ML/day after 6 years. This minimum flow of one quarter to one half the recommended value seems inconsistent with a stated objective of the plan to maintain self-sustaining populations of locally occurring native fish species. At present, it is likely that Macquarie Perch are surviving only because they are stocked (52,700 stocked since 1987–1988) (Yea River SFMP Consultative Committee, 2001).

On the Kiewa River, the environmental flows study recommended a minimum of 150 ML/day at Kiewa but the committee set the allowable minimum at 80 ML/day. The scientists involved in the Kiewa study found there was a critical minimum below which 'environmental conditions exceed tolerance levels and become dangerous to stream dependent biota'. This critical minimum was 130 ML/day, yet the plan allows pumping to continue down to 80 ML/day (Zampatti & Close, 2000). The Kiewa plan also specifies an 'operational tolerance' of 20 ML/day which means flows could go as low as 60 ML/day before a ban was applied; less than half the critical minimum (Kiewa River SFMP Consultative Committee, 2002).

On the Upper Ovens River, the recommended environmental minimum was 200 ML/day at Myrtleford (Sinclair Knight Mertz, 2001) yet the draft stream flow management plan allows diversions to occur down to less than 1 ML/day or 0.5% of the required flow. Rostering isn't proposed until a flow of 60 ML/day. (Upper Ovens River Consultative Committee, 2003).

Perhaps fortunately, none of these draft streamflow management plans were ever gazetted and as of 2008 they were being revised.

Clearly, this is public policy gone awry. In these three rivers, the Yea, Kiewa and Ovens it will be legal for diverters to dry the river to beyond critical levels much more often than occurs naturally. On the Ovens, the draft plan would allow almost the entire flow to be extracted. The Stream flow management plan process allows the water users to decide how much water they want, without requiring them to accept responsibility for protecting environmental values. The streamflow management plan process is also expensive and time consuming with cost around $250,000 per plan (Howell & Bennett, 2003). Perhaps spending this money on buying back water allocations would have resulted in better outcomes. The plans have also established a right for irrigators to take this water so they could then justifiably ask for compensation if, in the likely event, some of it is required to be returned to these rivers when their degraded state becomes apparent. The National Water Initiative specifies that governments are to bear the risks of any reduction in volumes or reliability arising from changes in policy, for example, if there are new environmental objectives (National Water Initiative, 2004, Section 50).

The problems associated with historical over-allocation of water are well known yet these streamflow management plans continue to do this in guise of community decision making.

3. What's 'left over'

If the volume of water available for consumptive use is completely specified, and it is less than the total volume available, then the remainder could be considered to be the environmental flow. (This is, more or less, the historic situation on many of the rivers of southern Australia).

This is a high risk strategy for environmental protection for four reasons. *First* there are a variety of natural events that can reduce the total volume of water available in a river. These include climate change and bushfire. Without protection, the environment's share of the resource will absorb most of the decreases in supply. This is the scenario that has occurred over the last decade in south eastern Australia.

Environmental flows have been delayed or cancelled in response to reduced inflows. Examples include suspension of environmental flows in the following NSW creeks and rivers: Hunter, Wybong, Ourimbah Creek, Macquarie, Cudgegong, Murray, Murrumbidgee, Lachlan, and the Hawkesbury-Nepean system (these river names were obtained by searching the NSW government gazette).

In Victoria, environmental releases to the Wimmera River were cancelled with water held in case of emergencies related to fires or algal blooms (Victoria, Parliament, 2006). Environmental flow to Hattah Lakes was ceased on 12 Nov 2006 to redirect the water for human consumption. The environmental flow was later resumed following pressure from irrigators and environmental groups (Weekes, 2006). There has also been 'qualification of rights' i.e. decrease in environmental flows in the Yarra and Thomson Rivers (Melbourne Water, 2008).

For rivers in northern Victoria, modelling the impact of a sequence of dry years suggest a dramatic decrease in the availability of water for both consumptive use and the environment but the proportional decrease in environmental flows is much greater (Table 14.1).

CSIRO modelling suggests similar scenarios on the Murrumbidgee River (CSIRO, 2008). The average annual runoff over the period 1997–2006 is 31% less than the long-term average (1895–2006). If this drier climate continues the average annual flooding volume to Murrumbidgee Wetlands would be reduced by a further 69% compared to long-term values to be only 16% of predevelopment levels. The average period between flooding events would double. The maximum period between flood events on the Lowbidgee floodplain would increase to 16 years which is longer than the reproductive life stage of all waterbirds. This is an area where there has already been extensive flow alteration and severe ecological consequences of high water use. For example, waterbird numbers decreased by 90% between 1983 and 2001 (Kingsford & Thomas, 2004).

Second, gaps in the entitlement system mean that exploitation is likely to be larger than planned which will reduce environmental flows. This problem arises because current water property rights and regulations do not cover all aspects of the resource. Over time, exploitation may shift to those aspects which are not controlled which will affect the availability of the components of the resource where the rules are clear. Over exploitation is likely.

For example, there may be a property rights regime associated with diverting water from a river along with clear and enforced rules, but if groundwater use is not regulated, extraction of groundwater is likely to increase which will reduce river flows, most critically in the low flow periods when most of the streamflow is derived from groundwater (Evans, 2007).

In fact, information on groundwater use in the Murray-Darling Basin shows this is happening. Surface water diversions were capped in 1997 but there is no basin wide groundwater cap and there has been a rapid increase in groundwater use (Murray-Darling Basin Commission, 2007). This is expected to reduce streamflows

Table 14.1 Forecast water availability; comparison of long term average conditions (1891/1892–2006/2007) to recent low inflows (1997/1998–2006/2007) (DSE, 2008)

River	Reduction in consumptive use (%)	Reduction in environmental flows (%)
Ovens	−2	−31
Broken	−10	−70
Goulburn	−30	−69
Campaspe	−49	−86
Loddon	−67	−84

in the Murray-Darling Basin by 840 GL/year by 2030 which is 8% of the long term average surface water diversions or 16% of surface water diversions in low flow years (CSIRO, 2008). There are also likely to be additional impacts from farm dams and land use change (van Dijk et al., 2006).

Third, some consumptive entitlements are not limited to a particular volume but expand with the amount of water that is available. An example is the entitlements to 'off-allocation' water that irrigators can pump from a river during high flows. This water is in addition to other entitlements that they may have (Stewart & Jones, 2003).

On regulated rivers, providing water for the environment reduces the volume in storage and can affect the availability of off-allocation water in future years. It probably also decreases the likelihood of off-allocation water because reservoirs are less likely to spill if some water has been released for environmental purposes. This was confirmed on the Goulburn River, where computer modelling showed that *any* water allocated to the environment reduced the security of supplies to irrigators. This implies there really is no water 'left over' in this system (Fitzpatrick & Bennett, 1994).

The *fourth* reason why environmental flows need to be clearly specified is that it is not just the volume of consumptive diversions that need to be controlled if environmental values are to be protected. The timing of diversions can also be critically important. This argument applies to the Yea, Kiewa and Ovens Rivers discussed above. On these rivers, the total volume of diversions is small compared to the annual average volume of water available. The threat to the environment comes because on these unregulated rivers, most of the irrigation demand comes at the time when natural flow is at its lowest. Most of the year, there will be ample water 'left over' but during critical periods diverters may take all the water, placing environmental values at risk.

The upshot is, that despite, what seems to be appropriate policy statements about providing water to protect the environment, when we look at what is actually happening with water allocation in rivers, there are few examples of environmental flows and they are at risk.

Wither Adaptive Management?

Given this background, is there a role of adaptive management of environmental flows?

As noted in previous chapters, and outlined in Chapter 2 of this volume, Walters and Holling (1990) describe adaptive management as a structured process of learning by doing. The aim is to:

1. Work with stakeholders to develop a shared understanding of the system to be managed, and the desirable outcomes, by developing a system model that can be used for policy screening

2. Use this model to identify policies that are likely to succeed or that probe key uncertainties
3. Implement policies
4. Monitor and evaluate outcomes; and apply the learning to develop a better understanding of the system

Adaptive management treats policies and management interventions as experimental probes designed to learn more about the system; they are not confident prescriptions (Lee, 1993). Monitoring before, and during, the intervention, enables the natural system response to be determined and thereby allows managers to learn from past experience. This learning is then used to fine-tune the next round of interventions (Dovers & Mobbs, 1997).

The promise of adaptive management is that it can facilitate learning through a structured dialogue between scientists and managers and allows meaningful participation of stakeholders (Lee 1993; Ladson & Argent, 2002; Poff et al., 2003).

Adaptive management is likely to be an appropriate response where uncertainty plagues management action and this is clearly the case with environmental flows. Uncertainty occurs at three levels. Firstly there is uncertainty in setting environmental objectives for the state of our rivers. We may wish to restore streams to a better condition but the costs can be high. In Australia, the federal government has committed $3 billion over the next 10-years to address over-allocation through buying water entitlements back from willing sellers with the intension of returning 1,500 GL to the Murray. However, independent analysis suggests that the government would need to acquire 100% of the water entitlement market for the next 14 years to achieve these volumes, with the specified funding only providing perhaps one third of the required volume (Waterfind, 2008). The first $50 million of purchases have now taken place but are only expected to return 5,500 ML to the river because allocations in 2007/2008 were so low i.e. the entitlements are to a share of a pool of water for consumptive use and that pool is small at present following a decade of low inflows.

Second there is uncertainly around designing environmental releases to achieve particular environmental outcomes. That is, we are not yet able to design an environmental water allocation to achieve a particular result with a great deal of certainty (Stewardson & Rutherfurd, 2006). Thirdly there is uncertainly about future availability of water given the likely impacts of climate change. Water Resources analysis usually requires the assumption that climate is stationary but that may no longer be true (Milly et al., 2008).

In Australian we seem to be struggling with getting the balance right between values around consumption or conservation. Our policies espouse certain values of environmental protection but our actions favour consumption. An adaptive management process set up to refine uncertainty around achieving certain environmental outcomes from a particular flow release will be undermined if the desirability of the environmental outcomes are questioned.

In practice, the use of adaptive management in large riparian systems has been challenging. Carl Walters (1997) acknowledges that experimental management planning has floundered in complex institutional settings and that adaptive management approaches in riparian ecosystems have often failed to produce useful models for policy comparison or good experimental management plans for resolving key uncertainties. The reasons he identifies are: management stakeholders seeing adaptive policy development as a threat to existing programs; research stakeholders threatened by the outcomes of adaptive management; difficulties in modelling processes that occur across a large range of time and space scales; lack of data to guide modelling of key processes; insufficient data to validate models; and concerns by management agencies that experimental policies are too costly or risky.

Ladson and Argent (2002) reviewed the use of adaptive management in three large rivers in North America – the Columbia, Colorado and Mississippi – and identified seven factors that were likely to indicate success: few jurisdictions, few points of intervention, credible science, an appropriate level of modelling, early success of experimental management, a sense of community amongst stakeholders, and barriers to stakeholders leaving the adaptive management process.

Perhaps we just need to make careful choice of a case study where adaptive management is likely to succeed and go through with the process. We need to be careful to avoid the danger of half-efforts. An experiment with a small effect size might be less risky but a poor response is difficult to interpret and there is likely to be waning enthusiasm for further experimentation (Ladson & Argent, 2002). The experimental manipulations of environmental flows that have been tried in Australia, e.g. on the Campaspe River (Humphries & Lake, 1996), have suffered from this problem.

Getting institution arrangements correct will be important (Ladson et al., 2008). Institutional conditions that favour adaptive management of environmental flows include (Lee, 1993):

- There is a mandate to take action in the face of uncertainty.
- Decision makers are aware that they are experimenting.
- Decision makers care about improving outcomes over biological timescales.
- Human interventions cannot produce desired outcomes predictably.
- Preservation of pristine environments is no longer an option.
- Resources are sufficient to measure ecosystem-scale behaviour.
- Theory, models and field methods are available to estimate and infer ecosystem scale behaviour.
- Hypotheses can be formulated.
- Organisation culture encourages learning from experience.
- There is sufficient stability to measure long-term outcomes.

Recently, the National Water Commission (NWC) completed an update of progress in implementing the National Water Initiative (National Water Commission, 2008). They noted that, in relation to environmental management, while progress has been made in establishing better institutional arrangements, the NWI outcomes for integrated management of environmental water are not yet being achieved and there is significant community concern about the lack of tangible results on the ground. They also found that achieving improvements is made difficult by the inadequate specification of the desired environmental and public benefit outcomes because of a lack of transparency of trade-offs between environmental and consumptive water. The inability to provide adequate specification of desired outcomes leads to difficulty in monitoring and consequent lower levels of accountability that undermine adaptive management processes.

I see the problem is more fundamental. We need to establish the legitimacy of water for the environment. There are many people in Australia who think that water that flows to the sea is wasted. Some consider that it is simply not legitimate to consider allocating water to the environment at the expense of production. For example this quote is from the Herald Sun (Australia's largest selling newspaper):

> *How is water "used... by the environment"? Who can tell if a river really is "using" water, or just wasting it? And if a river really is "using" water, who says I can't take it anyway? (Bolt, 2006).*

There have been similar sentiments expressed in the Victorian Parliament (Victoria, Parliament, 1992) and in a book published by CSIRO (Pigram, 2006, p. 157).

Unless we agree as a society that some water should remain in rivers for environmental purposes, processes such as adaptive management of environmental flows, are unlikely to produce useful outcomes. They will be just a short term distraction until we build the next dam and drill the next bore…

References

Australian Academy of Science. (1963). Water Resources Use and Management. *Proceedings of a National Symposium on Water Resources*. Melbourne: Melbourne University Press.
Bolt, A. (2006). Gods must be crazy. Herald Sun, Melbourne, 26 July 2006.
Colby, B. G. (2000). Cap-and-trade policy challenges: A tale of three markets. *Land Economics*, 76(4), 638–658.
CSIRO. (2008). Water availability in the Murrumbidgee. http://www.csiro.au/files/files/plee.pdf
Dovers, S. R. & Mobbs, C. D. (1997). An alluring prospect? Ecology, and the requirements of adaptive management. In N. Klomp & I. Lunt (Eds.), *Frontiers in Ecology: Building the Links* (pp. 39–52). Albury, NSW: Elsevier.
Department of Primary Industries and Energy. (1986). Proceedings of the specialist workshop on instream needs and water users. Department of Primary Industries and Energy and Australian Water Resources Council.
Department of Sustainability and Environment. (2006). Securing our water future together: Victorian Government White Paper. State Government of Victoria.
Erskine, W. D., Terrazzolo, N., & Warner, R. F. (1999). River rehabilitation from the hydrogeomorphic impacts of a large hydro-electric power project: Snowy River, Australia. *Regulated Rivers: Research and Management 15*, 3–24.
Evans, R. (2007). The impact of groundwater use on Australia's rivers. Land and Water Australia Technical Report, Canberra.
Firth, H. J. & Sawer, G. (Eds.). (1974). The Murray Waters: Man, nature and a river system. Sydney: Angus & Robertson.
Fitzpatrick, C. R. & Bennett, P. A. (1994). *Environmental Flows – Jumping the Implementation Hurdle*. AWWA Environmental Flows Seminar, Canberra, 25–26 August 1994. Australian Water and Waste Water Association: 103–110.

Goulburn-Murray Water. (2003) Streamflow management plans: What is a streamflow management plan? http://www.gmwater. com.au/browse.asp?ContainerID = streamflow_management_plans (Accessed 9 Nov 2004).

Howell, G. & Bennett, P. (2003) Managing river flows in Victoria: Successes and challenges. *River Basin News.*

Humphries, P. & Lake, P. S. (1996). Environmental flows in lowland rivers: Experimental flow manipulation in the Campaspe River, northern Victoria. *23rd Hydrology and Water Resources Symposium*, Hobart, Institution of Engineers, Australia 197–202.

Kiewa River SFMP Consultative Committee. (2002). Kiewa River Streamflow Management Plan Report.

Kingsford, R. T. & Thomas, R. F. (2004). Destruction of wetlands and waterbird populations by dams and irrigation on the Murrumbidgee River in Arid Australia. *Environmental Management, 34*(3), 383–396.

Ladson, A. R. & Argent, R. M. (2002). Adaptive management of environmental flows: Lessons for the Murray Darling Basin from three Large North American rivers. *Australian Journal of Water Resources, 5*(1), 89–102.

Ladson, A. R. &. Finlayson, B. L. (2002). Rhetoric and reality in the allocation of water to the environment: A case study of the Goulburn River, Victoria, Australia. *River Research and Applications, 18*(6), 555–568.

Ladson, A. R. & Finlayson, B. L. (2004). Specifying the environment's right to water: Lessons from Victoria. *Dialogue, 23,* 19–28.

Ladson, A. R., Schofield, N., Sih, K., Sanderson, A. & Pawley, S. (2008). Improving governance of environmental water. *OzWater09.* 16–18 March, Melbourne. Australian Water Association.

Lee, K. N. (1993). *Compass and Gyroscope: Integrating Science and Politics for the Environment.* Washington, DC: Island Press.

Melbourne Water (2008) Annual report. http://www.melbournewater.com.au/applications/annual_report_2008/stat_bulk.htm.

Milly, P. C. D., Betancourt, J., Falkenmark, M., Hirsch, R. M., Kundzewicz, Z. W., Lettenmaier, D. P., & Stouffer, R. J. (2008). Climate change: Stationarity is dead: Whither water management? *Science,* 319(5863), 573–574.

Murray-Darling Basin Commission. (2007). Risks to shared water resources. Updated summary of estimated impact of groundwater extraction on streamflow in the Murray-Darling Basin. Murray-Darling Basin Commission, Canberra. http://www.mdbc.gov.au/__data/page/2257/Groundwater_Extraction_Impact_-_Overview_Report.pdf.

Murray-Darling Basin Commission. (2008). Sustainable Rivers Audit: Murray-Darling Basin Rivers: Ecosystem Health Check, 2004–2007.

National Water Commission. (2008). National Water Commission update on progress in water reform. National Water Commission, Canberra.

National Water Initiative. (2004) Intergovernmental agreement on a National Water Initiative. National Water Commission, Canberra. http://www.nwc.gov.au/resources/documents/Intergovernmental-Agreement-on-a-national-water-initiative.pdf.

Overton, I. & Saintilan, N. (2008) Ecosystem response modelling in the Murray-Darling Basin. *Journal of the Australian Water Association, 35*(8), 34–36.

Pigram, J. J. (2006). *Australia's Water Resources.* Collingwood, Australia: CSIRO.

Poff, N. L., Allan, J. D., Palmer, M. A., Hart, D. D., Richter, B. D., Arthington, A. H., et al. (2003). River flows and water wars: Emerging science for environmental decision making. *Frontiers in Ecology and the Environment, 1*(6), 298–306.

Ridgeway, A. (2004). Rural water resource. Senate Rural and Regional Affairs and Transport Committee. Commonwealth of Australia. http://www.aph.gov.au/Senate/committee/RRAT_CTTE/completed_inquiries/2002-04/water/report/.

Sinclair Knight Merz. (2001). Environmental flows studies for the Upper Ovens River. Prepared for Goulburn-Murray Water.

Stewardson, M. & Rutherfurd, I. (2006). Quantifying uncertainty in environmental flow assessments. *Australian Journal of Water Resources, 10*(2), 151–160.

Stewart, J. & Jones, G. (2003). Renegotiating the environment: The power of politics in managing the environment. Annandale, N.S.W.: Federation Press.

Tan, P. L. (2001). Irrigators come first: Conversion of existing allocations to bulk entitlements in the Goulburn and Murray Catchments, Victoria. *Environmental Planning and Law Journal*, *18*(2), 154–187.

Upper Ovens River Consultative Committee. (2003). Upper Ovens River Streamflow Management Plan Report (Draft). Prepared for Goulburn-Murray Water.

Van Dijk, A., Evans, R., Hairsine, P., Khan, S., Nathan, R. J., Paydar, Z., Viney, N. & Zhang, L. (2006). Risks to the shared water resources of the Murray-Darling Basin. Canberra: CSIRO.

Victoria, Parliament. (1992). Legislative Council, 4 June 1992. Water (Rural Water Corporation) Bill, second reading. Hon. W. R. Baxter, p. 1423.

Victoria, Parliament. (2006). Water: Wimmera-Mallee. Legislative Assembly 13 September 2006. Mr Thwaites (Minister for Water), p. 3250.

Victorian Government. (2004). Victorian Government White Paper. Securing our water future together. Victorian Government Department of Sustainability and Environment. http://www.ourwater.vic.gov.au/programs/owof.

Victorian Government. (2008). Draft Northern Region Sustainable Water Strategy. Victorian Government Department of Sustainability and Environment. http://www.ourwater.vic.gov.au/programs/sws/northern/draft.

Walker, K. F. (1985). A review of the ecological effects of river regulation in Australia. *Hydrobiologia*, *125*, 111–129.

Walters, C. (1997). Challenges in adaptive management of riparian and coastal ecosystems. Conservation Ecology [online], 1(2): 1. http:// www.consecol.org/vol1/iss2/art3.

Walters, C. J. & Holling, C. S. (1990). Large-scale management experiments and learning by doing. *Ecology*, *71*(6), 2060–2068.

Water Act, 1989 (as amended) Parliament of Victoria. http://www.legislation.vic.gov.au/Domino/Web_Notes/LDMS/PubLawToday.nsf/a12f6f60fbd56800ca256de500201e54/E078BDE9A7497B16CA257520001AA413/$FILE/89-80a092.pdf.

Waterfind. (2008). Waterfind analysis of the federal government buyback. Waterfind.com.au.

Weekes, P. (2006). Greens stick to their gums over water. Sunday Age, Melbourne. November 12.

Wentworth Group of Concerned Scientists. (2003). *Blueprint for a National Water Plan*, 31 July 2003. Sydney:World Wide Fund for Nature (WWF).

Working Group on Water Resource Policy. (1994). Report of the Working Group on Water Resource Policy to the Council of Australian Governments, Canberra.

Yea River SFMP Consultative Committee. (2001). *Draft Yea River Catchment Streamflow Management Plan*. http://www.g-mwater.com.au/browse.asp?ContainerID = yea_.

Zampatti, B.P. & Close, P. G. (2000). An Assessment of Environmental Flow Requirements for the Kiewa River: A Component of the Kiewa River Streamflow Management Plan for Goulburn-Murray Water. Melbourne: Arthur Rylah Institute.w

Chapter 15
Collaborative Learning as Part of Adaptive Management of Forests Affected by Deer

Chris Jacobson, Will Allen, Clare Veltman, Dave Ramsey, David M. Forsyth, Simon Nicol, Rob Allen, Charles Todd, and Richard Barker

Abstract Adaptive management requires the merger of management with science to provide robust knowledge about the effect of management actions. It can also be applied as a model of collaborative learning to support effective resource management. Using the example of adaptive management of native forests affected by introduced deer in New Zealand, we set out to identify some of the tensions that become apparent when adaptive management is applied in this way. We describe the process of adaptive management as it was applied in this case study. Drawing from project documentation and participant reflections on the learning process, we highlight three key lessons: (1) the need to create 'space' – i.e. a permissive environment that allows for an evolving process rather than a formalised and legalistic one; (2) that adaptive management cannot be expected to progress in a standardised way but instead, role clarity will emerge over time and this will contribute to an emerging vision of contribution that participants see for their project; and (3) the collaborative learning component of adaptive management poses a new challenge for science as rather than providing solutions to management issues, scientists contribute technical expertise and methods *as part* of the management

---

C. Jacobson
School of Natural and Rural Systems Management, The University of Queensland, Gatton, Queensland, Australia

W. Allen and R. Allen
Landcare Research Ltd., Lincoln, New Zealand

C. Veltman
Research and Development Group, Department of Conservation,
c/o Landcare Research NZ Ltd., Palmerston North, New Zealand

D. Ramsey, D.M. Forsyth, and C. Todd
Arthur Rylah Institute for Environmental Research, Department of Sustainability and Environment, Heidelberg, Victoria, Australia

S. Nicol
Secretariat of the Pacific Community, Noumea CEDEX, New Caledonia

R. Barker
Department of Mathematics and Statistics, University of Otago, Dunedin, New Zealand

of an issue or situation of interest. We show that these tensions decrease with time and that the collaborative learning process in this project lead to new understanding of forests for most participants. Moreover, the inclusion of shared learning as a primary objective of the project improved the relationships between participants.

Introduction

The use of large-scale experimentation and quantitative models as part of adaptive management provides a way to optimize management performance where ecological uncertainties exist (Walters, Korman et al., 2000; Walters & Holling, 1990). Equally, the application of collaborative learning processes as part of adaptive management provides a way to manage conflict and increase the pool of contributions to potential management solutions, emphasizing discussion on 'what' to learn about and 'who' ought to be learning (Buck et al., 2001; Keen et al., 2005; Walker et al., 2002). Mendis-Millard and Reed (2007) and others (Shindler & Cheek, 1999; Walkerden, 2006) argue that the incorporations of local needs, interests and circumstances as part of collaborative approaches requires science to be willing to relinquish some control of the direction of adaptive management. However, a solely collaborative emphasis can draw attention from the need for management decisions to be based on robust knowledge. The process of collaborative learning in adaptive management necessarily involves a juxtaposition of management and science. This is well illustrated in this research project where adaptive management was applied to increase understanding about the effects of introduced deer on indigenous forests in New Zealand.

Although deer have been present in New Zealand since the 1850s, there is uncertainty about the extent and consequences of their impacts on native forests (Allen et al., 1984; Caughley, 1989; Nugent et al., 2001a; Wardle et al., 2001). There is evidence that deer have modified the abundance of palatable and unpalatable understorey species (e.g. Allen et al., 1984; Nugent et al., 2001b; Wardle et al., 2001), and can sometimes irreversibly alter successional pathways (e.g. Coomes et al., 2003). Concerns about the impacts of deer on rural and forest lands were first recognised in 1921 under the Animal Protection and Game Act, and again in 1956 under the Noxious Animals Act (Eggleston et al., 2003). After a reduction in deer numbers during the 1960s and 1970s (Challies, 1985), the 'noxious' label was removed in 1977, when the Wild Animal Control Act replaced earlier legislation. Further deer control is proposed to conserve New Zealand forests (Department of Conservation, 2001). However, the benefits for forest conservation of additional deer control have seldom been tested.

The New Zealand Department of Conservation, the government agency with responsibility for managing public conservation lands, undertook an extensive period of consultation on the future management of deer between 1997 and 2001. The department invited comment on different management options and approaches whilst clearly indicating its statutory role in maintaining the biodiversity and structure of indigenous forests. Adaptive management was suggested as an approach to reducing the uncertain benefits of additional deer control. As stated in the 2001 Policy Statement (2001), "an adaptive management approach will be needed to allow control to be varied in response to observed effects of management". Although cases exist where the impacts of deer on forests are well-documented, many of these are based on paired samples rather than monitoring response to management over time. For example, forests on islands without deer have been compared with forests with deer (Veblen & Stewart, 1980) and smaller scale studies have involved the use of fenced areas (e.g. Veblen & Stewart, 1982; Wardle et al., 2001). Some studies (e.g. Husheer & Robertson, 2005) have monitored the response to management over time, however, significant uncertainties remain about how reducing the abundance of deer will influence forests, especially at the forest scale.

In 2003, a project was initiated to determine whether the removal of deer leads to anticipated benefits by applying the principles of adaptive management at four study sites in New Zealand. The 8 years "Adaptive Management to Restore Forests Affected by Deer" (AM-FAD) project was established through the Research and Development Group of the Department of Conservation (DOC). The issue of 'deer' (as opposed to other pests such as possums) was identified through discussion between the project leader, operational managers and scientists. The project has two aims:

- Reduce deer abundance and test predictions made about the reduced effects of deer herbivory on forests and
- Document all steps of the adaptive management process as part of the development of an operational guide for adaptive management

Adaptive management in this project combined collaborative and experimental elements. As outlined by Jacobson et al. (In press) adaptive management emulates a process of collaborative learning between stakeholders and scientists interested in the management of a particular area or issue. Collaboration provides opportunity to engage institutions and communities affected by management. In this case study, managers, scientists and stakeholders were involved in 'learning groups' that developed 'rich pictures and made predictions about the effect of deer control on parts of the deer-forest ecosystem before monitoring the influence of decreased deer abundance to determine whether forests respond as expected. Drawing from process documentation and evaluations conducted with groups (as a collective) and individuals involved, we examine the process of adaptive management as applied in this project. This enables distillation of lessons about the process of collaborative learning in the project.

The chapter is co-authored by many of the project research team. The lead authors (Jacobson and Allen) are applied social scientists interested in collaborative learning and engagement processes and the perspective offered here emphasises these interests. We acknowledge at the outset that this brings a particular 'lens' to the interpretation presented. Our theoretical background in individual and organisational learning and adaptation results in a set of beliefs that influence how we have approached and analysed this case study. This gives rise to the premise that effective collaborative learning is central to the success of adaptive management

(Jacobson et al., In press). Collaborative and social learning processes recognise that individuals have different perspectives on situations relevant to their different knowledge, experience and involvement in managing an area (Keen et al., 2005). As such, we were cognisant of these factors when designing and facilitating project process, especially the need for inclusiveness and respect for a diversity of knowledge sources.

The Adaptive Management Process

The AM-FAD project is very much a model of 'active' adaptive management (See box in Chapter 2 of this volume). Existing management at four different forest sites was modified through the introduction of deer control above and beyond the previous control (all sites have a history of aerial shooting, recreational hunting and culling that have intermittently and at various levels since the introduction of deer). A learning group was established for each site comprised of the science team, local DOC/land managers and stakeholders (primarily those involved in deer hunting). Conceptual models (rich pictures) of the deer-forest ecosystem were constructed for each site. The use of multiple forest sites, each with paired deer management and 'control' (no deer management) recognised the fact that the effects of deer management may vary between places and should enable strong inferences to be made about the effects of deer control on forest ecosystems.

The project has potential to influence deer management in multiple ways. Depending on evidence from the experiment, managers of sites involved in the project will be able to continue deer control across all or part of the forests they manage or justify their choice not to. Increased understanding about the ecological relationship between deer control on parts of the deer-forest ecosystem and conservation outcomes may directly influence management action at other sites in an attempt to replicate results. The experiences gained in adaptive management might also be applied in other forests where the outcome of reducing deer abundance remains uncertain or to other socio-ecological situations where uncertainty about management outcomes is high. In this sense, the project outcomes may contribute to the 'mainstreaming' of adaptive management as an acceptable operational approach. We anticipate that stakeholder involvement in selecting indicators and in scrutinising the experimental process will contribute to increased relevance of the results for management. We also anticipate that the project will provide lessons of more general interest about adaptive management.

Process Overview

A research team was established that included animal ecologists, forest ecologists, modellers, statisticians and applied social scientists from six organisations. These scientists interacted with forest managers and stakeholders who were interested in learning about the management of a forest by way of site-based 'learning groups'. Together, they contributed their knowledge about each site, recorded in the form of rich pictures, timelines of events and spatial diagrams.

The specific research questions for each site emerged as the learning group process developed. Each learning group was devised a set of indicators they felt would be appropriate to measure the effects of deer control on parts of the deer-forest ecosystem. A mathematical model was then developed that could be adapted to the specific interests of each group. Although there was freedom to develop individual hypotheses, a central hypothesis emerged: seedling growth rates are dependent on the relative vulnerability of seedling species to deer browsing. Because the learning groups associated with each site had freedom to develop individual hypotheses about the deer-forest ecosystem, the sites might be considered independent units rather than being part of a spatially replicated experiment (for other examples see Parkes et al., 2006). However, given that a central hypothesis emerged and sites had multiple indicators in common, they might equally be considered to represent experimental replicates measuring the effect of a similar management treatment (decreased deer abundance). During the process of reviewing this chapter, it became apparent that mixed views existed among the research team members about the extent to which sites should be treated as either 'independent units' (and therefore reflect the autonomy of learning groups), or as 'replicates' (and therefore increase the inference that can be drawn from results at individual sites). Whether they are viewed as 'units' or 'replicates' by scientists depends on whether components of interest are considered similar or dissimilar across sites.

Step by Step

Before the project began in earnest (i.e. after it was funded), the project leader met with DOC managers and scientists to scope the project in sufficient detail for funders. After funding was obtained, the process in this project has involved ten key steps conceptualised in Fig. 15.1:

1. Identification of potential study areas according to ecological and social criteria
2. Meetings with local managers
3. Individual meetings with learning group participants
4. Formulation of a collective understanding of the study area, including the interactions between deer and local forests and the driving influences upon this
5. Development of rich pictures, their refinement and verification by learning groups
6. dentification of indicators, appraisal of research methods and predictions about the effect of deer control on parts of the deer-forest ecosystem
7. Baseline monitoring of indicators
8. Decisions about appropriate modelling approaches, mathematical model development and predictions
9. Ongoing management action (sustained deer control) and
10. Ongoing indicator monitoring

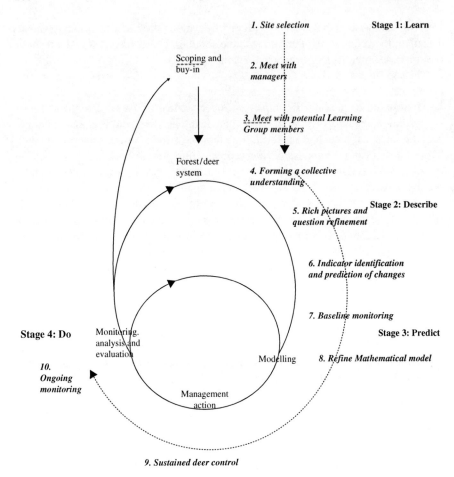

Fig. 15.1 The process of adaptive management. Italicised shadow diagram shows the actual project process thus far. Stages refer to those outlined in Chapter 2

The steps in the process reflect the model of adaptive management used in this specific project. We have aligned the steps taken in this project with the stages outlined in Chapter 2 of this volume. Steps 1 to 4 are equivalent to the stage of *'learn'* outlined in Chapter 2 where the negotiation process occurs around site selection and appropriate learning group composition and project goal setting. Steps 5 to 7 are equivalent to the stage of *'describe'*. Step 8 is equivalent to the stage of *'predict'*. Given that there was lack of agreement on the modelling approach at the outset, the simulation exercises described in Chapter 2 did not occur. These uncertainties (including uncertainty about the impact of deer abundance on the inter-relationships between system components) led to the development of a qualitative model of the system using fuzzy-cognitive maps (Ramsey & Veltman, 2005) a form of modelling that is uncommon in adaptive management. These models permit a variety of knowledge about the relationships between system components (e.g., deer abundance and hunting effort) including expert or consensus opinion. Components in the model are represented as states (e.g., 'high' or 'low' deer abundance) and qualitative descriptors are used to define interactions between components (e.g. the effect of deer on seedling growth is 'moderate'). 'Fuzzy logic' is then used to map the qualitative descriptor to an underlying numeric value in the model. As the model can be built using descriptions that are drawn from discussions, fuzzy cognitive map models are suited to transforming the 'rich picture' graphical models of the deer-forest ecosystem developed by the learning groups into a predictive model that can be used in the adaptive management framework.

As will be demonstrated in subsequent description of the project, learning occurred through the process of engaging in sharing ideas and perspectives on the topic of interest and reflecting on those in a structured way (e.g., model building) – i.e. it occurred throughout the process of adaptive management. A more apt description of the stages is therefore engage, describe, do, predict and evaluate.

In **Step 1**, sites were selected from a pool of those fitting the criteria outlined in the research investigation form (i.e. sites with ongoing pest possum control and lacking feral goats). To determine the effects of deer suppression on indicators, ecologists from the science team assessed forest areas within study sites that could be used as 'treatment' (i.e. deer control) and 'non-treatment' (i.e. no deer control) sites. Considerations were also made from a social research perspective. We sought local managers who were open-minded about the effects of suppressing deer abundance and interested in learning more about it. Whilst we have little information on the decision-making processes of managers in supporting the project, their understanding of the project and its potential outcomes are likely to have influenced their decisions to participate.

At **Step 2**, the project manager approached managers of forest sites of interest. The nature of the project and the commitment required (i.e. time) were explained. Given that the project manager was a scientist from with DOC but without operational management responsibilities, approvals from more senior operational managers were also obtained.

At **Step 3**, we provided local managers with a list of criteria for selecting potential learning group members. Criteria were based on experiences of applied social scientists, and targeted tow ensuring working relationships were as effective as possible. These included being known to the manager, open-minded about the impacts of deer on forests, familiar with the site, interested in working in a learning group alongside scientists, managers and other interest groups, and knowledgeable of other stakeholder perspectives on the management of deer and forests. We avoided selecting learning group members to represent different stakeholder groups, on the assumption that representatives would be more inclined to emphasise their stakeholder group perspective, rather than focus on interests, and that this could lead to increased potential for conflict. As noted by Walkerden (2006), a focus on interests and not on positions is one way that adaptive management can act to reduce conflict among stakeholders. As it happens, learning group members have a broad range of interests including hunting, community, Indigenous perspectives and other environmental concerns such as preservation of indigenous biodiversity. They also considered how other interests not represented by the groups might respond to the indicators selected.

Before bringing people together to meet as a group, some scientists visited prospective learning group members and local managers to introduce them to the project and gain insights (i.e. local knowledge?) about the sites. This step provided a crucial first interaction, giving potential learning group members an opportunity to familiarise themselves with the project aims without having committed themselves to ongoing project involvement, and to demonstrate our commitment to working in a collaborative way. Notes from interviews were circulated to other scientists not present so that they could familiarise themselves with this background.

Step 4 involved the first workshop. Individuals shared their reasons for being involved in learning groups and their expectations of the project. During interviews, workshop welcomes and follow-up interviews, we asked members to share their aspirations about the project so they could be incorporated into ongoing project management. These included:

- Improved working relations between local managers and hunters, including building relationships to improve future management planning and outcomes and the legitimisation of different interests through their participation in the process
- Learning about the local ecology, including learning from each others' different knowledge and experiences
- Quality science outcomes including ensuring that the project was able to detect a difference in relative deer abundance between treatment and non-treatment areas if one existed, that the effects of deer in comparison with other herbivores could be distinguished and that we could draw conclusions about abundance of deer that would be appropriate for different conservation outcomes and
- Learning about how institutions would use the information, including the need to act earlier rather than later with respect to building national-level relationships between DOC and interest groups so that results would be less likely to be treated with suspicion

Information from interviews was presented to the learning groups in two ways. The first was in the form of a chronology of observed events and change at the site. The second was as a map that was annotated to locate changes spatially (e.g. seasonal habitat use by deer, differences in perceived abundance of deer in different areas and changes in hunter pressure in different areas over time). This ensured we acknowledged value in the information provided. The discussion on links among forest components, human factors and deer prepared learning group members for the introduction of rich pictures that would later lend themselves to starting the mathematical modelling. Scientists then presented a summary of their knowledge of the effect of deer on forests. Focussed discussion allowed the incorporation of different knowledge about people, deer and plants, and on the connections between them. The workshop ended with a summary of potential research questions. Between meetings, workshop notes were circulated with opportunity for adjustment at the following meeting.

The second workshop (**Step 5**) involved a field trip to a study site followed by presentation of 'rich picture' models (see Britt, 1997, for examples of qualitative modelling approaches). These trips provided opportunity for checking plant identification, increased site-focused discussion and further opportunities for learning about the ecology of the area. They also provided an opportunity for the scientists to talk in a much more engaged and grounded way with the rest of the group.

Different types of rich pictures were developed with each learning group depending on the information provided. These included box and arrow models linking system components and their influence on one another, and more detailed models for each subsystem (e.g. for deer, forests and people) (Fig. 15.2). Discussion of rich pictures enabled identification of indicators and provided an opportunity for the scientists to present a considered response to investigation methods. Rich pictures or conceptual models (see Checkland, 1985; Eden, 1988) serve the purpose of 'principled negotiation' (Walkerden, 2006) by enabling representation of different knowledge about a system, identifying areas of common agreement on the interactions between system components and identifying areas where there

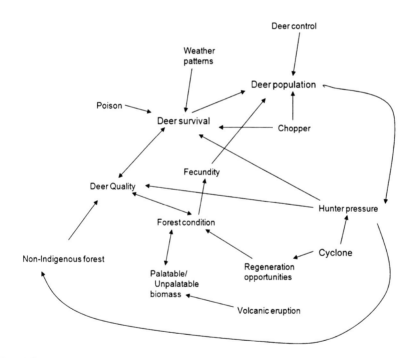

Fig. 15.2 Example of a rich picture used to convey ideas about relationships between system variables. Many of the components of this picture are idiosyncratic to the context of theses forests; 'chopper' relates to helicopter hunting of deer, volcanic eruption relates to the influence of volcanic ash on soil fertility, cyclone relates to a large wind and rain event that created light openings in forests and poison refers to the fact that toxins used in pest control also affect deer. Note that the diagram presented is in a very raw form hence there are four aspects related to deer population represented separately

is uncertainty about the behaviour of relationships between different components (Walters, 1986). In our case, discussion of rich pictures confirmed to learning group members that their view of interactions in the forest had been correctly recorded. It also provided a framework for ecologists to provide additional prompts and extra ecological detail for consideration. As a result of the additional input, learning group members rapidly identified tree species of interest at their study sites and modelling was able to commence.

In **Step 6,** learning groups made final decisions about indicators and how they would be measured. There was rapid consensus on indicators within groups, although the extent of input from the scientists varied between groups. Groups also identified additional unanticipated indicators; for example, measurement of changes in deer condition (e.g., jaw size, antler size, tissue depth above bone, body fat and fecundity) and measurement of carbon costs and benefits associated with deer control. In addition to making decisions about research questions, learning groups were involved in making predictions about research outcomes and stating their confidence in them. In this sense, whilst we had not at this point used mathematical simulation modelling techniques evident in other descriptions of adaptive management (Walters, 1986; Wollenberg et al., 2000) we did make predictions of change and assessed confidence in them, all of which can be revisited when results are analysed.

Step 7 involved baseline monitoring of deer, using a Faecal Pellet Index (Forsyth, 2005), and baseline measurements of the vegetation. Any required amendments were made to methods after consultation with learning groups. For example, in one site we considered planting additional seedlings of one indicator species due to low natural occurrence rates. From this step onwards, key decisions had been made and the focus shifted from regular face-to-face interaction to other forms of communication. Meetings were held less regularly, with the purpose of providing results from monitoring and deer control. One such meeting (**Step 8**) was held to review the fuzzy-cognitive mathematical model, focusing on the indicators, relationships between them and the certainty about strengths of relationships. The stepwise nature of working with learning groups at different sites means that at sustained deer control (**Step 9**) has been undertaken for different lengths of time (so far) at each site. Knowledge about the non-linear relationship between deer and forest response for palatable species led to the decision to maximise the difference in deer numbers between the 'control' and 'treatment' areas at each site.

Participation has varied throughout the project. Science team members participated in all four learning groups and hence had a multi-site perspective. This group has been added to when the need for additional specialist skills was identified (e.g. the inclusion of a fuzzy logic specialist) and members have been involved in different ways as the project has progressed (e.g., less intensive participation by applied social scientists as the project shifted emphasis from workshops to ongoing monitoring and deer culling). While there has been some interaction between learning groups, opportunity has been limited due to the distances between sites. New team members were able to gain a full understanding of project progress to this point through detailed project notes, working discussion documents and discussions with other team members. This pattern of participation is viewed as natural in the process of adaptive management as roles establish and different tasks are attended to (Stringer et al., 2006).

Lessons from the Project

Three key reflections have emerged from this project: 'Space' must be created for relationships in adaptive management to develop and grow; there is a need to be cognisant of group development and role assertion when planning collaborative processes; and there is an inevitable tension about the role of science vis a vis management authorities and responsibilities. The reflections draw from a combination of project documentation and from interviews conducted with individuals after each group meeting during the first three stages of the project (Jacobson, 2007). The use of interviews was based on the assumption that individuals might not feel comfortable sharing their perspectives on group process within a group setting.

Create Space for Building Relationships

Adaptive management often incorporates formalised processes of engagement with a range of stakeholders. The statutory and policy requirements of government agencies may require formal consultation processes (including deciding whom to consult) and memoranda of understanding with stakeholder groups that can influence the processes used for participation, the roles of stakeholder groups and the extent to which their perspectives are incorporated into management decision-making. As noted by Kootnz and Bodine (2008), this level of formalisation can act as a barrier to participation in adaptive management given the potential for issues to be managed in an adversarial way, and for positions taken on issues by stakeholder groups to become ingrained. Collaborative learning therefore requires 'space'. This 'space' enables the evolution of a project by creating a working environment that is permissive to participants, allowing them to clarify the project purpose, consider how they feel it will evolve, determine what their role in it is, and reflect on how they feel about their relationships with others who are involved. This enables groups to develop a joint understanding on an issue and how it could be managed without focussing on positions that can stall progress of collaborative projects (Walkerden, 2006). While learning groups were developing working relations and progressing ideas about appropriate indicators, discussions about learning groups were largely limited to DOC. Detailed and prolonged formal consultation processes that could lead to scrutiny from other interest groups were avoided given that the project was led by the Research and Development Group of DOC rather than by operational managers.

As groups develop, the way in which they make decisions also develops. Management of the project at each site has been based on joint decision-making.

A range of budget-appropriate choices for indicators and measurement methods was always sought, in addition to detailing supporting and limiting factors associated with these choices. Any amendments to methods were presented to learning groups for their acceptance. In our example, the situation never arose where there was fundamental disagreement among the learning group, scientists and DOC managers and therefore the significance of gaining this 'acceptance' was not tested. The involvement of senior managers in learning groups could have led to learning group members feeling less comfortable with the process and interpreting their role as 'placation' (Arnstein, 1979), and consequently choosing to participate less fully. Although senior managers were not involved directly, learning groups recognised the need for their support. In recognition of this, the learning groups subsequently took steps to incorporate concerns about research questions raised by mid-level managers. Whilst 'rules' for engagement were similar across groups, each developed on their own pathway.

Collaborative adaptive management requires effective working relationships among a community, managers and scientists if learning is to occur. In the case study described, the selection of learning group members was made to maximise opportunity for discussion and minimise the likelihood of members reverting to ingrained interest group positions. Indeed, learning group members commented on the potential for the process to be disrupted if debates about the status of deer in conservation lands ensued. Eleven of 20 learning group members involved in individual follow-up interviews commented, without prompting, that whilst individuals within groups had diverse interests, they were open-minded and this contributed to a good process. Well-facilitated processes that illuminated different perspectives of all participants and provided the flexibility to work at the pace of non-specialists were also viewed as important factors contributing to this. One learning group member commented on the positive way in which local knowledge was valued and represented in the project compared with others in which they had been involved. Thus, even where the participatory process avoids conflict through minimising opportunities for entrenched 'positions' to emerge, careful attention to process during workshops is beneficial.

Group Development and Role Assertion

Flexibility is required if adaptive management is to be responsive to the concerns and needs of the learning group and its membership. The progress of each learning group around the steps outlined varied: groups took between three and five workshops to reach step 7. Each learning group progressed differently through the modelling and indicator steps. Reasons for this include differences in the complexity of interactions between variables identified as important at individual sites, differences in the level of engagement of individuals with the process and differences in the contributions that learning group members felt they could make to the process. The lesson here is that each group will develop and attend to tasks at different rates. Practitioners of adaptive management should be wary of expecting progress to occur in a standardised way.

Adaptive management requires people to have a range of understandings from science, management and local knowledge. The roles of scientists and conservation managers are normally well-defined, as is their mandate for being involved. Other participants can take on a range of roles, depending on how their roles are defined (if a narrower role such as knowledge contribution is identified), or how they self-define their role. In our project, members began to assert different roles as the groups progressed through the different steps of adaptive management. Identified roles included contributing local knowledge (4 of 12), overseeing process and ensuring the work is practical (7 of 12) and ensuring credibility through quality science so that people would accept results (8 of 12). Learning group participants who defined their role as 'ensuring credibility' were less interested in indicator selection and more interested in measurement design and methods. In seven instances, participants did not feel comfortable with their role until later stages. Thus, role clarity emerges as projects progress.

The participation of different stakeholders in adaptive management and the roles they define for themselves in the process contribute to an emerging vision for the project. In response to the identified need to ensure project credibility with those not directly involved in it, a website was developed, and popular pieces were written for interest group magazines. For one of the learning groups, a diagram was also developed to explain indicator choices so that others not involved in the process would be able to understand the rationale. Some learning group members raised the need to articulate a vision so they had something to work towards. Without a self-identified vision, the reason for participation can be unclear, causing the momentum required for success in collaborative learning to lag. Whilst never explicitly articulated as such, the vision that emerged for many participants in follow-up interviews was that "the project contributes to knowledge about conservation management of forests with deer in a way that should be accepted by a broad range of stakeholders engaged in discussion about deer management in New Zealand."

Science Management Tensions

Adaptive management challenges the role of science and its dominance in solving management problems. Traditionally, science has been used in a way that provides technical solutions to management problems (see systems hierarchy box). However, as highlighted in the introductory chapter of this volume, problems addressed by adaptive management are often 'wicked' in nature, including uncertainty and difference in opinion about values, the nature of the problem and the acceptance of actions taken to manage it. In such systems, the act of discussing problems serves to change individuals' perceptions about them, meaning that a purely technical solution to a problem assumed true may not address the underlying uncertainties sufficiently. Nonetheless, the role of science is still significant.

On the one hand, the use of science provides robust knowledge about whether a management intervention can deliver expected benefits in a way that meets quality standards identified by society (Walters & Holling, 1990). On the other hand, it comes with a set of language (e.g., Latin species names), technical skills (e.g., experimental design, mathematical analysis and modelling) and in-depth knowledge (e.g., one person studying the ecology of a particular species for their working life) that can privilege the ideas of one group over another within multi-stakeholder situations (Ravetz, 1990). Funtowicz and Ravetz (1994) call for the application of 'post-normal science' where there is high uncertainty and outcomes that can affect many people in significant ways (i.e. situations in which adaptive management is commonly applied). Under post-normal science, broader contributions to problem delineation (from managers, interest groups and scientists) are considered appropriate given that a problem might never be fully resolved and the quality of the outcome is judged by a range of individuals with a stake in the issue. A recent Delphi study by Plummer and Armitage (2007) identified core issues in collaborative adaptive management. The learning process, shared authority, collectively developed roles and responsibilities, community and capacity to evolve were identified as the five most significant elements in collaborative adaptive management. In the same study, power asymmetries (e.g., where there are rules or processes that result in substantial differences in the ability of participants to influence decision-making) were identified as the most significant challenge to adaptive management, in addition to the reliance on and imposition of western science information, structures and management models. This study highlights the significance of science-management tensions associated with collaborative learning in adaptive management.

Tensions between the traditional role for science experts and their role in post-normal science were evident in this project. Members of one learning group exhibited anxiety about their contribution to formulating research questions, identifying indicators and developing methods to measure indicators, and asked for increased direction from the science team. This provides an example where scientists were expected to act in a more traditional way e.g. they were being asked to respond in a way that reflected their scientific interests. In other situations, learning group members were comfortable in a range of tasks including making decisions about indicator selection. Some scientists involved in the project expressed apprehension about their role in the project, noting that they felt they were expected to act as both 'experts' and as 'learning group members'. This was described by one scientist as "an almost schizophrenic experience". Although scientists were introduced to learning groups as 'experts', we were adamant about wanting to capture the different knowledge and viewpoints of non-expert learning group members and not devalue their expertise. A tension exists in adaptive management between 'traditional' and 'post normal' operational paradigms of science. One explanation for this is that adaptive management and collaborative processes were a novel approach that may have differed from their expectations. Most participants noted that this experience was a new way of working for them. An alternative or complementary explanation is that scientists are interested in addressing questions of scientific interest, whereas adaptive management requires understanding about responses to management actions that pose limitations on the way science is conducted. The tension that existed could therefore by equally explained by the fact that science is not typically accustomed to asking and answering questions about management (Walters & Holling, 1990). As community and scientists gain experience in working in this way, we would expect concerns about the role of science to decrease.

Feedback from individuals often led to adaptations in the process. Learning groups were established in a stepwise fashion. Feedback ensured we continued practices that learning group members found valuable. This included facilitation of the process and circulation of notes between meetings and site visits, ensuring that individuals were able to identify their ideas in the decisions made, and that they felt valued. Feedback also enabled refinements to the process based on feedback from individuals within the learning groups. Examples included:

- Clarifying the role of learning group members, although this concern decreased over time
- Providing more science input but minimising technical language
- Managing differences in plant identification skills by bringing books and/or samples to meetings
- Managing anxiety of learning group members about not knowing which of their interests to represent by making it clear that the group is interested in a range of perspectives, and they are welcome to identify which they are presenting when they raise issues or make comments during discussions
- Providing briefing papers for discussion before meetings and
- Avoiding over-loading the agenda for any given meeting so as that there is adequate time to consider all issues

At our fourth study site, measurement showed that deer abundance was so low it would not be feasible to further reduce numbers by hunting. As we write this chapter, an answer has yet to be found to the question of what can be learnt by suppressing deer abundance at that site. This has required scientists to acknowledge that it might not be possible to gain buy-in to measuring the same types of variables at each site and that more modelling may be necessary.

What Was Learnt?

An inherent assumption in the name 'learning group' is that learning occurs. Social learning theorists have argued that involvement in processes such as adaptive management has the potential to lead to profound changes in individuals' perceptions about a situation (Keen et al., 2005). Overall, the project was considered a successful exercise in relation to learning. Participants indicated that the process challenged their assumptions about the effects of deer on forests; one added that the site visit revealed "the exact opposite to what they expected to find." For two others, learning about forests and forest processes had changed their experience of being in the forests, saying that "it makes me look harder" and that it "has changed my horizons." Another participant noted that it had led to a realisation of how little they know, and a desire to learn more.

Table 15.1 (Jacobson, 2007) summarises things people say they learnt and the factors that contributed to them, grouping scientists separately from DOC and other learning group members given differences in their pre-existing ecological knowledge. Two key standout factors are modelling and knowledge about forests and forest processes.

Learning about physical things (e.g., forests) is just one aspect of learning in adaptive management. In addition, more abstract learning can occur (e.g. about group process) where there are likely to be multiple perspectives on what happened, the factors that were significant in contributing to that and the implication and meaning of it. The process of participation can also lead to learning about another's perspectives that subsequently leads to changes in relationships between learning group members. These relationship changes have the potential to continue beyond the life of the project. Reflections on relationships by individual learning group members indicated that some anti-Department of Conservation sentiment existed when the project began. These included comments that the views of DOC are "big time pre-determined", "out of touch with reality", that forest users have "absolutely no faith in them", and that the agency is "less than the sum of its parts". These views appeared to change substantially through the process of engagement. Rather than commenting about DOC as a whole, remaining criticisms were focussed on engagement at the national policy level. Due to the project's existence, the agency was perceived to be "fighting fires at the bottom rather than letting them get too big." This suggests potential for the project to improve relationships on multiple levels. Reflections by local managers indicate that improvements in local relations were a desired outcome in all cases. One manager stated that they wanted learning group participants to "walk away thinking they could stake their reputations on it [the project]." A DOC

Table 15.1 What was learnt by learning group members

What was learnt?	Specified contributing factors
Scientists	
Project differs to other adaptive management in New Zealand and Australia	Interdisciplinary team
The model building process used has been different to what was expected.	A broad topic and a desire to be inter-disciplinary
Can get to the modelling phase in a less-abrupt way	Desire to incorporate and demonstrate progression from rich picture to model
Whole system processes are messy	Research team and learning group relations
Other learning group members	Field visits and discussion by plant scientists
About forest ecology	
About the bigger picture	Listening
How to work with communities, including the value of listening	Observation
About range of perspectives (sometimes conflicting) within the hunting community	Discussions with other learning group members

manager involved in the project commented that this sentiment was evident during the first workshop. After the second workshop (Step 5) the same manager said that this tension had subsided and that trust was "60% there".

Concluding Comments

Using the example of adaptive management of forests affected by introduced deer in New Zealand, we set out to identify some of the tensions that become apparent when adaptive management is applied as a model of collaborative learning. We identified lessons about creating space for building relationships in adaptive management, that adaptive management cannot be expected to progress in a standardised way (and therefore that attention must be given to process), that role clarity emerges over time, and that collaborative learning in adaptive management poses challenges for scientists. Further, we have demonstrated that participation in this project led to different understanding of forests for most participants, and that relationships between participants improved. In our example of adaptive management, a more apt description of the stages is engage, describe, predict, do and evaluate with learning as a central component.

Chapter 2 of this book noted that attention to participation in earlier stages of adaptive management can lead to increased success in the 'doing' phase. While the participatory or collaborative aspects of adaptive management are often considered separately or at best as a 'graft' before experimentation (Jacobson et al., In press), this project sought to integrate them explicitly throughout the adaptive management process.

In the 'learning' phase of adaptive management, collaboration first begins through the identification of participants. There are many reasons for participation that can influence selection of participants. Unlike many planning processes, the reason for participation in adaptive management is pragmatic (as opposed to ideological), leading to the need to identify the 'right people' to involve. Reflections of learning group members highlighted the importance of trying to select the right people for learning groups. Proponents of collaborative approaches must also recognise that relationships take time to develop before a 'shared understanding' can be reached. These relationships enable 'trust' in one another and increase the likelihood that results will be accepted at the end of the project. Further, collaboration does not necessarily progress in a standardised way, resulting in variability in both project tasks and project timeframes. Although roles often are not clear, and this can lead to anxiety in this stage of adaptive management, our experience demonstrated that they generally became clearer as the project progressed.

In the 'describe' stage, modelling becomes a focus as knowledge from descriptions is synthesised. Models can then be used to identify indicators for management and test scenarios about what might happen. In the case study presented, models began from a conceptual basis in an attempt to engage learning group members in the indicator selection process.

Modelling falls in the traditional realm of science that provides technical solutions (e.g., prediction) to complex issues. However, adaptive management is more than a purely 'technical' ways to resolve problems; it also provides opportunity to discuss the framing of the problem under consideration. Under collaboration, multiple processes occur concurrent to model building: discussion about the nature of the issue or the system components under consideration, emergence of individual self-identified roles within groups, and the identification and framing of a 'vision' in order to enable individual 'ownership' of the project. These processes have the capacity to influence the traditional role of scientists as 'experts'. Depending on the outcomes of role definition and visioning, scientists may be expected to take on more or less of a lead role at this stage. If individuals' reflections presented in this chapter are anything to go by, the 'describe' stage presents a challenge for all involved but that the challenge diminishes with the increasing role clarification that occurs naturally over time.

In the current 'doing' phase of this case study, the nature of participation adapted to meet the ongoing needs of the project. At the last check, participants were comfortable with the group process and excited about potential results. On completion in 2011, the project in its formal sense will be ready for review. Our aim in writing this case study was not to provide a recipe for how to do adaptive management, but rather to share our experiences about the learning process. There have been substantial process wins in terms of relationship development. There has also been 'fuzziness' about how the project would progress. Through the process of reviewing this manuscript, scientists noted other lessons they felt were significant and instances where there were different perspectives on the project. We have been fortunate that the open working environment created by the project leader has enabled this to be explored. We anticipate there is much to learn yet!

Acknowledgments This project has involved contributions from numerous community members and Department of Conservation staff who have contributed to a vibrant working environment. We acknowledge their input into the project. Although two research team members (Peter Bellingham and Sarah Richardson) were unable to contribute to the preparation of this chapter, we acknowledge their important contributions to the project described in this chapter. We also acknowledge the role of the New Zealand Department of Conservation in supporting this project. Lastly, we thank Bill Carter for reviewing an earlier draft of this chapter.

References

Allen, R. B., Payton, I. J., & Knowlton, J. E. (1984). Effects of ungulates on structure and species composition in the Urewera forests as shown by exclosures. *New Zealand Journal of Ecology, 7*, 119–130.

Arnstein, S. (1979). A ladder of citizen participation. *Journal of the American Institute of Planners, 35*, 216–224.

Britt, D. W. (1997). *A Conceptual Introduction to Modelling: Qualitative and Quantitative Perspectives*. New Jersey: Erlbaum.

Buck, L. E., Geisler, C. C., Schellas, J., & Wollenberg, E. (2001). *Biological Diversity: Balancing Interests Through Adaptive Collaborative Management*. Florida: CRC Press.

Caughley, G. (1989). New Zealand plant-herbivore ecosystems past and present. *New Zealand Journal of Ecology, 12*, 3–10.
Challies, C. N. (1985). Establishment, control, and commercial exploitation of wild deer in New Zealand. *Royal Society of New Zealand Bulliten, 22*, 23–26.
Checkland, P. (1985). From optimizing to learning: A development of systems thinking for the 1990's. *Journal of the Operation Research Society, 36*(9), 757–767.
Coomes, D. A., Allen, R. B., Forsyth, D. M., & Lee, B. (2003). Factors preventing the recovery of New Zealand forests following control of invasive deer. *Conservation Biology, 17*(2), 450–459.
Department of Conservation. (2001). Department of Conservation Policy Statement on Deer Control. Retrieved 5 March, 2002, from http://www.doc.govt.nz/Conservation/002~Animal-Pests/Policy-Statement-on-Deer-Control/index.asp.
Eden, C. (1988). Cognitive mapping. *European Journal of Operational Research, 36*, 1–13.
Eggleston, J. E., Rixecker, S. S., & Hickling, G. J. (2003). The role of ethics on the management of New Zealand's wild mammals. *New Zealand Journal of Zoology, 30*, 361–376.
Forsyth, D. M. (2005). *Protocol for Estimating Changes in the Relative Abundance of Deer in New Zealand Forests Using the Faecal Pellet Index (FPI)*. Lincoln, New Zealand: Landcare Research contract report LC0506/027 to the Department of Conservation.
Funtowicz, S. O. & Ravetz, J. R. (1994). Uncertainty, complexity and post-normal science. *Environmental Toxicology and Chemistry, 13*(12), 1881–1885.
Husheer, S. W. & Robertson, A. W. (2005). High-intensity deer culling increases growth of mountain beech seedlings in New Zealand. *Wildlife Research, 32*, 273–280.
Jacobson, C., Hughey, K. F. D., Allen, W. A., Rixecker, S., & Carter, R. W. (In press). Towards more reflexive use of adaptive management. *Society and Natural Resources*, In press.
Jacobson, C. L. (2007). *Towards Improving the Practice of Adaptive Management in the New Zealand Conservation Sector*. Unpublished Ph.D. thesis, Lincoln University, Christchurch, New Zealand.
Keen, M., Brown, V. A., & Dyball, R. (2005). *Social Learning in Environmental Management: Building a Sustainable Future*. London: Earthscan.
Kootnz, T. M. & Bodine, J. (2008). Implementing ecosystem management in public agencies: Lessons from the U.S. Bureau of Land Management and the Forest Service. *Conservation Biology, 22*(1), 60–69.
Mendis-Millard, S. & Reed, M. G. (2007). Understanding community capacity using adaptive and reflexive research practices: Lessons from two Canadian Biosphere Reserves. *Society and Natural Resources, 20*, 543–559.
Nugent, G., Fraser, K. W., Asher, G. W., & Tustin, K. G. (2001a). Advances in New Zealand Mammalogy 1990–2000: Deer. *Journal of the Royal Society of New Zealand, 31*(1), 263–298.
Nugent, G., Fraser, W., & Sweetapple, P. (2001b). Top down or bottom up? Comparing the impacts of introduced arboreal possums and 'terrestrial' ruminants on native forests in New Zealand. *Biological Conservation, 99*, 65–79.
Parkes, J., Robley, A., Forsyth, D., & Choquenot, D. (2006). Adaptive management experiments in vertebrate pest control in New Zealand and Australia. *Wildlife Society Bulletin, 34*(1), 229–236.
Plummer, L. & Armitage, D. R. (2007). Charting the new territory of adaptive co-management: A Delphi study. *Ecology and Society, 12*(2), 10.
Ramsey, D. & Veltman, C. J. (2005). Predicting the effects of perturbations on ecological communities: What can qualitative models offer? *Journal of Animal Ecology, 74*, 905–916.
Ravetz, J. (1990). *The Merger of Knowledge with Power: Essays in Critical Science*. London: Mansell.
Shindler, B. & Cheek, A. (1999). Integrating citizens in adaptive management: A propositional analysis. *Conservation Ecology, 3*(1), 9.
Stringer, L. C., Dougill, A. J., Fraser, E., Hubacek, K., Prell, C., & Reed, M. S. (2006). Unpacking "participation" in adaptive management of socio-ecological systems: A critical review. *Ecology and Society, 11*(2), 39.
Veblen, T. T., & Stewart, G. H. (1980). Comparison of forest structure and regeneration on Bench and Stewart Islands, New Zealand. *New Zealand Journal of Ecology, 3*, 50–68.

Veblen, T. T. & Stewart, G. H. (1982). The effects of introduced wild animals on New Zealand forests. *Annals of the Association of American Geographers, 72*(3), 372–397.
Walker, B., Carpenter, S., Anderies, J., Abel, N., Cumming, G., Janssen, M., et al. (2002). Resilience management in socio-ecological systems: A working hypothesis for a participatory approach. *Conservation Ecology, 6*(1), 14.
Walkerden, G. (2006). Adaptive management planning projects as conflict resolution processes. *Ecology and Society, 11*(1), 48.
Walters, C. (1986). *Adaptive Management of Renewable Resources*. New York: McMillan.
Walters, C., Korman, J., Stevens, L. E., & Gold, B. (2000). Ecosystem modelling for evaluation of adaptive management policies in the Grand Canyon. *Conservation Ecology, 4*(2), Article 1.
Walters, C. J., & Holling, C. S. (1990). Large-scale management experiments and learning by doing? *Ecology, 71*(6), 2060–2068.
Wardle, D. A., Barker, R. J., Yeates, G. W., Bonner, K. I., & Ghani, A. (2001). Introduced browsing mammals in New Zealand natural forests: Aboveground and belowground consequences. *Ecological Monographs, 71*, 587–614.
Wollenberg, E., Edmunds, D., & Buck, L. (2000). Using scenarios to make decisions about the future: Anticipatory learning for the adaptive co-management of community forests. *Landscape and Urban Planning, 47*, 65–77.

Chapter 16
Effective Leadership for Adaptive Management

Lisen Schultz and Ioan Fazey

Abstract This chapter is about *making it happen*. Adaptive management, that is. How do you transform a conventional management regime into one of adaptive management? How do you make it safe and rewarding to fail? And how do you sustain the processes of adaptive management over time? More specifically, we focus on leadership types, leadership processes, leadership skills and characteristics that seem to help catalyzing and maintaining adaptive management. We discuss the role of adaptive leadership, administrative leadership and enabling leadership. We highlight the importance of managing through, managing out, managing in and managing up. Facilitative leadership is described briefly and then we provide a case study where transformational leadership was instrumental in shifting management regimes. Although the chapter does not provide any blue print solutions, we try to illuminate some of the processes that need to be taken into account when leading for adaptive management.

Introduction

"While many managers claim to be practicing adaptive management, most practice some variant of trial and error management. (…) One key distinction between these approaches is that adaptive management assumes policy failures will occur and that they provide a valuable contribution for learning, while other approaches seek to avoid policy failure." (Gunderson & Light, 2006)

As reflected in the quote above, implementing adaptive management in natural resource agencies often requires a shift in mindset, from one that seeks to avoid

---

L. Schultz
Stockholm Resilience Centre and Department of Systems Ecology,
Stockholm University, Stockholm, Sweden

I. Fazey
Institute of Biological, Environmental and Rural Sciences,
Aberystwyth University, Aberystwyth, Ceredigion, UK

failures to one that embraces them. Adaptive management is about continuous learning, not with the objective of finding the perfect final solution to a problem, but to navigate complexity while keeping a direction towards improved environmental conditions, increased human wellbeing or resilient generation of ecosystem goods and services. To some natural resource managers and organizations, the ideas of adaptive management are easily embraced, perhaps because they have always had to deal with uncertainty and rapid change. But to the many that have worked in stable conditions where routines have been refined over time, outcomes are reasonably predictable and roles are set, the notion of adaptive management and the continuous change it brings can be uncomfortable or even threatening (Parson & Clark, 1995). The potential benefits of adaptive management may be outweighed by the risks that learning also involves. What if this experiment shows that my specific role or activity is unnecessary or even destructive? Such worries may even stop people from participating in adaptive management efforts (Lee, 1999). It is in these situations that active facilitation of the shift in management regimes is most needed. But how does such a shift come about? And how it is sustained over time? This chapter is about the process of *making it happen*, or in other ways, the human agency that drives the cycle of learning and doing described in Chapter 2 of this volume. Reflecting on some relevant literature and a case study of a real world example, it explores the role of leadership in catalyzing and maintaining the process of adaptive management.

A Focus on Leadership...

We probably all know of at least one success story in relation to biodiversity conservation or natural resource management. There are always some projects that overcome the limited funding, the rigid institutions, the interest conflicts and the incomplete knowledge that cause so many other projects to fail. When we find these success stories, we search for success factors and lessons to be learned, and we may find explanations in management practices, institutional design, in social capital, in the ecosystem processes, in the mental models of human-nature relationships, or in the timing of the project. All of these factors and more can affect whether adaptive management works or not. But here, we will focus on a factor that seems to always be present in any success story, at least in the beginning: The Key Individual. He or she can be a champion (e.g. Gilmour et al., 1999; Howell & Boies, 2004; Stankey et al., 2005), a change agent (e.g. Bahamon et al., 2006; Crawford et al., 2006), an organizational entrepreneur (Hahn et al., 2006), a policy entrepreneur (Shannon, 1991; Kingdon, 1995), a local steward (Schultz et al., 2007), a key steward (Olsson et al., 2004), a facilitator (Vasseur et al., 1997), a broker (Bebbington, 1997), or a leader (e.g. Leach & Pelkey, 2001; Westley, 2002; Olsson et al., 2007). Although these terms refer to different roles in a change process, they all refer to a person who was instrumental in making something happen. The existence of such a person can, to some extent, compensate for the fact that many organizations and institutions of today are not built for adaptive management. But what is it that these people do? What kind of leadership do they provide?

...But Not in the Traditional Sense

The word 'leadership' often ignites images of a charismatic or authoritarian person who is in control, holds a formal position, and decides how his/her employees are to work to achieve the objectives of the organization. This image seems far away from the adaptive management approach, which accepts that no single person will ever have the full picture of what exactly needs to be done. However, in the vast field of leadership studies there is a growing body of examples of more "low key" leaders that facilitate and stimulate collaborative processes, supporting self-organization and experimentation rather than dictating every step towards a set goal. There is a parallel shift in focus, from the individual leader at the top, to the process of leadership, which can occur throughout an organization and is not restricted to the person in the formal role. For firms in the "new economy", just as for managers of complex adaptive ecosystems, the challenge is to create an environment in which knowledge accumulates and is shared at a low cost. Rather than leading for efficiency and control, organizations find themselves leading for adaptability, knowledge and learning (Volberda, 1996).

Leading for Adaptability and Learning

Complexity leadership theory suggests that three forms of leadership are needed to address this challenge (Uhl-Bien et al., 2007). Table 16.1 provides a summary of these leadership forms and their associated processes. *Adaptive leadership* refers to the dynamic that emerges from interactions between people of conflicting needs and different knowledge, skills and beliefs, resulting in adaptive outcomes, such as a new understanding of a problem. Adaptive leadership emerges when "expertise and creativity coincide in an adaptive moment between two individuals". *Administrative leadership* refers to the bureaucratic function, and involves processes such as planning, resource allocation, crisis management and organizational strategy. *Enabling leadership* provides and protects the conditions that catalyze adaptive leadership and allow for emergence. Enabling leadership involves both ensuring that processes of experimentation and creativity are protected from the rationalization and standardization driven by administrative leadership, and ensuring that the experimentation is consistent with the organization's mission and strategy (Uhl-Bien et al., 2007). In other words, enabling leadership for adaptive management needs to strike a balance between keeping the organization on track and making sure its goals are

Table 16.1 A summary of leadership forms and their associated processes

Leadership types and associated processes
Adaptive leadership
Nurturing creativity and diversity in views and knowledge
Interacting with people of different expertise
Reflecting on practice and learning
Managing through – taking an experimentation approach to problem-solving
Managing out – building relationships with stakeholders outside the organization
Administrative leadership
Keeping track of progress and communicating it
Managing up – gaining political and financial support for adaptive management
Enabling leadership
Setting aside time and providing space during work days for reflection and learning
Setting example – facing mistakes and learning from them through critical and open self-evaluation
Involving and empowering co-workers – building their capacity and confidence
Managing in – building support and involvement from within the organization
Managing up – gaining political and financial support for adaptive management
Facilitative leadership
Trust-building
Securing commitment of people involved – e.g. through 'small wins'
Building a shared understanding of the problem and the way forward
Transformational leadership
Connecting groups and key individuals
Building a knowledge base
Establishing a shared vision
Motivating people to invest in an alternative approach
Identifying and seizing windows of opportunity
Leadership skills and characteristics
Ability to match leadership style with context
Ability to integrate, understand and communicate a wide set of technical, social and political perspectives regarding the particular resource issues at hand
Ability to fabricate new and vital meanings
Ability to overcome contradictions
Ability to identify and build personal contacts with key individuals
Interpersonal skills such as conflict management and active listening
Strong values
Emotional control

met on the one hand, and leaving space for creativity and following unforeseen side paths on the other. In the context of adaptive management, this balance has been referred to as framed creativity (Folke et al., 2005).

Enabling leadership in the context of adaptive management involves removing blockages as well as creating opportunities for adaptive management, such as setting aside times and places for learning and reflection on experiments and practices (cf. Rushmer et al., 2004). Leaders also need to set example, by facing mistakes and learning from them through critical and open self-examination. Furthermore, leadership means involving and empowering co-workers to become adaptive learners, building their capacity and confidence in the adaptive management process (Rushmer et al., 2004; Chapter 18, Fazey and Schultz, this volume).

Leading for Collaboration and Participation

The full utilization and continuous updating of the knowledge base put forward in adaptive management often requires collaboration with groups outside the management organization itself (Charles, 2007). Adaptive management depends on a diversity of knowledge, skills and beliefs, and adaptive managers need to enable collaboration in this diverse setting. In a thorough review of 137 cases of collaborative governance, Ansell and Gash (2007) found that institutional design and *facilitative* leadership are two key variables in making collaboration successful. Such leadership can emerge from the people involved, or be provided by a neutral facilitator from outside. A particularly important process is trust-building and securing commitment to the project from the people involved (Ansell & Gash, 2006). This is especially relevant when aiming to embrace failures rather than avoid them. Communicating the rationale behind taking an experimental approach rather than promising a particular outcome can be challenging. One way of overcoming this challenge is to secure some intermediate outcomes, in the forms of "small wins". Such early results can create a virtuous cycle of trust-building and commitment, eventually leading to a shared understanding of the problems at hand and the way forward. Throughout the cases reviewed, face-to-face dialogue was indispensable (Ansell & Gash, 2006).

In an in-depth study of one adaptive manager, Westley (2002) identified four processes that must be handled simultaneously in adaptive management projects: *managing through* refers to the experimentation and testing approach to learning about the ecosystem, *managing out* refers to building relationships with local stakeholders outside the management agency, *managing up* refers to gaining political and financial support for the adaptive management projects, and *managing in* refers to building and maintaining support from within the organization. When any of these processes are lacking, the project becomes vulnerable.

Leading for Transformation

Initiating and sustaining smaller adaptive management projects as described above is useful enough, but what type of leadership is needed for shifting the approach to natural resource management on a larger scale? Can adaptive management somehow be institutionalized, so that it becomes less dependent on the existence of dedicated, energetic individuals?

Although this is an area that needs further exploration, we would like to share one case study of such a facilitated shift, which took place in a wetland in Southern Sweden: Kristianstads Vattenrike Biosphere Reserve. The wetland is a result of annual flooding and millennia of grazing by cattle, and it is a habitat of high biodiversity that produces a range of ecosystem services (Olsson et al., 2004). It is managed and monitored by a loosely connected network of official managers and local stewards (Fig. 16.1), including farmers, bird watchers, conservation associations, hunting associations, school children, scientists and angler societies (Schultz et al., 2007). Stimulated and supported by a municipal organization called the Biosphere office, these local stewards have transformed their approach to ecosystem management, as well as the area itself, from one of conventional management, unresolved conflicts and biodiversity decline, to one of adaptive, collaborative management and increasing biodiversity (Olsson et al., 2004). The case was examined during the Millennium Ecosystem Assessment, and has been described in several articles (e.g. Olsson et al., 2004; Hahn et al., 2006; Schultz et al., 2007; Olsson et al., 2007) but in short the story is as follows: In the 1970s, the wetland's biological and aesthetical values were declining because of cessation of cattle grazing on the wet grasslands. Conservation efforts, such as protecting the area under the RAMSAR international wetland agreement, had no profound effect on the negative development. Then, a key individual who perceived

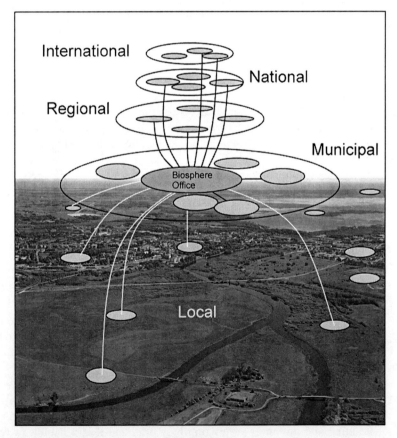

Fig. 16.1 The network of actors involved in managing the wetlands of Kristianstads Vattenrike. Networks are tailored to each project and coordinated by the bridging organization Biosphere Office (Modified from Hahn et al., 2006)

the pending crisis took action. More specifically, he discussed the wetland with key individuals representing the groups mentioned above, and from these discussions and his own experience, he built an attractive vision of the area – to create a water realm where the biological values would be both conserved and used sustainably. During the early discussions he also built a knowledge base about what caused the loss of valued species, and what could help reverse the trend. Through skillful communication with politicians, entrepreneurs and conservation funds, he was able to develop an acceptable and fundable proposal. When a window of opportunity opened in 1989, as environmental issues were high on the agenda, and local politicians were looking for a new image for Kristianstad municipality, the Ecomuseum Kristianstads Vattenrike was launched. The Ecomuseum was set up as a small and flexible municipality organization with the mission to initiate, facilitate and maintain ecosystem management of the wetland. Since then, the area has been designated a Man and the Biosphere reserve by UNESCO (United Nations Educational, Scientific and Cultural Organization) and the Ecomuseum has turned into a Biosphere office. Habitats have been restored, practices of cattle grazing and hay-making have been revived, rare species have returned, access to the area has improved and local decision-makers' perception of the area has shifted from a "swamp" to a "water realm". The approach to management involves experimentation, adhocracy, continuous learning and strategic communication and collaboration (Hahn et al., 2006).

The Kristianstads Vattenrike case illuminates some of the most important leadership processes that are needed to prepare a conventional management system for change, pushing it through a window of opportunity, and to sustain the new direction of adaptive management (Olsson et al., 2004). In a comparison between this case and four other cases of transformation towards adaptive governance, Olsson et al. (2007) identified transformational leadership (Kotter, 1995) and bridging organizations as essential features of such transformations. During the *preparation phase*, leadership involves connecting groups and key individuals, building a knowledge base, establishing a shared vision and motivating people to invest in an alternative approach. For *navigating the transition*, leadership involves identifying and seizing, or even creating a window of opportunity. Sometimes, a governance shift towards adaptive management is induced by an ecological crisis, but windows of opportunity can also be opened by a change in staff, a new funding program, a change in laws, or a change in perceptions.

Leadership Skills and Characteristics

Effective leaders in adaptive management are able to span multiple arenas of discourse and they are able to integrate, understand and communicate a wide set of technical, social, and political perspectives regarding the particular resource issues at hand (Olsson et al., 2007). Visionary leaders fabricate new and vital meanings, overcome contradictions, create new syntheses, and forge new alliances between knowledge and action (Westley, 1995).

Westley (2002) suggests that strong values, emotional control and interpersonal skills are critical to adaptive managers. Hahn et al. (2006) add that personal contacts with other key individuals are essential for building local, political and financial support. Together, these characteristics enable the adaptive manager to create the right links at the right time, around the right issues, overcoming the impossibility of identifying best practices or institutional arrangements in complex adaptive systems (Westley, 2002; Olsson et al., 2006, 2007).

Key Messages

As noted in the beginning of this chapter, success stories tend to point to at least one key individual as being critical. Considering the diversity of skills and characteristics necessary for leading adaptive management described in this chapter (Table 16.1), it seems that such a key individual would need super-natural powers. However, these kinds of skills can, at least to some extent, be taught, practiced and learned (e.g. see Chapter 18, Fazey and Schultz, this volume). If we accept that collaborative leadership skills such as active listening, conflict management, the ability to build coalitions and knowledge management skills are as important as ecological knowledge for adaptive management, we can start paying more attention to these skills and develop them. Furthermore, by identifying these key skills and characteristics, we have the option of building "adaptive teams" (Westley, 2002), making best use of the different skills of people to drive the process of adaptive management forward. Such a strategy also reduces the vulnerability involved in relying in one single individual, and emphasizes the importance of facilitating the development of the general adaptive capabilities of all involved.

In conclusion, we believe that most human beings have the potential to be creative, innovative, collaborative and adaptive. The role of leaders is, to put it simply, to provide a safe and rewarding environment to realize this potential.

References

Ansell, C. & Gash, A. (2004). Collaborative Governance in Theory and Practice. *Journal of Public Administration Research and Theory*, Nov. 13: 1–28.

Bahamon, C., Dwyer, J., & Buxbaum, A. (2006). Leading a change process to improve health service delivery. *Bulletin of the World Health Organization*, 84(8): 658–661.

Bebbington, A. (1997). Social capital and rural intensification: local organizations and islands of sustainability in the rural Andes. *The Geographical Journal*, 163: 189–197.

Charles, A. (2007). Adaptive co-management for resilient resource systems: Some ingredients and the implications of their absence. In D. Armitage, F. Berkes, & N. Doubleday (eds). Adaptive Co-Management: Collaboration, Learning and Multi-Level Governance (pp. 83–102). Vancouver/Toronto: UBC Press.

Crawford, B., Kasmidi, M., Korompis, F., & Pollnac, R.B. (2006). Factors influencing progress in establishing community-based marine protected areas in Indonesia. *Coastal Management*, 34(1): 39–64.

Folke, C., Hahn, T., Olsson, P., & Norberg, J. (2005). Adaptive governance of social-ecological systems. *Annual Review of Environment and Resources*, 30: 441–473.

Gilmour, A., Walkerden, G., & Scandol, J. (1999). Adaptive management of the water cycle on the urban fringe: three Australian case studies. *Conservation Ecology*, 3(1): 11. http://www.consecol.org/vol3/iss1/art11.

Gunderson, L. & Light, S. (2006). Adaptive management and adaptive governance in the everglades ecosystem, *Policy Sciences*, 39(4): 323–334

Hahn, T., Olsson, P., Folke, C., & Johansson, K. (2006). Trust-building, knowledge generation and organizational innovations: the role of a bridging organization for adaptive co-management of a wetland landscape around Kristianstad, Sweden. *Human Ecology*, 34: 573–592.

Howell, J.M. & Boies, K. (2004). Champions of technological innovation: The influence of contextual knowledge, role orientation, idea generation, and idea promotion on champion emergence. *The Leadership Quarterly*, 15(1): 123–143.

Kingdon, J.W. (1995). *Agendas, alternatives, and public policies*. Harper Collins, New York, USA.

Kotter, J.P. (1995). Leading change: Why transformational efforts fail. *Harvard Business Review*, March–April: 59–67.

Leach, W.D. & Pelkey, N.W. (2001). Making watershed partnerships work: a review of the empirical literature. *Journal of Water Resources Planning & Management*, 127: 378–385.

Lee, K.N. (1999) Appraising Adaptive Management. *Conservation Ecology*, 3: 3

Olsson, P., Gunderson, L. H., Carpenter, S. R., Ryan, P., Lebel, L., Folke, C., & Holling, C.S. (2006). Shooting the Rapids: Navigating Transitions to Adaptive Governance of Social-Ecological Systems. *Ecology and Society*, 11(1): 18.

Olsson, P., Folke, C., Galaz, V., Hahn, T., &. Schultz, L. (2007). Enhancing the fit through Adaptive Co-management: Creating and maintaining bridging functions for matching scales in the Kristianstads Vattenrike Biosphere Reserve, Sweden. *Ecology and Society*, 12(1): 28.

Olsson, P., Folke, C., & Hahn, T.(2004). Social-Ecological Transformation for Ecosystem Management: The Development of Adaptive Co-management of a Wetland Landscape in Southern Sweden. *Ecology and Society*, 9(4): 2.

Parson, E.A. & Clark, W.C. (1995). Sustainable development as social learning: Theoretical perspectives and practical challenges for the design of a research program. In L.H. Gunderson, C.S. Holling, & S.S. Light (eds). Barriers and Bridges to the Renewal of Ecosystems and Institutions (pp 428–60). New York: Columbia University Press.

Rushmer, R., Kelly, D., Lough, M., & Wikinson, J.E. (2004). Introducing the learning practice-1. The characteristics of learning organizations in primary care. *Journal of Evaluation in Clinical Practice*, 10(3): 375–386.

Schultz, L., Folke, C., & Olsson, P. (2007). Enhancing Ecosystem Management Through Social-Ecological Inventories: Lessons from Kristianstads Vattenrike, Sweden. *Environmental Conservation*, 34: 140–152.

Shannon M.A. (1991). Resource managers as policy entrepreneurs. *Journal of Forestry*, 89: 27–30

Stankey, G.H., Clark, R.N., & Bormann, B.T. (2005). *Adaptive management of natural resources: Theory, concepts and management institutions*. US Department of Agriculture, Portland, OR

Uhl-Bien, M., Marion, R., & McKelvey, B. (2007) Complexity leadership theory: Shifting leadership from the industrial age to the knowledge era. *The Leadership Quarterly*, 18(4): 298–318.

Vasseur, L., LaFrance, L., Ansseau, C., Renaud, D., Morin, D., & Audet, T. (1997). Advisory committee: A powerful tool for helping decision makers in environmental issues. *Environmental Management*, 21(3): 359–365.

Volberda H.W. (1996). Toward the flexible form: How to remain vital in hypercompetitive environments, *Organization Science*, 7 (4): 359.

Westley, F. (1995). Governing design: The management of social systems and ecosystem management. In L.H. Gunderson, C.S. Holling, & S.S. Light (eds.). *Barriers and Bridges to the Renewal of Ecosystems and Institutions* (pp. 391–427). New York: Columbia University Press.

Westley F. (2002). The devil in the dynamics: Adaptive management on the front lines. In: L.H. Gunderson, & C.S. Holling (eds.). *Panarchy: Understanding Transformations in Human and Natural Systems*. Island Press, Washington, DC.

Chapter 17
Institutionalising Adaptive Management: Creating a Culture of Learning in New South Wales Parks and Wildlife Service

Peter Stathis and Chris Jacobson

Abstract 'Learn by doing' is the mantra of adaptive management. Organisations that undertake conservation management are often challenged by high levels of uncertainty and a multiplicity of competing priorities leading to more doing than learning. Adaptive management provides a sound approach for these organisations to effectively manage uncertainty and ambiguity. However, institutional characteristics can impede the development of a learning culture and thus the uptake of adaptive management. Following on from a major review of the organisation's performance, the New South Wales National Parks and Wildlife Service (NPWS), responsible for managing over 6,5000,000 ha and over 750 protected areas, embarked on an ambitious program to introduce a performance management program based on adaptive management principles and to institutionalise it so that it became an indelible part of the way NPWS undertakes conservation. Through the combination of an adaptive management framework, a comprehensive performance evaluation program and set of common denominators defining the services provided in the organisation, NPWS has evolved its approach to ensure maximum penetration and uptake of the adaptive management ethos, by actively influencing key institutional facets such as policy, planning regimes, programs, projects and systems to link and align them, and ultimately to help close the adaptive management loop. While adaptive management is becoming normalised in NPWS, future efforts will be geared towards making the institutionalisation of adaptive management more robust and permanent.

P. Stathis
Management Effectiveness Unit, Parks and Wildlife Division, Department of Environment and Climate Change, Hurstville, New South Wales, Australia

C. Jacobson
School of Natural and Rural Systems Management, The University of Queensland, Gatton, Queensland, Australia

Introduction

Adaptive management is increasingly being identified by large organisations responsible for conservation management as an effective operational approach. Conservation agencies such as Parks Victoria (Australia), Parks Canada and Metsahallitus (Finland) have each established adaptive management frameworks and programs to assess their management effectiveness. Others have adaptive management frameworks that emphasise experimental management (e.g. Innes & Barker, 1999), modelling (e.g. Parkes et al., 2006) and collaboration with communities (Uychiaoco et al., 2005; Wollenberg et al., 2000) Whether this equates to adaptive management is dependent on criteria used for assessing 'what' adaptive management is (Jacobson et al. In press). It is axiomatic in accepting the ethos of adaptive management as a continuous learning process that the related implementation or development of adaptive management programs is also on a continuum. In short, you have to start somewhere: preferably with a clearly stated intention (or objective) to move towards adaptive management.

This is unsurprising given the multiplicity of objectives and pressures on such organisations and the varieties of uncertainty inherent in conservation, including uncertainty about ways to achieve particular goals, uncertainty about the appropriateness of goals, and uncertainty about the effect of management trade-offs between competing values and political priorities. Compromises between the laboratory precision of science and the risks of trial and error make adaptive management an attractive prospect (Lee, 1999). Adaptive management has been used in conservation planning, in evaluating performance (e.g. Uychiaoco et al., 2005), in determining the best approach to managing threats to endangered species (e.g. Bearlin et al., 2002) and in the management of protected areas in a developmental context (e.g. Agrawal, 2000; Buck et al., 2001).

The history of conservation in developed countries is regularly characterised by changes in land tenure that result in public agency mandates to preserve values of significance, including ecosystems and species in addition to sites of cultural significance. Such situations often lead to conflict about the appropriateness of activities such as recreation, ecotourism, harvesting of indigenous species and the use of sites for spiritual reasons. In cases where competing values exist, fear of political repercussions resulting from refuting previous management practices and policies, compounded with other factors that inhibit learning (e.g. procedural and resource rigidity and limitations on learning new ways of doing things (Argyris, 1990; Weiss, 1998)) can lead agencies to choose not to learn management. The institutional context of adaptive management can also affect the way in which adaptive management is applied. Groups within organisations can act to inhibit learning by limiting information access from levels above or below (what Siggelkow & Rivkin, 2006 refer to as 'screening'), by creating organisational structures that limit the capacity of individuals to affect change (Bodin et al., 2006) by limiting the influence of science experts on management decision-making (Jacobson, 2007) and by using risk avoidance as a justification for inflexible policy design that prevents its adaptation and renewal (Mertsky et al., 2000; Volkman & Lee, 1994). Building organisational capacity for adaptive management is therefore significant for managing uncertainty, ambiguity and competing priorities in conservation. However, there is wide spread concern (e.g. Lee, 1999; Walters, 2007) that institutional factors such as lack of leadership and political and or management ease of not changing behaviour can act to limit the success of adaptive management.

Adaptive management has been identified as an approach to management in the New South Wales National Parks and Wildlife Service (NPWS), Australia. In 2004, a review of park management in New South Wales by the State Audit Office indicated that the agency needed to "develop an adequate information base to measure its success" and to "implement a comprehensive system to measure and evaluate its results" (New South Wales Audit Office, 2004). These comments establish the need for managers to understand their performance and learn from their actions, adapting as they go to achieve 'continuous improvement'. The review acted as a catalyst, or 'crisis' as in the Biscuit Fire case study, to motivate the organisation to introduce adaptive management as an operational approach. In response, a periodic performance evaluation program was established to assess the performance of all parks in the park network. The performance evaluation program intended to address the dual goals of (1) performance evaluation, and (2) facilitating management adaptation and performance improvement. Thus, it was vital to instil a corporate culture amenable to adaptive management where the context for assessing performance of individual parks as components of a park network was confounded by complexity, uncertainty and ambiguity.

This case study focuses on the role of performance evaluation as part of an adaptive management approach for the park network, and the use of the Park Management Framework and Park Management Program in developing capacity for it. Results of the evaluation, and more specific detail about it, are readily available elsewhere (Department of Environment and Conservation, 2005; Hockings et al., 2006, pp. 85–89). The experiences shared here are drawn from our roles as manager with responsibility for the design and delivery of the management effectiveness evaluation system and the capacity for supporting adaptive management within the agency (Stathis) and as an external researcher specialising in capacity development for adaptive management (Jacobson). Along with others, we are involved in a 3 year research program to build capacity for adaptive management in the NPWS through improved systems for monitoring and evaluation.

Case Study Background

When the Audit Office conducted its review of the park system, it recommended the application of adaptive management as outlined in the review of the Tasmanian World Heritage Area (Tasmanian Parks and Wildlife Service, 2004). This approach required further development given the Audit Office's interest in evaluating performance across a network that included hundred of reserves as compared with seven in the Tasmanian example. To support adaptive management at this scale an evaluation framework was required that could be applied in vastly different contexts across a range of work areas. Thus, the agency chose to evaluate against a generic model of park management, and to facilitate capacity development in line with this. The intention was adaptive management as described in Chapter 2 of this volume, that is, to look forward in an effective and structured way while continuing with management.

The foundation components of the NSW State of the Parks approach include (1) a performance evaluation system (State of the Parks Program) based on a network wide survey every 3 years; the main (but not sole) information gathering process about performance to support adaptive management; (2) a Park Management Framework that establishes the structure and process for Adaptive Management within NPWS; (3) Service Themes that provide a set of common denominators based on Services provided within NPWS and act as points of congruence for linking and aligning programs, projects and approaches (see Table 17.1); and (4) the Park Management Program, a mechanism for actively linking and aligning organisational processes and projects that support continuous improvement by relating them to the Park Management Framework, the evaluation data obtained from the State of the Parks Program (and other sources) and/or Service Themes.

Management evaluation was conducted in 2004, 2005 and 2007. The most recent iteration included more than 750 reserves. The size of the evaluation limited involvement of assessors in its design. Eighteen staff from a range of management levels were involved in the original design and pilot testing. Regular meetings with the agency head ensured continued support and legitimised data collection with managers who might otherwise avoid performance assessment initiatives. Part of the process of cultivating the right conditions for learning to occur involved emphasising the risk and consequences of not engaging in the process, which involved reiterating the critical findings of the Audit Office report and the likelihood of a follow-up audit occurring. On the other hand, emphasis was also given

Table 17.1 Service themes for park management activities

Assessments acquisition and establishment
Fire
Pest animals
Weeds
On-park ecological conservation
Off-park ecological conservation
Aboriginal cultural heritage
Historic heritage
Threatened species
Visitor services
Visitor infrastructure
General infrastructure maintenance
General planning and policy
Community programs and education
Customer services and administration

to the opportunity to address some long held concerns by both staff and management (such as concern about NPWS' evaluation processes and corporate direction on key threats) and for staff to participate in resolving these problems. The internal delivery of the evaluation tool included staff training attended by over 60% of staff. A mid-level management sign-off process raised awareness of the evaluation and has supported use of the data at this level. A strategy of continuous engagement with key staff and experts has contributed to refinement of the process.

The evaluation was based on the World Commission on Protected Areas framework for assessing management effectiveness (Hockings et al., 2000). Managers made assessments using a four point ordinal scale and were asked to identify supporting evidence to justify their assessment; e.g., monitoring information. The evaluation was broad in scope, consisting of 36 questions plus written justification for key questions. For each Service Theme, key components include:

1. Providing contextual information such as reserve category, reserve design
2. Identifying key values and their significance and assessing their condition, the condition of threats to them and the adequacy of information for identifying them
3. Assessing the presence of clear and documented management directions
4. Identifying the relevant plans that exist, assessing their influence on management and assessing the comprehensiveness of planning
5. Providing quantitative information on operational costs, staff time and volunteer time
6. Assessing processes such as relationships with communities and indigenous peoples
7. Assessing management actions designed to ameliorate negative impacts of threats on values
8. Assessing the extent to which proposed actions were achieved and
9. Assessing the extent of monitoring and evaluation

Whilst the network level evaluation of management has created increased opportunity for adaptive management in NPWS, it is not sufficient to ensure it. In order for learning to occur, in an institution in a structured, programmed way a range of conditions must be met, including that managers want to learn (Argyris, 1990). Of course learning can occur in any organisation randomly without the aid of an adaptive management framework. Experience shows that without a model to help guide and direct the processes, knowledge is unlikely to be secured in the institutional memory, systems and processes to ensure it is available, valued and useful. As noted by Chapter 18, Fazey and Schultz, in this volume, adaptive people have particular skills that support learning. The role of the facilitator, in this case, agency staff that developed the evaluation system, is to develop tools and processes to support learning. As experience with evaluation and interest in its application grew, so did efforts to institutionalise the vision for adaptive management. The first step involved the adaptation of the evaluation model to form the basis of the Park Management Framework (Fig. 17.1). This framework was used to link and align existing management processes and identify areas for improvement. Tools to support the application of this framework as part of operational management became the focus of the Park Management Program.

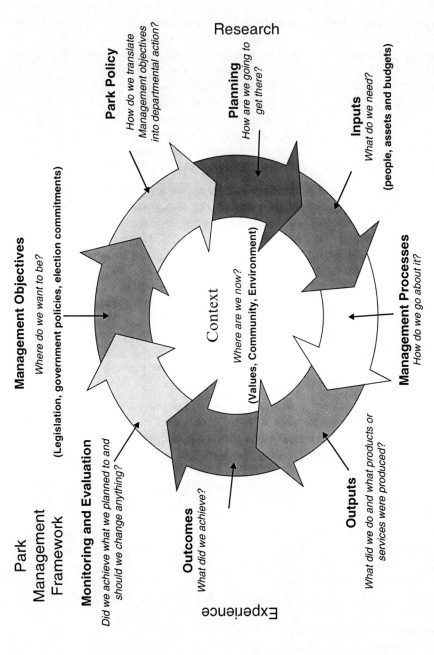

Fig. 17.1 Park management framework. This framework as used as a way to link and align management processes in a way that enabled adaptive manage-

The park management framework is linear. It represents a technically simple, sequential process, moving from one point to another within the cycle. This should not be misconstrued as being simply a technical-rational model or an over simplification of the challenges involved in implementing the adaptive management program (which are many). Clearly, it is not enough just to have plans or policies that support or reference adaptive management. However, the purpose of the framework is to create capacity for adaptive management to exist. While the framework is presented as a linear process for ease of presentation, by no means is it suggested that management progress through the steps will be linear. The aim of it is to help achieve a synchronicity of all the component parts. So the criteria for successful implementation of adaptive management would be evidence of the linkage of the components of the adaptive management model and of learning or reflection taking place in interactions between the component parts.

The Application of Adaptive Management – Beyond Ecosystems and Single Issues

The use of evaluation information as part of adaptive management has been prioritised to senior and middle levels of management but also provides for access and use by field based staff. This is largely due to the fact that these levels are accountable for, and more influential through corporate processes for directing, a multitude of lower level operations. This does not mean that it is not useful at the lower level. In fact, the evaluation information captured in corporate processes originates from this level upwards to be distilled back into 'products' for use by all staff. This 'looping' provides the opportunity for dialogue, reflection and adaptation at corporately significant places and times (e.g. during strategic reviews) or on a as needed basis. The process of completing the evaluation and the justification provision requirement provides opportunity for all staff involved in the management of a reserve to reflect on and consider performance. It provides opportunity that might not otherwise exist to consider the links between different management components (e.g. sufficiency of information and planning) and integrate perspectives across service areas in order to form a holistic understanding of management. Taking action based on the results of reflection typically occurs in adaptation of policies (e.g. to achieve different outcomes when working with stakeholders), priorities (e.g. setting resourcing target to address critical threats) and plans or strategies (e.g. to consolidate numerous individual local research needs into a regional plan).

At the mid-level management, operational planning balances the applicability of policy and strategy within the regional reserve context. For example, if analysis of evaluation data indicates that changes in impacts of invasive animals (e.g., dogs) are positively correlated with comprehensiveness of planning, then a resulting operational action could be to develop comprehensive plans for all parks with these invasive animals. Whether or not an action is relevant to a particular park will depend on the existence of the dogs, the significance and extent of threat posed by them in comparison to other threats to values (e.g., foxes), and whether there are appropriate resources and skills to respond. At this management level, emphasis has been given to Regional level planning operations given that these are the least senior management tier with a consistent and formalised planning process that evaluation information can be linked to and subsidiary strategy is developed at this level. Evaluation information and analyses have been aligned with operational outlook planning (based on a 3-year time horizon) in order to support its use.

Broad strategic directions are provided for in the statutory creation of protected areas. Public sector agencies interpret these in the form of policy and strategy. This involves a process of contrasting feedback from external influences (e.g., changes in policy at the State level, pressure exerted by the public and stakeholder groups, and landscape level ecological changes) with feedback from internal influences (e.g., ways in which management can realistically be improved within a given set of constraints such as funding, information access, individual park values and the relative performance of management). The depth of the evaluation means that it could be combined in different ways to inform specific reporting requirements; e.g., reporting on Ecologically Sustainable Forest Management and on outcomes for World Heritage Areas. Combinations of evaluation data with landscape wide information about the distribution and abundance of invasive species has enabled NPWS to demonstrate effort and achievement in dealing with whole-of-landscape issues. This fosters better understanding and more realistic expectations from other agencies and stakeholder groups.

It is difficult to categorise this case study in the evolutionary-active-passive paradigm. The approach taken by NPWS yields potentially multiple adaptations to policy, operations and plans. Arguably it is not passive either because the outputs of the evaluation are used in explicit and dynamic ways that are not appropriately categorised as simple or tame management situations. The modelling component of adaptive management is encapsulated in the evaluation framework. The types of information collected are sufficiently broad to facilitate diagnostic analysis of factors contributing to performance. Analysis of evaluation information enables managers to formulate hypotheses about changes to management that are relevant to their own context. For example, the influence of staff time, visitor information and numbers can be viewed in relation to visitor impacts. A hypothesis might be that when visitors are at high levels and information about them is insufficient, increased staff time in parks will lead to a decrease in their impacts. Thus, it clearly incorporates the modelling component of active adaptive management. While not experimental in the form of a large-scale field experiment (Walters & Holling, 1990), it does enable experimentation with policy, process and strategy. The evaluation program is itself a form of experiment. Adaptations are made both to the design of the program (changes to survey, data delivery methods, definitions etc.) as well as to the operations based on findings from the evaluations. Moving round the adaptive management cycle from objectives through policy, planning, operations to monitoring and evaluation also provides opportunities for policy experimentation in, and between, each of these elements. And it is at this juncture that the NSW program is arguably at its most active – that is, in the dynamic process of moving around the adaptive management cycle and between each element of that cycle, looking for ways to link and align learning to close the loop. So the NSW approach sits on the continuum described in Chapter 2, but also oscillates between the categories described above, depending on the context of issues and factors in play at anyone time. It encourages characteristics of evolutionary, passive and active approaches depending on the scale and scope of uncertainties evident in the evaluation.

The development of mechanisms to facilitate access to, and interpretation of, evaluation results are a more recent development in the history of the project. Capacity for utilisation is achieved through a number of mechanisms associated with the conduct of the evaluation and the delivery of results:

- Comprehensive guidelines for interpretation of data and results, including defining base levels of performance around the precautionary principle (see Box on precautionary principle)
- On-line assessment completion with electronic review (of each question) of evaluations by assessors and their managers
- Automated database queries for different components of the evaluation
- Internal reporting and support tool designed specifically to meet manager needs, as determined by workshops held with staff and
- Correlations of different evaluation components to address information needs specified by managers (e.g., comparison of outcomes for weed and pest management in relation to inputs)

Adaptation using evaluation data is promoted by considering a range of pre-defined categories for each Service Theme using on-line tools specifically designed for this purpose (see Table 17.2). This analysis does not provide results that offer definitive analyses of good or bad performance. Instead, it provides high-level analyses of performance and issues at a range of levels across the park system and through this provides a systematic way of considering indicative risks or problems.

An aligned research project is exploring additional ways in which to maximise information utility. It is focussed on improving understanding about the decisions made at each management level and the information used to inform them. From this information, the set of possible evaluation analyses is narrowed. The project has demonstrated that the analyses, interpretation and presentation of results in spatial format are linked to improvements in perceived usefulness of evaluation information for decision making. Additional uses for evaluation identified by managers included monitoring visitor impacts in different reserves to determine relative impacts and management priorities, demonstrating that evaluation information can encourage the initiation of site-based experimentation. Evaluation information is viewed as a 'support tool' rather than a decision making tool, and its use by managers at this level varies depending on their experience in an area, the number of parks they manage and the iconic nature of different parts of parks.

Table 17.2 Prompts to help guide consideration of performance evaluation data

Category of analyses for each service theme	Prompt
Condition	Consider investigating parks reporting the condition of values is at risk without corrective action.
	Consider investigating parks reporting insufficient information to answer.
	Look at justifications provided in answer. To do this use the evaluation tool – online data query: 'Assessment of Management Approaches (with consideration to management effect)' and select the relevant question as the filter.
Sufficiency of information	Consider parks indicating that they have insufficient information to support planning or there is little or no information available.
	Look at justifications provided in answer. To do this use the evaluation tool – online data query 'Assessment of Management Approaches (with consideration to management effect)' and select the relevant question as the filter.
Management directions and plans	Consider parks that indicate that they have insufficient or no clear management directions.
	Look at justifications provided in answer. evaluation tool – online data query: 'Assessment of Management Approaches (with consideration to management effect)' and select the relevant question as the filter.
	Consider the plans available for management and their influence on management. To do this use the evaluation tool – online data query 'What plans are available for park management within each PWD region/area?'.
Extent of threat	Consider which parks are reporting severe threats that are widespread or throughout.
	Use the evaluation tool – online data query 'What are the size and extent of threats facing each region/area?'.
Inputs	Consider whether the level of staff, volunteer and financial input is unexpected (higher or lower). To do this use the evaluation tool – Part B review query. This will show you data for each park grouped by area.
Management effectiveness (planned approach and impact on values)	Consider parks that indicate that they have only reactive or little or no management for an issue.
	Consider parks where assessments show that negative impacts are assessed as increasing.
	Look at justifications provided in answer. To do this use the evaluation tool – online data query: 'Assessment of Management Approaches (with consideration to management effect)'.
	Look at justifications provided in answer. To do this use the evaluation tool – online data query: 'Assessment of Management Effect (with consideration to management approach)'.
Change over time	Consider unexpected negative changes and unchanged responses where it is known considerable effort has been made to improve results.
	Looking at the justifications provided in the answer in the latest survey. To do this use evaluation tool – online data query Part D (Management Effectiveness) Review.

Building Support for Adaptive Management

Support for evaluating management effectiveness and an adaptive approach to conservation management is evident at the federal level in Australia in direction statements of the National Reserve System, and in review of park management and indigenous protected areas commissioned by them (Commonwealth of Australia, 2005; Gilligan, 2006a, b). Some State-based protected area management agencies have detailed systems for monitoring a comprehensive range of outcomes, while other have very detailed knowledge on only a more limited set of outcomes (Jacobson, Carter, & Hockings, In press). Management effectiveness evaluation in other protected area tenures (including Indigenous Protected Area Management agencies) appears to be in its infancy (Gilligan, 2006a, b).

Reviews of regional scale Natural Resource Management in Australia have highlighted that evaluation is a problematic area for adaptive management given biases in the way people evaluate in different sectors, the rarity of evaluation of socio-political components of projects and that there is often a lack of appropriate measures especially for socio-political aspects of projects (Bellamy et al., 2001). In 2008, the Federal government introduced guidelines for the Monitoring, Evaluation Reporting and Improvement approach. Several NRM bodies have additional evaluation requirements at the State level; e.g., the NSW Standard for Quality Natural Resource Management (Natural Resources Commission, 2005). The NPWS case study should therefore be viewed in light of an emerging impetus for management evaluation and performance improvement through adaptive management.

The initiation of performance evaluation and network level adaptive management in this case study initially was driven by the head of the agency. The significance of support from senior executive should not be down played (Walters, 2007). However, while committed to improved accountability and adaptive management, decision makers were initially sensitive to potential adverse responses from public reporting of performance data. In response to this sensitivity extensive preparations were made for media release and high level briefings throughout the agency.

Performance evaluation is clearly useful as a corporate reporting tool. The conduct of assessment and reporting alone does not necessarily lead to adaptive management. The most apparent lesson from this case study is that the institutionalisation of adaptive management is dependent on building capacity for a culture of learning and systems that enable institutional memory of that learning as others have identified (Bormann et al., 2007; Hagmann et al., 2002). In saying that, we recognise the difficulties faced in overcoming these and other barriers associated with institutionalising learning. Multiple conditions must be met for learning to occur, including that individuals can identify why they need to learn, that they operate in an environment conducive to learning, that they want to learn and have the skills to do so, that the don't perceive the demands of learning as being too high and they can identify an alternative (Argyris, 1990; Moon, 2004).

Efforts to enhance institutionalisation of adaptive management were synchronised through an overarching Park Management Program. The objectives of the Park Management Program are to achieve management excellence through adaptive management by (1) clearly defining organisational values and objectives (2) providing clear and consistent operational procedures and standards (3) linking evaluation to decision making at the strategic and operational levels and (4) improving ability to demonstrate and report on performance.

The intention of the program has been to co-ordinate, link, align and support a range of agency initiatives to improve park management. The program has included reviews of policies, standards and procedures at each step of the park management framework. These policies, standards and procedures enable broad agency directions to be translated into a format that is relevant for different management levels within the agency. For example, consider visitor facilities. Strategic planning priorities are geared towards expanding visitation opportunities. Policy for visitation implies the provision of suitable standards for park facilities. These are incorporated into planning objectives. Facilities are built and managed in accordance with the operational procedures guide and the park facilities manual. Works conducted are recorded in the Asset Maintenance System. The overall appropriateness of facilities, the maintenance of built infrastructure and whether actions are conducted in accordance with plans is assessed during performance evaluation. The latter informs reviews of mid-level management directives and the preparation of the next round of plans. In this way, capacity for utilisation of evaluation information can be developed and plans and strategies can be adapted. Ultimately, the longer term linking of operational planning processes with the evaluation process provides an effective means of applying adaptive management across the whole park system, thus driving the institutionalisation of adaptive management.

Prior to the Park Management Program there was no corporate capacity for system wide evaluation of performance, no systemic processes linking evaluation with management decision making or corporate projects to any framework for capturing knowledge and applying learning at the broad scale necessary to respond to the Audit Office findings. Further, there was no fostering of dialogue between park managers at all levels about management effectiveness, less capacity to provide comprehensive reporting and analysis of management effectiveness to stakeholders and finally, the corporate culture was not so readily receptive to discussion about park management. All these factors have improved with the Park Management Program. The Park Management Framework embedded a learning culture (or at least the precursors for it) in an attempt to overcome individual and organisational defences to structured approaches to learning.

Apart from the normal attributes that all significant change management programs require (such as clear leadership, ongoing communications and resourcing), five key lessons have emerged regarding the institutionalisation of adaptive management in NPWS. Firstly, it is essential to establish the imperatives and communicate the objectives for the 'doing' and the 'learning' parts of the adaptive management cycle. The continued existence of the Park Management Program has been dependent on building internal support for it at all management levels in response to a clearly defined need, motivated by an external stimulus but augmented by an internal commitment to improve and recognition of the opportunity at hand. This occurred in a staggered but quick process, with the initial introduction of the performance evaluation system, followed by the Park Management Framework, the Service Themes and then the Park Management Program. Whilst this may seem back to front, it provided opportunity for benefits to be demonstrated, for criticisms to be considered and for momentum to gather. In order to ensure continued support for the Park Management Program, both 'carrot' and 'stick' approaches have been used. The 'carrot' has come in the form of showing how it can benefit staff at different levels. The 'stick' is that aspects of it must be conformed with. In combination, the use of carrot and stick approaches has led to an increase in institutional appetite for using evaluation information to support adaptive management.

Secondly, broad scale adaptive management needs to be built on a clear framework and supported by corporate structures and systems. To address the goals of the Park Management Program, a number of existing and new strategic projects were brought together and aligned around the Park Management Framework. The framework provided a means of unifying these projects towards common objectives using an agreed model. It also provided the means for ensuring corporate systems use and make space for adaptive management processes.

Thirdly, it is vitally important to provide a means to demonstrate how core services delivered by the organisation relate to these higher order strategic objectives. This ensured an understanding of how an individual's responsibilities and efforts aggregate to help achieve the objectives of the whole organisation and to overcome the disengagement of field based staff when implementing strategic measures in an operational context. The Service Themes (Table 17.1) represent the broad areas of operations undertaken in the agency. Service Themes are used to categorise actions in operational and strategic plans, to organise financial reporting, to assess staff and volunteer time, in identifying information gaps, in developing new information systems and designing new projects and plans and to organise policies and operational procedures.

The fourth key to ensuring institutionalisation of adaptive management approaches is to prove the worth of the approach by providing tools and products that link to decision making processes. This has been done by designing a tool kit of products that support strategic planning, regional operational planning and communication of performance evaluation to relevant stakeholders. These tools help managers to adapt operational level management based on performance evaluation data by using data specific to their sphere of management. In this way, the adaptive management approach becomes part of the operational planning process and the institutional management systems.

Fifth, it is important to respond to organisational changes, sensitivities and concerns the institution has about the adaptive management program. In other words, the adaptive management program must itself be adaptable. Multiple adaptations have been made in the NSW example. A list of response strategies was developed in order to deal with staff concerns about the interpretation and reporting of negative evaluation data. These emphasis approaches to data contextualisation, messaging (colour and wording usages) and the benefits that can arise from agency transparency. To address the desire to 'group' parks, staff conducted a trial comparison of park evaluation results based on suggested groupings that found significant differences between parks perceived to be homogenous in management. Emphasis was then placed on addressing the reason for the concern (i.e., inputs and time required to undertake the evaluation) by developing systems to capture information more efficiently (e.g., online completion and pre-population of the survey). Concerns about information reliability have been addressed through associated research involving ground-truthing of some evaluation components (i.e., comparison to quantitative assessments).

The next phase of development will further emphasise the institutionalisation of adaptive management. A critical review of the design and product delivery will be undertaken to test whether the Park Management Program and performance evaluation is meeting its stated aims and objectives. Linkages and alignment of programs and systems within and outside the agency will be extended. The intent is to cultivate multiple drivers and supporters (internal and external) at a range of scales (local to national), to ensure the program continues to be supported, and that learning as an output from the program continues across a broad range of spatial, temporal and institutional scales and thus the program is seen as the preferred way of doing business. More closely coupling performance evaluation systems with site and species specific monitoring programs will evolve the approach to provide a logical continuum between qualitative and quantitative monitoring and evaluation from the ground up to the strategic level, but focussed on priority areas to ensure programs are affordable, practical and therefore sustainable. There is interest in improving organisational capacity to define standards of management for key service themes linked to the evaluation system that in turn informs operational and strategic planning. Lastly, a desire to maximise the potential of the evaluation component of the Park Management Program led to research collaboration with the University of Queensland and other Australian park management agencies undertaking similar evaluation. The project aims to support capacity building for adaptive management by improving understanding about the reliability of the evaluation, conducting more technical data analyses and supporting information integration into decision making.

Concluding Comments

Adaptive management offers conservation managers a way to balance the risks of trial and error management with the costs of laboratory precision science. One problematic area for adaptive management has been in building institutional capacity (Allan & Curtis, 2003; Stankey et al., 2006).With this in mind, this case study set out to explore the application of adaptive management as an approach to conservation management in the New South Wales Parks and Wildlife Service, focussing on the ways in which institutional capacity was developed. The approach began with evaluation and was supported by a management framework and strategic approach to its reinforcement in all levels of management through the application of the Park Management Program. In this case, a broader climate for evaluation existed and this was seized upon and driven within the organisation by its head. Critical to the institutionalisation of a learning culture in NPWS has been building the management framework and evaluation process into corporate systems and processes, demonstrating the practicality of adaptive management, and being responsive to staff ideas and concerns.

Adaptive management in this case study appears at odds with academic descriptions of the approach. The evaluation system supports the identification of areas where experimentation can reduce uncertainties associated with management at the park level. It also supports policy and process experimentation and feedback across the agency as a whole. Further, learning also occurs in the institutionalisation of the approach within the agency. Thus while the approach does not fit the quantitative modelling and scientific experimentation interpretation apparent in other examples of adaptive management (e.g., Chapter 15, Jacobson et al., this volume), the notion of models, delineating assumptions about expected outcomes, experimentation, monitoring and evaluation are clearly evident. This contrast highlights the multi-dimensional nature of adaptive management, and the fact that adaptive management may be multidimensional in application within any individual case study.

Fundamental to the ongoing commitment to this program in NPWS is the understanding that there is still much to learn and a desire to improve. Equally important is the understanding that the most effective institutional wide learning is not extemporised but is actively facilitated through a framework that provides structure, rigour and seeks continuous improvement. Ideally, there should be a permanent commitment and capacity to realise the objectives of an adaptive management approach, irrespective of the changes in 'fashionable' conservation linguistics or policy options.

Acknowledgments Support for this collaboration was provided by an Australian Research Council linkage grant LP0667672, partnered by The University of Queensland, NSW Department of Environment and Climate Change, Parks Victoria and Parks Australia (including the National Heritage Trust and the Director of National Parks).

References

Agrawal, A. (2000). Adaptive Management in Transboundary Protected Areas: The Bialowieza National Park and Biosphere Reserve as a Case Study. *Environmental Conservation, 27*(4), 326–333.
Allan, C., & Curtis, A. (2003). Regional Scale Adaptive Management: Lessons from the North East Salinity Strategy (NESS). *Australasian Journal of Environmental Management, 10*, 76–84.
Argyris, C. (1990). Overcoming Organizational Defences: *Facilitating Organizational Learning.* Boston, MA: Allyn & Bacon.
Bearlin, A. R., Schreiber, E. S. G., Nicol, S. J., Starfield, A. M., & Todd, C. R. (2002). Identifying the Weakest Link: *Simulating Adaptive Management of the Reintroduction of a Threatened Fish. Canadian Journal of Fisheries and Aquatic Science, 59*(11), 1709–1716.

Bellamy, J. A., Walker, D. H., McDonald, G. T., & Syme, G. J. (2001). *A systems approach to the evaluation of natural resource management initiatives.* Journal of Environmental Management, 63, 407–423.

Bodin, O., Crona, B., & Ernstson, H. (2006). Social Networks in Natural Resource Management: *What is there to Learn from a Structural Perspective.* Ecology and Society, 11(2), 42.

Bormann, B. T., Haynes, R. W., & Martin, J. R. (2007). Adaptive management of forest ecosystems: *Did some rubber hit the road?* BioScience, 57(2), 186–191.

Buck, L. E., Geisler, C. C., Schellas, J., & Wollenberg, E. (2001). *Biological Diversity: Balancing Interests Through Adaptive Collaborative Management.* Florida, U.S.A.: CRC Press.

Commonwealth of Australia. (2005). Directions for the Natural Reserve System - A Partnership Approach. Canberra, Australia.

Department of Environment and Conservation. (2005). State of the Parks 2004. Sydney.

Gilligan, B. (2006a). The Indigenous Protected Areas Programme: 2006 Evaluation by Brian Gilligan. Canberra, Australia: Commonwealth of Australia.

Gilligan, B. (2006b). The National Reserve System Programme: 2006 Evaluation. Canberra, Australia: Australian Government Department of Environment and Heritage.

Hagmann, J., Chuma, E., Murwira, K., Connolly, M., & Ficarelli, P. (2002). *Enhancing the adaptive capacity of the resource users in natural resource management.* Agricultural Systems, 73, 23–29.

Hockings, M., Stolton, S., Leverington, F., Dudley, N., Courrau, J., & Valentine, P. (2006). Evaluating effectiveness: *a framework for assessing management effectiveness of protected areas.* 2nd edition.. Gland, Switzerland/Cambridge: IUCN.

Innes, J., & Barker, G. (1999). Ecological consequences of toxin use for mammalian pest control in New Zealand - an overview. *New Zealand Journal of Ecology,* 23(2), 111–127.

Jacobson, C., Carter, R. W., & Hockings, M. (In press). The status of protected area evaluation in Australia and implications for its future. *Australasian Journal of Environmental Management.*

Jacobson, C., Hughey, K. F. D., Allen, W. A., Rixecker, S., & Carter, R. W. (In press). Towards more reflexive use of adaptive management. *Society and Natural Resources,* In press.

Jacobson, C. L. (2007). Towards Improving the Practice of Adaptive Management in the New Zealand Conservation Sector. Unpublished Ph.D. thesis, *Lincoln University,* Christchurch, New Zealand.

Lee, K. H. (1999). Appraising Adaptive Management. Conservation Ecology, 3(2), issue 3.

Mertsky, V. J., Wegner, D. L., & Stevens, L. E. (2000). Balancing Endangered Species and Ecosystems: A Case Study of Adaptive Management in Grand Canyon. *Environmental Management,* 25(6), 579–586.

Moon, J. A. (2004). A Handbook of Reflective and Experiential Learning: *Theory and Practice.* London: Routledge-Falmer.

New South Wales Audit Office (2004). Auditor-General's Report: Performance Audit. Managing Natural and Cultural Heritage in Parks and Reserves: *National Parks and Wildlife Service.* Sydney.

Parkes, J., Robley, A., Forsyth, D., & Choquenot, D. (2006). Adaptive management experiments in vertebrate pest control in New Zealand and Australia. *Wildlife Society Bulletin,* 34(1), 229–236.

Service P. a. W. (2004). State of the Tasmanian Wilderness World Heritage Area: An evaluation of management effectiveness. Hobart, Australia: Department of Tourism Parks Heritage and the Arts.

Siggelkow, N., & Rivkin, J. W. (2006). When Exploitation Backfires: *Unintended Consequences of Multi-Level Organisational Search.* Academy of Management Journal, 49(4), 779–795.

Stankey, G. H., Clark, R. N., & Bornmann, B. T. (2006). *Learning to manage a complex ecosystem:* adaptive management and the Northwest Forest Plan Oregon: United States Department of Agriculture Forest Service.

Uychiaoco, A. J., Arceo, H. O., Green, S. J., De La Cruz, M. T., Gaite, P. A., & Alino, P. M. (2005). Monitoring and Evaluation of Reef Protected Areas by Local Fishers in the

Philippines: *Tightening the Adaptive Management Cycle. Biodiversity and Conservation, 14(11),* 2775–2794.

Volkman, J. M., & Lee, K. N. (1994). The Owl and Minerva: Ecosystem Lessons from the Columbia. *Journal of Forestry,* 92(4), 48–53.

Walters, C. (2007). Is adaptive management helping to solve fisheries problems? *Ambio, 36(4),* 304–307.

Walters, C. J., & Holling, C. S. (1990). Large-scale management experiments and learning by doing? *Ecology, 71(6),* 2060–2068.

Weiss, C. H. (1998). Have We Learned Anything New About the Use of Evaluation? American *Journal of Evaluation,* 19(1), 21–33.

Wollenberg, E., Edmunds, D., & Buck, L. (2000). Using scenarios to make decisions about the future: *Anticipatory learning for the adaptive co-management of community forests. Landscape and Urban Planning, 47,* 65–77.

Chapter 18
Adaptive People for Adaptive Management

Ioan Fazey and Lisen Schultz

Abstract Adaptive management needs people within organizations that can learn flexibly and be adaptive. Unfortunately, people are not generally very good at changing thinking or understanding or translating such change into doing things differently. Insights into the sorts of characteristics that make people adaptive can be found in educational psychology, including work on how people improve performance and the personal beliefs they hold about the nature of knowledge and how they come to know something. These fields of research help understand how adaptive expertise can be developed and how people can deal more effectively with uncertain and messy real world problems. Doing adaptive management provides the kinds of circumstances highlighted in educational psychology that are likely to help develop adaptability of individuals. These contexts, however, are only likely to assist development of the ability of people to learn flexibly if appropriate attention is given to the structure and culture of the organizations in which adaptive managers are embedded.

Introduction

Adaptive management is an important way of thinking about managing and dealing with uncertainty. Adaptive management involves actively seeking new ways of doing management, actively trying to work out what happened through that management, and actively evaluating how things might be done differently next time, either to improve management outcomes or to improve what can be learnt about the system being managed.

I. Fazey
Institute of Biological, Environmental and Rural Sciences, Aberystwyth University, Aberystwyth, Ceredigion, UK

L. Schultz
Stockholm Resilience Centre and Department of Systems Ecology, Stockholm University, Stockholm, Sweden

Doing adaptive management well is not easy. A major challenge lies in managing the structure and culture of institutions in ways that enable them to continuously refine how they operate. Ultimately, however, it is individuals who do the learning, not the organization in which they are embedded. The practice of adaptive management is therefore also particularly challenging because it depends on the extent to which the individuals within an organization can continually re-evaluate their understanding, be open to changing that understanding, and translate that change into appropriate management behaviour and outcomes. That is, successful adaptive management requires adaptive people.

Unfortunately, most people rarely engage beyond surface level thinking about problems (King & Kitchener, 2002) and while history appears to show that human beings can respond creatively as problems arise, much of the adaptive behavior in response to changing conditions results in continuation or reinforcement of the issue it is meant to address (Fazey et al., 2007). This is particularly the case for learning in complex social–ecological systems. For example, engineers working in India in the nineteenth century learnt a considerable amount about building irrigation systems and structures, including understanding the contribution their activities made to creating serious environmental problems by raising groundwater that mobilized salt in the soil. When the same engineers later worked in Australia however, they failed to translate their experiences into effective behavioural change, and they ended up constructing the same sorts of structures, resulting in the same sorts of salinity problems (Proust, 2004).

In this chapter we consider some of the factors that promote adaptability in people. We refer to two areas of educational psychology that have previously received little attention in the adaptive management of environmental systems. These are: (1) research on how people can develop adaptive expertise; and (2) the influence of personal beliefs of the nature of knowledge and knowing on capacity to deal with messy, real world problems. The chapter is illustrated by examples of the facilitation of reflective practice by on-ground conservation managers and of teaching practices that influence students' understanding of knowledge and knowing. Finally we briefly discuss the kinds and structures and cultures required in organizations that are most likely to promote learning.

Adaptive Experts

Adaptive Expertise

One of the most useful areas of research for understanding adaptability in people stems from the study of adaptive expertise in educational psychology. Expertise can be defined by a person's capability for skillful physical, cognitive and meta-cognitive behaviours; deep and contextualized understanding of a body of knowledge; ability to retrieve and apply that knowledge to familiar problems; and/or an ability to notice patterns of information in a novel situation (Bransford et al., 2000). Experts generally tend to do extremely well in a particular domain because of their extensive experience (Ericsson, 1996; Gobet & Waters, 2003; Taylor, 2006). However, experts of the same skill or ability can display very different degrees of flexibility in being able to adapt to novel situations. A hypothetical example is a trapper who demonstrates expertise in keeping a site free of rabbits. In this context, the specific trapping skill may be sufficient to achieve the desired outcome. However, if the desired outcome is to maintain the rabbit population for optimum grazing to conserve flora, more flexibility in their skill is required. Further, the deep and conceptual understanding of a more flexible manager may mean that they can quickly adapt to working in a completely new situation, such as a grazing related issue in another country with different flora and fauna. Experts who are highly competent in flexibly dealing with new situations are described as having 'adaptive expertise'.

The term 'adaptive expertise' was first used to highlight the flexibility of expert Japanese sushi chefs who demonstrated creativity and adaptability to external demands, as opposed to other experts who were technically very proficient, but were relatively routinised (Hatano & Inagaki, 1986). That is, certain chefs were more able to produce excellent food despite lack of an important ingredient whereas other experts were not able to deal with the novel situations because they were used to following fixed recipes. Other studies have described adaptive experts as being 'highly competent' rather than 'merely skilled', or being 'virtuosos' rather than 'artisans' (Bransford et al., 2000). Such differences exist across a range of professions, from historians to information system designers (Bransford et al., 2000), and across a range of physical, social, and intellectual skills, including the ability of conservation managers to learn about and manage complex dynamic ecosystems (Fazey et al., 2005). Table 18.1 outlines the key features and outcomes of adaptive expertise.

Table 18.1 Features and outcomes of adaptive expertise (From Bransford et al., 2000; Feltovich et al., 1997; Woods et al., 1994; Gott et al., 1992; Holyoak, 1991)

Features of adaptive expertise
- Applying knowledge/skills/capabilities effectively to novel problems or atypical situations
- Inventing new procedures for solving unique or fresh problems rather than simply applying procedures that have already been mastered
- Continuous refinement of understanding through problem-solving experiences
- Application of strong conceptual foundations to help make sense of complexity

Outcomes of adaptive expertise
- Improved performance of dealing with technical or complex problems
- Greater avoidance of errors
- Greater transferability of a skill/ability to completely new contexts
- More accurate diagnoses of problems

Developing Adaptive Expertise

Understanding how adaptive expertise is developed provides important insights into understanding how to become better adaptive managers. Insights into the process of developing adaptive expertise are found in the cognitive psychology literature related to how people improve performance of a skill or ability. This includes the importance of: (1) practicing the skill/ability; (2) reflecting on practice; and (3) varying the way that something is practiced or reflected on.

Practicing the Skill/Ability

In simple terms, learning how to do something better, whether it is a cognitive thinking ability, or a physical skill, requires regular practice. This practice can be enhanced by imagined or detailed mental rehearsal and review (Feltz & Landers, 1983; Malouin et al., 2004) and practicing making judgments about the performance of a task before and after receiving external feedback (Wulf & Shea, 2003).

Reflecting on Practice

Reflecting on the practice of something is a key factor in effective learning (King & Kitchener, 2002). That is, learning requires exposure to new circumstances, experiences and ideas (i.e. those things that can be practiced), but also requires personal consideration of what and why something happened, the way in which an event influenced thinking, and the role the person doing the reflection played in the process (Fazey et al., 2005). Effective reflection requires the development of a range of different thinking dispositions, such as sometimes being broad and adventurous and at other times being intellectually careful (Perkins et al., unpublished). Effective reflection also requires motivation and intention to improve understanding and is therefore considered to be much more than just a thinking skill or ability, (Perkins et al., 2000). Effective, critical and reflective thinking can be characterised as involving three components- inclination, sensitivity in choice and ability, acting on seven different thinking dispositions (Table 18.2).

Varying the Way That Something Is Practiced and Experienced

The crucial element to developing adaptability is varying the way that something is practiced. That is, those who have experienced variation are more able to use that which is learned in new contexts and deal with new circumstances more effectively (Fazey & Fazey, 1989; Schwartz et al., 2005). In general, introducing variation in the practice of something assists development of adaptability because it helps break habitual assumptions that what we experience is reality rather than

Table 18.2 Developing an appropriate learning attitude is influenced by how we think. Good thinking has seven broad dispositions, each with three components (From Pekins et al., unpublished)

Disposition	Component		
	Inclination (examples)	Sensitivity (examples)	Ability (examples)
(1) To be broad and adventurous	Tendency to be open-minded, impulse to probe assumptions, desire to tinker with boundaries	Alertness to binariness, dogmatism, sweeping generalities, narrow thinking	Identify assumptions, empathic and flexible thinking, to look at things from other points of view
(2) Toward sustained intellectual curiosity	Zest for inquiry, urge to find and pose problems, tendency to wonder	Alertness to unasked questions, anomalies, hidden facets, detecting gaps in knowledge	To observe closely, focus and persist in a line of inquiry
(3) To clarify and seek understanding	Desire to grasp the essence of things, impulse to anchor ideas to experience and seek connections to prior knowledge	Alertness to unclarity, discomfort with vagueness, a leaning towards hard questions	Ability to ask pointed questions and build complex conceptualisations, ability to make analogies and comparisons
(4) To plan and be strategic	Urge to set goals, make and execute plans, a desire to think ahead	Alertness to lack of direction, lack of orientation, sprawling thinking	Ability to formulate goals, evaluate alternative modes of approach, make plans and forecast possible outcomes
(5) To be intellectually careful	Urge for precision, a desire for mental orderliness, organisation, and thoroughness	Alertness to possibility of error, disorder and disorganisation, inaccuracy and inconsistency	Ability to process information precisely, to recognize and apply intellectual standards
(6) To seek and evaluate reasons	A leaning towards healthy scepticism, the drive to pursue and demand justification, the urge to discover grounds and sources	Alertness to evidential foundations, responsiveness to superficiality and over generalization	Ability to distinguish cause and effect, to identify logical structure, reason inductively
(7) To be metacognitive	Urge to be cognitively self-aware and to monitor the flow of one's thinking, desire to be self challenging	Alertness to loss of control of one's thinking, detection of complex thinking situations requiring self monitoring	Ability to exercise control of mental processes, to conceive of the mind as active and interpretive, to be self evaluative, to reflect on prior thinking

reality experienced in a particular way (Fazey & Marton, 2002). Trying to look at a problem from different perspectives is, therefore, possibly one of the most crucial elements of variation that needs to be practiced (Marton & Wenestam, 1988). This can sometimes be achieved in simple ways, such as discussing experiences with others, roleplaying (e.g. Lynam et al., 2002), or building relatively simple conceptual models such as spidergrams which can be used to share personal theories of what may be happening in a complex system that managers are working in.

Developing Adaptive Expertise in Learning

The key to becoming more adaptive is learning how to learn more flexibly, or developing adaptive expertise in learning. That is, adding variation to the practice or reflection on anything that we do. As with learning anything, once we get used to doing this, the process becomes natural, unconscious and automated in the same way that learning to drive a car initially requires deep concentration but then becomes a largely subconscious action.

Practicing and reflecting on learning in variable ways is therefore the key to developing adaptability. To practice learning about complex and dynamic systems (e.g. socio-economic, politic and bio-physical aspects of a wetland or forest) the principles of variable practice and reflection need to be applied to three main areas (Fazey et al., 2005):

1. Whenever we use any technique, display skill or demonstrate ability, such as when building a fence for stock management, designing an experiment, evaluating the effectiveness of policy, or conducting an environmental impact assessment.
2. Regularly going out into real ecological settings and reflecting on causes and linkages between systems to ensure that thinking is sufficiently grounded and maintains relevance.
3. Developing expertise in exploring feedback in systems either through simple conceptual models or more complex quantitative ones.

An example of applying these principles to practicing thinking about feedback in systems is provided in Fig. 18.1.

Finally, adding variation to practice and reflection highlights two important aspects of adaptive expertise. First, greater exposure to variation means that adaptive experts are more flexible in responding to completely new contexts and circumstances:- rather than relying on having come across the same situation before they are accustomed to dealing with Second, although it might initially take longer to learn something by adding variation, performance can be enhanced through varying practice. This increases the retention of what is learned compared with constantly practicing the same thing repeatedly (Shea & Morgan, 1979; Magill, 1998). It also means that, in general, there does not have to be a trade-off between a person being a routinised or an adaptive expert.

18 Adaptive People for Adaptive Management

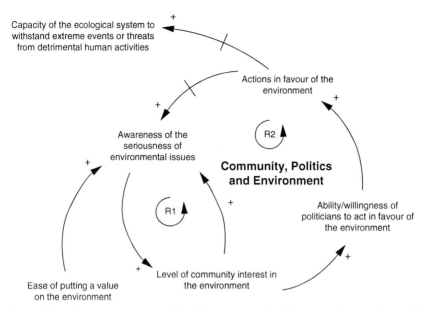

Part of the final conceptual model of the managers' collective understanding of why wetland conservation was not being achieved

Fig. 18.1 Applying principles of variable practice and reflection when eliciting the perceptions of on-ground managers of the Macquarie Marshes (from Fazey et al., 2005, 2006). Research primarily aimed to elicit the implicit knowledge of seven on-ground expert managers about the current conservation problems and issues facing a complex 220,000 ha wetland social–ecological system in southeast Australia. The experts were cattle grazers and Parks and Wildlife staff who between them had 140 years of experience of water management and 234 years of general experience in the Marshes. A secondary aim of the research was to conduct it in ways that provided repeated opportunities for the managers to practice and vary the way they explored their personal understanding of the complex social and ecological dynamics. This aimed to help them further develop their expertise in the complex dynamics. There were six research stages: (1) The researcher worked with the individuals for 2 months to build trust and familiarity of the context in which the participants were working; (2) Interviews were held separately with each participant to develop simple diagrams for initial examination of some of the feedback process occurring in the system; (3) A second interview with each participant was held that focused on a different topic; (4) A workshop with all participants was used to identify and discuss the significant historical changes to the environmental system that had contributed to current conservation problems; (5) A preliminary conceptual model describing the environmental system was analysed with each participant in a third interview; and (6) a meeting was held with all the participants to discuss the accuracy of the conceptual model and the presentation of their expert understanding. The process enabled variation to occur in the way the participants shared perspectives and articulated and reevaluated their understanding about the dynamics of the system. The result not only provided a more accurate articulation of the expert managers' understanding but also helped managers to consolidate their knowledge about why conservation was not being achieved

How Beliefs of Knowledge and Knowing Influence Capacity to Learn About Complex Systems

Personal Epistemological Beliefs

Another important area of research relevant to understanding how to promote adaptability in learning in complex systems involves the study of personal epistemological beliefs. These are the personal beliefs people hold about the nature of knowledge and how something is known, and how this affects perception, learning, and behaviour (Hofer & Pintrich, 1997; Buehl & Alexander, 2001; Hofer, 2001). Such beliefs operate at higher levels than many other forms of thinking and have a major impact on how people tackle ill-structured problems (Kitchener, 1983; Kuhn, 2000). They influence use of strategies (Schommer et al., 1992; Kitchener, 1983), thinking processes (Kardash & Howell, 2000), whether deeper conceptual change occurs during learning (Qian & Alvermann, 2000), and are related to the ability to make reasoned judgments (King, 1992). They are directly relevant to understanding how individuals deal with conflicts over what people claim to be 'knowledge' or 'evidence' and how they evaluate new information and make important decisions (King & Kitchener, 1994; Kuhn, 1991).

An example is the impact of these beliefs on data analysis and interpretation. Such processes require thinking about operation and immediate interpretation of physical actions (e.g. operating a computer, statistical package, or writing). Such processes also involve higher levels of thinking, including formulation of the strategies and monitoring involved in particular analytical approaches, and making sense of the outcomes. Both these levels of cognition are, however, significantly influenced by even higher levels of thinking such as personal epistemological beliefs. If, for example, knowledge is viewed as being tentative, evolving, or context dependent there will be a greater tendency to dig deeper into the data, look for hidden relationships, and consider the multiple possible interpretations than if knowledge is considered to be fixed, certain, or made of concrete facts.

Hofer (2000) identifies two consistent themes in existing models of personal epistemological beliefs. First, she suggests that beliefs can each be presented along a continuum from less sophisticated to more sophisticated and can develop over time. For example, seminal longitudinal studies in the 1950s and 1960s found that epistemological beliefs of Harvard students changed along a particular directional pattern. Individuals started with a dualistic perception of knowledge with a belief that knowledge is 'black or white' and could be known. Students then began to take increasingly relativistic perspectives where they first acknowledged the existence of multiple and diverse views and that uncertainty was possible, then increasingly began to recognize that some of these views were better than others. Finally, individuals developed greater ability to commit themselves to a particular view through careful judgment and evaluation of evidence and arguments (Perry, 1970).

The second consistent theme identified by Hofer (2000) in different models of epistemological beliefs is that beliefs are comprised of multiple dimensions (Schommer-Aikins, 2002; Hofer & Pintrich, 1997). For example, Hofer (2000, 2001) clusters these beliefs into two main areas: (1) The nature of knowledge (beliefs about what knowledge is); and (2) The nature or process of knowing (beliefs about how a person comes to know something). These areas each consist of two dimensions that can be expressed as continua (Table 18.3). While it is accepted that beliefs are comprised of multiple dimensions, some of the models suggest that different dimensions develop in tandem (e.g. Baxter Magolda, 1992; Perry, 1970) and others that they can develop independently (Schommer-Aikins, 2002). There is also some suggestion that an individual's beliefs can be different for different contexts and that broad generalizations (e.g., Table 18.2) might be limited in their capacity to explain the variability of beliefs a person has about knowledge (Elby & Hammer, 2001).

Personal epistemological beliefs are extremely important in influencing how a person learns in dynamic and complex settings. For example, educational research found that students who believed that learning occurs quickly tended to overestimate how much they understood, and to draw oversimplified conclusions (Schommer et al., 1992). Similarly, students who believed that knowledge was 'right or wrong' (dualistic) and considered themselves to have reached understanding when they could recite 'the facts' tended to have lower grades than students who believed that knowledge was context dependent (relativistic) and that understanding was only achieved when they could apply that knowledge to another situation (Ryan, 1984). In short, people who have more sophisticated epistemological beliefs are much more likely to be effective learners and more capable of being good adaptive managers. Unfortunately, while personal epistemological beliefs have a major impact on behaviour, sophisticated views of epistemology appear to be relatively rare (King & Kitchener, 1994; Kuhn, 1991),and education appears to have significantly less of an impact on its development than is often claimed (Hofer & Pintrich, 1997; Tsui, 1999; Hofer, 2001).

Facilitating Change in Beliefs of Knowledge and Knowing

There is considerable evidence and strong theoretical foundations for certain kinds of practices and environments in educational settings that are most likely to promote the development of more sophisticated epistemological beliefs. These include high levels of student participation, praise and use of student's ideas, peer-peer interaction, problem based learning, use of active reflection, class presentations, critical analysis of papers by tutors, and taking essay exams rather than answering multiple choice questions (Miri et al., 2007; Tsui, 1999; Terenzini et al., 1995). Such practices need to be conducted in environments that are both supportive and motivating and where the teaching of thinking is actively pursued. In addition, it is important to engage directly with messy-real world problems because it is in these situations that people are most likely to be forced to revaluate their thinking (Hofer, 2000). The uncertain context in which adaptive management usually occurs is ideal for this and therefore has considerable potential to encourage thinking in ways consistent with more sophisticated epistemological beliefs (Table 18.3).

Table 18.3 The different dimensions of epistemological beliefs (Hofer, 2000, 2001)

Dimension	Explanation	Dimensions expressed along a continuum	
Nature of knowledge			
Certainty of knowledge	The degree to which a person sees knowledge as fixed or fluid	Absolute truth and certainty exists	Knowledge is tentative and evolving
Simplicity of knowledge	The degree to which a person sees knowledge as accumulation of facts or as highly interrelated concepts	Knowledge consists of discrete, concrete, knowable facts	Knowledge is relative, contingent, and context dependent
Nature or process of knowing			
Source of knowledge	Where a person believes knowledge comes from	Knowledge originating from outside the self (e.g. an expert or external authority)	Knowledge is constructed by individuals through interaction with their environment and others
Justification for knowing	How individuals justify what they know and how they evaluate knowledge	Justification of a view through observation, authority, or on the basis of what feels right	Justification of a view through active evaluation or assessment of the evidence, expertise or authority involved

Development of epistemological thinking, however, will not simply happen just because people are working in a context with high degrees of uncertainty. This is because there is no guarantee that the context alone will get people actively engaged in thinking about their thinking. There are three crucial preconditions that will significantly assist managers to develop their thinking skills:

1. Responsibility for learning needs to be handed over to those who need to learn. For example, in hierarchical organizations people at lower levels are often not given the opportunity to make, and be responsible for, decisions with the result that they may be demotivated and less likely to engage deeply or care about the problem they are working on. If they are empowered to learn, then they are much more likely to do so (Rushmer et al., 2004b).
2. Adaptive managers need time, incentives, and a motivating environment to get them thinking about what they and their colleagues understand, and to share perspectives. For example, opportunities for sharing perspectives will be far fewer where there is little participation in discussions about how to deal with complex issues (e.g. because people don't have time or where they feel uncomfortable speaking up in front of senior managers).
3. Adaptive managers need to be encouraged to think about their beliefs of the nature of knowledge and knowing. Finally, direct discussion and thinking about knowledge and knowing can be severely constrained in organizations where certain worldviews dominate. In the case of the Macquarie Marshes (Fig. 18.1), the on-ground managers highlighted that the agency allocating water to different stakeholders was dominated by a traditional engineering worldview with perceptions that water management was something that was predictable and could be 'controlled'. At the time, the managers cited many examples of decisions that reflected this general view and suggested that it would be very difficult to change such beliefs and achieve management actions that better reflected the dynamism and uncertainty associated with the management of water flows and conservation of the wetland.

Figure 18.2 provides an example of how these three elements were included in the processes of teaching students to analyse complex systems. Such approaches could be applied in continuous professional development in organizations doing adaptive management.

In summary, while the dynamic and uncertain context in which adaptive management occurs provides ample opportunity for the development of more sophisticated epistemological beliefs, this can only happen when people have the space, incentive, and motivation to actively engage with learning. This in turn requires organizations with appropriate cultures and structures that both enable and facilitate learning.

Structure and Culture of Learning Organizations

Much has been written about how to develop greater adaptability and learning in organisations (Rushmer et al., 2004a, b) and will not be repeated here. However, it is worth briefly highlighting the sort of culture and structure that is necessary to empower people to learn and develop adaptability (Tables 18.4 and 18.5).

Key features
• Students chose and worked on one of four complex case studies to examine the inter-relationships between components of a social and ecological system;
• Classes were run as workshops where students were encouraged to develop their own understanding of the system;
• Students regularly worked in groups to promote discussion and reflection;
• Mini assignments to help students build conceptual models and apply concepts from resilience thinking;
• Students required to complete formal self assessments of their own work and that of their peers that required them to actively reflect on complexity;
• Students regularly compared their chosen case studies with others chosen by their peers;
• During class, students practiced building models and applying theory b different case studies;
• Students actively engaged in discussions about the nature of knowledge and knowing.

Fig. 18.2 Example of teaching that aimed to facilitate development of more sophisticated beliefs of knowledge and knowing

Table 18.4 Cultural and structural characteristics of adaptive organisations (From Rushmer et al., 2004a)

Culture	
Celebration of success	Reducing fear of failure when trying to create positive change. Requires finding time to find out and share the success of people.
Absence of complacency	Continually trying to find better ways of doing things that make life easier, simpler and more rewarding.
Tolerance of mistakes	Dealing with uncertainty means that mistakes will happen, so systems (mentoring, support, training etc.) need to be strong enough to make sure that simple mistakes don't create disasters. But there must be tolerance of honest mistakes to ensure that creativity, innovation and change doesn't stop.
Belief in human potential	Trying to create conditions where people believe they can make a real difference through their work. This can be constrained where the structure of the organisation does not allow people's ideas and efforts to be heard or noticed.
Recognition of tacit knowledge	Recognition that those who do the jobs and tasks will have the best knowledge about them.
Prioritizing the immeasurable	Collecting information about the quality of the work and its outcomes are usually forgotten while the facts and figures about 'meeting targets' are prioritized. This reduces potential for learning because it focuses attention to quantities rather than qualities or questions about the reasons why something may not have turned out as intended.

Openness	Openness in sharing knowledge so that everyone can learn from events. This usually occurs better through informal multiprofessional teams, staff rotations etc. rather than formal reports and communications.
Trust	Staff must be confident that managers and leaders will not be punished for making mistakes and leaders and managers need to know that staff will use time, space and resources given to them to facilitate learning wisely.
Outward looking	Active seeking up-to-date information from outside the organisation can ensure learning about valuable lessons from other agencies.
Structure	
Flatter hierarchies	These increase empowerment and increase information flow, trust and participation. In terms of learning and being a learner, all should be equal. Skillful use of delegation and the release of autonomy to show initiative does not need to expose an organization to risky decision-making.
Team work structures	Team work for tasks where people need to work together to achieve the outcome. The more often and greater the number of combinations that staff work together, the stronger the communication links and trusting behaviours will be.
Incentives and rewards for learning	If organisations are meant to learn staff need to have incentives and rewards for working collaboratively and cooperatively.
Information and communication networks	Informal flow (talking) is the key to fast learning, but to be permanent, records are required. Key skills held and reasons for decisions can be recorded so that important lessons are not lost as staff retire or move on.
Research and development budgets and programmes	Active adaptive management requires planning and support, While resources for staff development may be constrained, creative use can be made of existing staff skills and coaching.

Table 18.5 Activities to encourage learning (From Rushmer et al., 2004b)

Activities that help people learn in organisations	Activities that help organisations learn
• Learning is encouraged and not judged	• They strive to enhance the individuals and capabilities of their staff
• Others are learning too	• They allow staff to learn together in teams
• Learners have had a chance to practice new behaviours	• They update and challenge assumptions they hold
• There is not too much to learn at one time	• They develop and share a cohesive vision
• The learning is relevant and meaningful to the person	• They consider the bigger picture (open systems thinking)

These components demonstrate that an organisation's transition to becoming adaptive is not easy. It is, however, much more likely to be possible in the presence of skillful leadership, which is discussed in Chapter 16, Schultz and Fazey, in this volume.

Key Messages

This chapter has considered two areas of educational research, the study of adaptive expertise and personal epistemological beliefs, to help understand the sorts of practices and conditions that assist help people become more able to flexibly learn in complex social and ecological systems. There are five main conclusions from this chapter:

1. It is ultimately people that learn, not an organization. Adaptive people are therefore essential for adaptive management.
2. Adaptive people exhibit a strong proactive desire to continuously learn from their experiences and improve performance. They accept that their understanding will always change which reinforces tendencies for them to be more open to how change may influence their current understanding and behaviour.
3. Truly adaptive people are rare.
4. To be able to deal adaptively with complex systems people need to learn how to learn flexibly in different situations. This requires practicing learning and reflecting in different ways on thinking such as through regular exposure to different contexts and problems, being exposed and used to examining issues from different perspectives, and continuously questioning and reflecting on those perspectives. Becoming used to adding variation to the way you practice and reflect on anything they do is the key to developing adaptability.
5. Personal beliefs about knowledge and knowing have a major impact on how people tackle ill-structured problems. The context in which adaptive management usually occurs is ideal for the development of more sophisticated beliefs but this can only occur if a trainee of adaptability is given responsibility for their own learning, has the space, incentives, and motivation to think about their thinking, and actively encourages thinking about their beliefs of knowledge and knowing.
6. Traditional organizational settings rarely provide an appropriate culture and structure for people to develop adaptability. Focusing both on the practitioner as a learner of learning and on ensuring that the organization in which they are embedded has an appropriate structure and culture to promote learning is essential if adaptive management is to be successful.

References

Baxter Magolda, M.B. (1992). *Knowing and reasoning in college: Gender related patterns in students' intellectual development*. Jossey Bass, San Francisco, CA.

Bransford, J.D., Brown, A.L., Cocking R.L. (2000). *How people learn: Brain, mind, experience, and school*. National Academy Press, Washington, DC.

Buehl, M.M., & Alexander, P.A. (2001). Beliefs about academic knowledge. *Educational Psychology Review*, 13, 385–418.

Elby, A., & Hammer, D. (2001). On the substance of a sophisticated epistemology. *Science*, 85, 554–567.

Ericsson, K.A. (1996). *The road to excellence: The acquisition of expert performance in the arts and sciences, sports, and games*. Mahwah, NJ: Erlbaum.

Fazey, D.M.A., & Fazey, J.A. (1989). Modification of Transfer Effects in Different Practice Schedules - an Extension of the Variability Hypothesis. *Journal of Human Movement Studies*, 17, 239–258.

Fazey, I., Fazey, J.A., Fazey, D.M.A. (2005). Learning more effectively from experience. *Ecology and Society*, 10, art 4.

Fazey, I., Proust, K., Newell, B., Johnson, B., Fazey, J.A. (2006). Eliciting the implicit knowledge and perceptions of on-ground conservation managers of the Macquarie Marshes. *Ecology and Society*, 11, 28.

Fazey, I., Fazey, J.A., Fischer, J., Sherren, K., Warren, J., Noss, R.F., Dovers, S.R. (2007). Adaptive capacity and learning to learn as leverage for social-ecological resilience. *Frontiers in Ecology and the Environment*, 5, 375–380.

Fazey, J.A., & Marton, F. (2002). Understanding the space of experiential variation. *Active Learning in Higher Education*, 3, 234–250.

Feltovich, P.J., Spiro, R.J., & Coulson, R.L. (1997). Issues of expert flexibility in contexts characterized by complexity and change. In P.J. Feltovich, K.M. Ford, & R.R. Hoffman (Eds.), *Expertise in context* (pp. 126–146). Menlo Park, CA: AAAI Press/MIT.

Feltz, D. L., & Landers. D.M. (1983). The effects of mental practice on motor skill learning and performance - a meta-analysis. *Journal of Sport Psychology*, 5, 25–57.

Gobet, F., & Waters, A. J. (2003). The role of constraints in expert memory. Journal of Experimental Psychology: Learning, Memory & Cognition, 29: 1082–1094.

Gott, S., Hall, P., Pokorny, A., Dibble, E., & Glaser, R. (1992). A naturalistic study of transfer: Adaptive expertise in technical domains. In D. Detterman, & R. Sternberg (Eds.), *Transfer on trial: Intelligence, cognition, and instruction* (pp. 258–288). Norwood, NJ: Ablex.

Hatano, G., & Inagaki K. (1986). Two courses of expertise. In H. Stevenson, H. Azuma, & K. Hakuta (Eds.), *Child development and education in Japan* (pp. 262–272). New York: W.H. Freeman.

Hofer, B.K. (2000). Dimensionality and disciplinary differences in personal epistemology. *Contemporary Educational Psychology*, 25, 378–405.

Hofer, B.K. (2001). Personal epistemology research: Implications for learning and teaching. *Educational Psychology Review*, 13, 353–383.

Hofer, B.K., & Pintrich, P.R. (1997). The development of epistemological theories: Beliefs about knowledge and knowing and their relation to learning. *Review of Educational Research*, 67, 88–140.

Holyoak, K.J. (1991). Symbolic connectionism: Toward third-generation theories of expertise. In K.A. Ericsson, & J. Smith (Eds.), *Toward a general theory of expertise: Prospects and limits* (pp. 301–335). Cambridge, UK: Cambridge University Press.

Kardash, C. M., &. Howell, K.L. (2000). Effects of epistemological beliefs and topic-specific beliefs on undergraduates' cognitive and strategic processing of dual-positional text. *Journal of Educational Psychology*, 92, 524–535.

King, P.M. (1992). How do we know? Why do we believe? Learning to make reflective judgements. *Library Education*, 78, 2–9.

King, P., &. Kitchener, K.S. (1994). *Developing reflective judgement*. Jossey-Bass Publishers, San Francisco, CA.

King, P. M., & Kitchener, K.S. (2002). The reflective judgment model: Twenty years of research on epistemic cognition. In B.K. Hofer, & P.R. Pintrich (Eds.), *Personal epistemology: The psychology of beliefs about knowledge and knowing*. Erlbaum, Mahwah, NJ.

Kitchener, K. S. (1983). Cognition, metacognition, and epistemic cognition - a 3 level model of cognitive processing. *Human Development*, 26, 222–232.

Kuhn, D. (1991). *The skills of argument*. Cambridge University Press, Cambridge.

Kuhn, D. (2000). Theory of mind, metacognition, and reasoning: A life-span perspective. In P. Mitchell, & K.J. Riggs (Eds.), *Children's reasoning and the mind* (pp. 301–326). Psychology Press, Hove.

Lynam, T., Bousquet, F., Le Page, C., d'Aquino, P., Barreteau, O., Chinembiri, F., Mombeshora, B. (2002). Adapting science to adaptive managers: Spidergrams, belief models, and multi-agent systems modeling. *Conservation Ecology*, 5.

Magill, R.A. (1998). *Motor learning: Concepts and applications.* McGraw-Hill, Boston, MA.

Malouin, F., Belleville, S., Richards, C.L., Desrosiers, J., Doyon. J. (2004). Working memory and mental practice outcomes after stroke. *Archives of Physical Medicine and Rehabilitation,* 85, 177–183.

Marton, F., & Wenestam, C.G. (1988). Qualitative differences in retention when a text is read several times. —. In: M.M. Gruneberg, P.E. Morris, & R.N. Sykes (Eds.), *Practical Aspects of Memory: Current research and issues* (pp. 370–376). Wiley, Chichester.

Miri, B., David, B.C., Uri, Z. (2007). Purposely teaching for the promotion of higher-order thinking skills: A case of critical thinking. *Research in Science Education,* 37, 353–369.

Perkins, D., Jay, E., Tishman, S. (1993). A dispositional theory of learning. Access through: http://learnweb.harvard.edu/alps/thinking/docs/merrill.htm.

Perkins, D., Tishman, S., Ritchhart, R., Donis, K., Andrade, A. (2000). Intelligence in the wild: A dispositional view of intellectual traits. *Educational Psychology Review,* 12, 269–293.

Perry, W. G. (1970). *Forms of intellectual and ethical development in the college years: A scheme.* Holt, Rinehart & Winston, New York.

Proust (2004). Learning from the past for sustainability: towards an integrated approach. Ph.D. thesis. Australian National University, Canberra.

Qian, G. & Alvermann, D. (2000). Relationship between epistemological beliefs and conceptual change learning. *Reading Writing Q* 16.

Rushmer, R., Kelly, D., Lough, M., Wilkinson, J.E., Davies, H.T.O. (2004a). Introducing the learning practice - I. The characteristics of learning organizations in primary care. *Journal of Evaluation in Clinical Practice,* 10, 375–386.

Rushmer, R., Kelly, D., Lough, M., Wilkinson, J.E., Davies, H.T.O. (2004b). Introducing the Learning Practice - III. Leadership, empowerment, protected time and reflective practice as core contextual conditions. *Journal of Evaluation in Clinical Practice,* 10, 399–405.

Ryan, M.P. (1984). Monitoring text comprehension: Individual differences in epistemological standards. *Journal of Educational Psychology,* 76, 248–258.

Schommer, M., A. Crouse &. Rhodes, N. (1992). Epistemological beliefs and mathematical text comprehension - believing it is simple does not make it so. *Journal of Educational Psychology,* 84, 435–443.

Schommer-Aikins, M. (2002). An evolving theoretical framework for an epistemological belief system. In B.K. Hofer, & P.R. Pintrich (Eds.). *Personal epistemology: The psychology of beliefs about knowledge and knowing.* Erlbaum, Mahwah, NJ.

Schwartz, D. L., Bransford, J.D., Sears, D. (2005). Efficiency and innovation in transfer. In J.P. Mestre (Ed.). *Transfer of learning from a modern multidisciplinary perspective: Research and perspectives* (pp. 1–51). Portland, Or.: Information Age Publishing.

Shea, J. B. & Morgan, R.L. (1979). Contextual interference effects on the acquisition, retention, and transfer of a motor skill. *Journal of Experimental Psychology-Human Learning and Memory,* 5, 179–187.

Taylor, B. (2006). Coaching. In F. Ferrero (Ed.). *British canoe union coaching handbook* (pp. 7–48). Caernarfon, UK: Presda Press.

Terenzini, P., Springer, L., Pascarella, E.T., Nora, A. (1995). Influences affecting the development of students critical thinking skills. *Research in Higher Education,* 36, 23–39.

Tsui, L. (1999). Courses and instruction affecting critical thinking. *Research in Higher Education,* 40, 185–200.

Woods, D. D., Johannesen, L., Cook, R. I., & Sarter, N. B. (1994). Behind human error: Cognitive systems, computers, and hindsight. Dayton, OH.: Crew Systems Ergonomic Information and Analysis Center.

Wulf, G. & Shea, C.H. (2003). Feedback: The good, the bad, and the ugly. in M. Williams, N. Hodges, M. Scott, and M. Court, editors. Skill acquisition in sport: Research, theory and practice. London: Routledge.

Part V
Conclusion

Chapter 19
Synthesis of Lessons

Catherine Allan and George Stankey

Abstract In this final chapter we suggest how the lessons learned in the case studies could be used to inform other attempts to move from conventional to adaptive management. All forms of adaptive management are purposeful and deliberate, characterized by careful documentation processes and they are designed to promote learning that translates to action. The cases suggest that adaptive managers must understand and articulate their particular context, even as they seek to understand the tools and philosophies of adaptive approaches. Support for people- leaders, champions, and managers- and participatory approaches are also necessary to achieve significant moves from traditional to adaptive management.

Introduction

The authors of this book have presented stories from boreal forests to southern semi-arid rangelands, commenting on a variety of attempts to manage water, soils, plants and animals to achieve multiple goals. The common thread in these varied tales is a desire on the part of managers to use their current management as a means of learning how to improve future management. Notwithstanding the variety of disciplinary areas, physical conditions, resources and institutional arrangements, some clear patterns emerge when the stories are viewed as a whole. In this final chapter, we suggest how the lessons learned in the case studies could be used to inform other attempts to move from conventional to adaptive management, with a particular emphasis on understanding context, understanding adaptive approaches, and supporting adaptive people. We conclude by speculating on what future adaptive management could entail.

C. Allan
Institute for Land, Water and Society, Charles Sturt University, Albury, Australia

G.H. Stankey
Private consultant, Seal Rock, Oregon, USA (Retired research social scientist,
Pacific Northwest Research Station, USDA Forest Service, Corvallis, Oregon, USA

Key Lessons

Understanding Context Is Critical

A consistent quality in all the case studies is that the context within which the project unfolded was a key influence on what could and did happen. Proponents of an adaptive project must be aware of, and sensitive to, context. Context provides an important source of information about the processes that have been previously undertaken, the nature of participation (including who and why), issues and concerns, and the body of existing knowledge and experience. Understanding context means having a clear sense of the history and dynamics of any given situation, so the person or organisation undertaking the task of articulation is required to think deeply and clearly about the situation in which they will be acting. Contextual understanding enables the framing of appropriate strategies for the future.

The importance of context is underlined by the format of this book. The case studies were written with the intention of being useful to readers involved with planning, undertaking, or reviewing their own adaptive management. Each author provided a detailed story, complete with people, contingencies, events and surprises, emphasising that each attempt at adaptive management is unique. Having the context specified in a clear, available manner allows the reader to better assess the extent to which any given project might serve as a template for their own study. It is always tempting to think "that's just like us," but often the contextual history differs significantly, meaning the similarities are more apparent than real.

Acknowledging that context is critical reinforces the importance of a broadly-based, inclusive, and participatory structure as part of any adaptive management enterprise. This is because it is difficult, if not impossible, to have a full and comprehensive grasp of the scope and detail of context without such inclusive structures and processes. Inclusion also provides a demonstrative measure of legitimizing the knowledge and experience of group participants in fashioning the overall adaptive management strategy. Whether one is involved with framing appropriate and useful computer models, developing long term professional relationships, or achieving an institutionalised commitment to adaptive management, an inclusive and participatory *modus operandi* is essential for encouraging the kinds of trust-based relationships needed to operate effectively.

Adaptive management, with its need for inclusion and its acknowledgement of the importance of context, reminds us that all resource management is, at its heart, political. It is simply naïve to think that all interests will be supportive of adaptive management – indeed, the case studies reported here and in the wider literature point to the widespread resistance to such an approach – and this raises the importance of negotiation among interested parties.

Understanding Adaptive Approaches

Just as there is a need to understand and articulate the context in which questions arise and management occurs, there is an equally pressing imperative to understand the management approach(es) that may be used. The careful, honest, and public articulation of what it means to undertake **adaptive** management must go beyond the rhetorical assertion; words and concepts need to mean something. Adaptive management needs to be seen as something more than making it up as we go, or "business as usual," or the way we've always managed. Simply put, it isn't; it is a significant departure from past practice and it will require new and specific policies, skills, and resources to succeed. Policy makers and practitioners must have explicit discussions of what adaptive management means, and what it doesn't, before directing its use or embarking on a project. When organisational leaders fail to do this, or when they fail to understand the requirements, demands, and implications of an adaptive approach, practitioners face an uphill battle, with resistance likely from a variety of sources, including those very organisational leaders who initially urged them on.

So, what does adaptive management mean, or not mean in any given context? We took considerable care in Chapter 2 to present and articulate our accepted theoretical starting point for adaptive management, encompassing as it does the deceptively simple idea of learning from doing. That chapter stressed that there are multiple ways in which adaptive approaches can proceed, but irrespective of the particular form, all share three qualities; they are *purposeful* and *deliberate*, they are characterized by *careful documentation processes* and they are *designed to promote learning that translates to action*.

Purposeful and Deliberate

Effective adaptive management begins with the framing of good questions. As many of the cases in this book demonstrate, good question framing helps direct subsequent undertakings, guides the monitoring and evaluation processes, and emphasises the social and political nature of the adaptive process. Adaptive management must be anchored in a process that focuses on clarifying and framing the underlying problem in a way that ensures that subsequent management actions are relevant and useful. One of the reasons why adaptive management advocates have argued for sound modelling approaches is that such techniques help clarify what key variables affect the underlying system dynamics; what are the key factors that will shape the response of a system, how might different management interventions affect that system, what are the likely outcomes of any particular strategy, and what are the key uncertainties we face? The adaptive process needs to provide a clear rationale for why a particular problem focus has been chosen as opposed to other formulations.

Many of the case studies point to the need to recognize the existence of multiple "mini-loops" in the adaptive management process. The simple depiction of adaptive management as a "plan-act-monitor-evaluate" scheme should now be replaced by a more complex one in which there is a continuous process of feedback, leading to a reformulation of problems, tactics and strategies. Knowing where you are operating at a given time within that complex scheme helps with planning and evaluation and knowing where it might be best to be at another time.

Careful Documentation

Good documentation is transparent and open to scrutiny, it is designed to encourage thoughtful and constructive debate. Good documentation is necessary to facilitate examination and analysis of data, and for sharing the lessons and new knowledge with other practitioners, including those of the future. Documentation makes processes visible, transparent, and traceable; this facilitates review and evaluation by any interested party. It is an essential component of any future replication to test or verify programmatic outcomes. Documentation also is critical to ensuring that a permanent record of the outcomes of efforts is available; what was tried, what rationale underlay those efforts, what specific treatments were undertaken (and why), what outcomes resulted and how those outcomes compare with anticipated results. Such documentation processes ensure that it is possible to identify where mistakes or errors (of commission or omission) occurred; ironically, risk-averse organisations are often reluctant to do this.

Designed to Promote Learning That Translate into Action

Organisations that undertake adaptive management must acknowledge early on that it is a hard, time-consuming, expensive undertaking, requiring an ongoing investment and commitment to complete successfully. It also requires significant organizational capacity in terms of skills and abilities. This means that it is critical there be both an organisational commitment and the will to act, as well as a capacity to act. Organisational commitment suggests a willingness to acknowledge that the lack of knowledge constrains the ability to act in an informed manner. It acknowledges that previous organisational policies and programs might have been incorrect. And it also acknowledges that there needs to be a commitment of organisational resources and skills to sustain any adaptive effort. This begins with accepting that achieving the goals of an adaptive program might require significant time commitments, significant financial resources, and the patience and tolerance to allow on-the-ground applications to unfold before leaping to what might prove erroneous conclusions. In addition to an organisational will, there must be a capacity to act adaptively. This requires the internal resources to act, including time, money, and technical and social expertise and skills, or sufficient resources to commission them. Effective adaptive management implementation will require special abilities in areas such as sampling and research design. It will also be dependent upon the necessary legal and political licence to act; i.e., the statutory and administrative mechanisms that permit experimentation and an ability to act in the face of uncertainty. Finally, organisations and proponents of an adaptive approach must be prepared to accept that the findings of such approaches, particularly their policy implications, will not necessarily be acceptable to all interests, either within or external to the organisation. The *status quo* is often a comfortable state and any change – irrespective of the basis for such change – may be seen as threatening or disruptive. Building and sustaining a compelling rationale for such changes will always be important; again, this emphasises the need for both organisational commitment and will as well as champions and advocates for the adaptive process and its results.

Supporting the "Right" People Is Critical

Various authors, both in this book and elsewhere, have discussed the importance of choosing suitable participants in an adaptive project. This is especially important at the inception of a new adaptive enterprise. There can be strong organisational inertia that works to hamper efforts to employ an adaptive approach, and advocates must be careful and strategic in designing these early efforts. People who bring enthusiasm and energy, who have established respect and trust among their colleagues and other interests, and who have a commitment to change and a capacity to cope with ambiguity and uncertainty are essential. They must also have the ability and willingness to accept dissent and differences and the confidence to be both a strong advocate of their perspective and to change in response to articulate, reasoned dialogue.

Effective adaptive enterprises almost always reveal people who have taken on a strong leadership role. Such leadership is not always at the top of the organisational hierarchy. Advocates and champions can be found – indeed, need to be found – at multiple levels in any organisation as well as from internal and external sources. These individuals provide the energy, the initiative, and the enthusiasm for undertaking what might be seen as risky endeavors. However, without their efforts, adaptive management could be easily dismissed or, worse, treated as a mechanised and routinised protocol, rather than an activity requiring imagination and creative thinking. Perhaps the most important role organisational leaders – in a hierarchical sense – can provide is the assurance that practitioners have the latitude, organisational support and resources to undertake their work. Because adaptive enterprises often will take shape under conditions of high risk and uncertainty, there will be inevitable pressures for guarantees of success. However, it will seldom be possible to make such assurances and those who promote experimentation and risk-taking in the face of uncertainty must have the unqualified support of their political masters.

The Future of Adaptive Management

Many of the cases presented in this book are characterized by being "works in progress." That might be frustrating or disconcerting to the reader, but this is a common quality of such projects. That is, the adaptive process and the adaptive cycle typically will lead to new issues and concerns, they will often trigger a reframed problem, and they will redefine the types of knowledge required, perhaps even the scope of participation. This stands in contrast to the dominant model of resource management in which a focus on *problem solution* predominates. It also means that the typical distinction between planning and management, or even management and research fades in an adaptive setting; these traditional, compartmentalised ways of thinking and organising will begin to blur together, with consequent impacts on traditional roles and responsibilities. This further reinforces the notion discussed earlier about taking care in choosing people who are open and accepting of such changes.

Many tough, difficult questions and challenges remain before adaptive management is an accepted mainstream alternative to conventional management. There are complex technical questions about how adaptive management strategies are undertaken (e.g., sampling, analysis). How are results implemented in a complex social, institutional, and political environment (and as a corollary, how does one acquire ownership and political support across competing interests?) What specific types of organisational capacity are required and how does one acquire and maintain them? How does one deal with the inevitable tensions that arise from the time scales involved; e.g., political and budgetary time frames are typically short-term, ecological while social scales regarding implementation and measurement of results, determination of cause-and-effect relations can be significantly long-term? What criteria guide the determination of when we know if we have sufficient learning about complex, uncertain systems to act and in what ways? How do we encourage, support, and reward risk-taking behaviour in risk-averse social and political environments? How do we bridge the gap between traditional, reductionist scientific paradigms to more integrated models of thinking and behaving? Finally, given the persistent and critical role of institutional factors in adaptive management – reported both in the wider literature and in these chapters – what would be the requisite qualities of "satisfactory" adaptive management structures and processes? What qualities are necessary, which sufficient?

The cases presented here suggest that these challenges are not insurmountable. Within the myriad constraints detailed, learning has come from doing, and policies have been changed in response to the knowledge and learning achieved through adaptive processes. Think then, what **could** be achieved in environmental/natural resource management by adaptive people, working together in trusting relationships, and within supportive organisations and institutions. When these types of processes become the norm and the standard – rather than the exception or the noteworthy – we will have made significant process to the goal of sustainable environmental management.

Index

A
active adaptive management, 13, 33–34, 48, 56, 65, 91, 278, 312
adaptive expertise, 323–29, 336
adaptive leadership, 295, 297, 298
adaptive management
    components, 11–27
    definition, 228
    institutionalisation, 305–19
    learning, 4, 11, 13–16
    role, 3–7
administrative leadership, 295, 297, 298
agent-based models, 16, 173, 178, 180, 194
agriculture, Australia
    environmental management systems, 209–25
    Signposts for Australian Agriculture, 203–07
Anangu
    wildlife management, Australia, 117–40
assessment
    definition, 228
Australia
    agriculture, 203–07, 209–25
    Barrier Ranges, 73–94
    Indigenous management, 117–40
    Mitta Mitta River flows, 59–68
    National Parks and Wildlife Service, culture, 305–19
    Tasmanian Wilderness World Heritage Area, 227–56

B
Barrier Ranges, Australia, 73–94
Bayesian networks, 16, 46, 106, 173, 176, 178, 183, 184
biodiversity, 40, 44, 45, 81–82, 97, 98, 120, 123, 212, 218, 222, 299–300

Biscuit fire, Oregon USA, 143–66
bottom up, 52, 193
British Columbia, Canada
    fisheries, 175
    forest management, 39–54

C
camels, 125, 137, 138
Canada
    British Columbia, forest management, 39–54
    Species at Risk Act 46, 48
caribou, 39, 46–48
catchment management, 75, 174, 178, 182–83, 210, 224
champions, 22, 23, 66, 93, 186, 296, 341, 345
citizens, 60, 70, 143, 153, 154, 159, 163, 165
'clean and green' marketing, 215, 218, 220, 225
climate change, 60, 93, 148, 155, 190, 198, 236, 252, 266, 269
closing the loop, 39, 40, 48, 53, 313
Coast Forest Strategy, 40–44, 53
collaboration, 101, 106, 115, 153, 173, 193, 299
collaborative learning, 275–92
Columbia River Basin, USA, 4, 27, 175
component tree, 204, 205
conflict, 70, 101, 103, 110, 111, 115, 152, 165, 281, 302, 306, 330
consultation, 66, 124, 183, 232, 276, 284, 285
continuous learning, 67, 296, 301, 306
coordination, 22, 241, 242
corporate knowledge, 25
crisis
    Biscuit fire, Oregon, 143–66, 307
    leadership, 297, 301
culture — *see* organisational culture

347

D

Dartmouth Reservoir, 59–68
data, 20–21, 25, 44, 103, 221, 234, 241, 242
   analysis, 50, 330
   Australian agriculture, 203, 204, 206
   British Columbian forests, 50
   collection and capture, 15, 50, 181, 183, 206, 207
   grasslands, NZ, 103
   insufficient, 270
   modelling, 173–176, 178, 181, 182, 183, 185, 194, 270
   National Parks and Wildlife Service, 312, 313, 315, 317
   underutilisation, 203
   wildfires, Pacific Northwest forests, 144, 145, 146
   wildlife, Barrier Ranges (Australia), 82, 88, 89
databases
   coast forest strategy, 43
   grasslands, NZ, 103
   Indigenous management, 129, 131
decision trees, 106, 107, 108
decision-making, 4, 162–64, 183, 194, 198, 285
deer affected forests, 275–92
describing a system, 11, 15–17
description processes, 15, 16, 27
disjointed incrementalism, 4
divergent learning, 27
documentation, 25, 156, 215, 244, 341, 344

E

ecological modelling — *see* modelling
ecological monitoring, 62
ecologically sustainable development, 60, 76, 82, 83, 92
ecosystem health, 12, 24, 81–82, 186
ecosystem services, 12, 24, 81–82, 196, 299
ecotourism, 306
effectiveness evaluation, 227–56
enabling leadership, 295, 297, 298
England
   Peak District National Park, 189–200
environmental flows, 183, 261–71
environmental management systems, 209–25
environmental performance, 210, 212, 218, 219
epistemic uncertainty, 9
epistemological beliefs, 330–33, 336
evaluation
   definition, 228

Forest and Range Evaluation Program, 44–45, 52, 53
   Indigenous management, 138–39
   learning, 15
   management effectiveness, 227–56
   underutilisation of data, 203
Everglades, 4
evidence, 18, 87, 89, 145, 184, 199, 330
evidence-based management, 228, 235, 237, 251
experimental design, 18, 24, 34, 56, 175, 178, 180, 181, 288
experimental management, 175, 269, 270, 306
experiments, 17, 51, 92, 99, 118, 169, 206, 276, 278, 279, 297, 299, 312, 319
expertise, 323–29, 336
experts, 21, 292, 306, 323–29

F

feedback loops, 14, 101, 159, 164, 180
feral animals, 74, 131, 137, 139
fire management
   Biscuit fire, Oregon, 143, 144, 149, 151, 154, 160, 161
   Indigenous Australians, 119, 128, 139
   Pacific Northwest forests, 143–66
   Tasmanian Wilderness World Heritage Area, 248, 249
First Nations, 39
flow regime
   Mitta Mitta River, SE Australia, 59–68
Forest and Range Evaluation Program, 39, 44–45, 52
forest management
   Biscuit Fire, Oregon USA, 148
   British Columbia, Canada, 39–54
   deer affected forests, New Zealand, 275–92
Franklin River, Tasmania, 230
funding for management, 230, 231, 248

G

gaming activities, 11, 17
goats, monitoring, 49, 50, 53
Grand Canyon, 4
grasslands, New Zealand, 95–111
green marketing, 215, 218, 220, 225
greenhouse gases, 93, 119, 222
ground rules, 101, 146

Index

H
Hieracium management, 98, 99, 104, 105, 107
hierarchy theory, 35
Hume Reservoir, 59–68
hypotheses, 34, 35, 85, 92, 146, 154, 185, 279, 312

I
Indigenous Australians
  wildlife management, 117–40
information management systems, 236, 250
information sharing, 99, 106, 110, 115, 178, 246, 344
institutional inertia, 5, 143
institutionalising adaptive management, 305–19
institutions, 161–62, 197, 204, 277, 282, 296, 324, 346
integrated systems for knowledge management, 100, 101, 105, 110
interdependence, 70
iterative process, 180, 181, 182, 183, 189, 190, 191, 241

K
kangaroo management, 73–94
key players, 23, 66, 193
knowledge
  beliefs, 330–33, 336
  corporate, 25
  data, 20–21, 23, 174
  gaps, 11, 14, 25, 173, 175, 182
  integrated system, 99, 100–12
  limitations, 5
  scientific, 155
  sharing, 95, 96, 177, 297, 344
  structuring, 14, 17
  traditional, 120–40
knowledge base, 180, 193, 222, 299, 301
Kuka Kanyini, 117–40

L
Landcare, 74, 75, 102, 105, 217
landscape, 60, 75, 109, 118, 123, 155, 189, 196, 199, 312
landscape function analysis, 82, 87, 92
leadership
  Biscuit fire, 153, 154, 156, 162, 165
  collaboration, 299
  effective, 295–302, 345
  forest management, 39, 44, 52
  Indigenous management, 139
  participation, 299
  skills, 301–02
  transformation, 299–301
learning, 26, 33, 65, 66, 67, 86, 218, 269, 296, 298, 315, 324, 326
  adaptive, 328
  Australian agriculture, 203
  Biscuit fire, 151, 159
  collaborative, 275–92
  developing expertise, 328–29
  grassland management (NZ), 95–112
  limitations, 306
  loop learning, 169–70
  monitoring, 11
  organisational, 240, 241, 277, 333–35, 336
  social — *see* social learning
  synthesis 341–46
learning by doing, 5, 13, 14, 27, 62, 268, 305, 316
learning culture, 305–319
learning from adaptive management, 4, 11, 13–16
learning groups, 275–91
lichen woodlands, 39, 40, 46–48, 52
linguistic uncertainty, 9
litigation, 146, 165
local knowledge, 99, 123, 127, 195, 282, 286, 287
loop learning, 169–70

M
management effectiveness, 227–56
Maori land rights, 97
metadata, 43
Mitta Mitta River, SE Australia, 59–68
modelling
  adaptive management tool, 16–17, 173–86
  aims, 174
  barriers, 185–86
  climate change, 252
  deer management, NZ, 279, 280, 282, 283, 284, 286, 290, 291, 292
  development, 173
  environmental flows, 267, 268, 270
  eucalypt study, 183–85
  National Parks and Wildlife Service, 312, 319
  Ospika Mountain Goat Project, 48
  participatory processes, 106, 174, 177, 190

modelling (*cont.*)
  Peak District National Park, England, 189, 193, 196, 197, 199, 200
  selection of approach, 177–80
  sensitivity assessment, 180–81
  standards, 176
  usefulness, 173, 175
  water quality study, 182–83
monitoring
  Barrier Ranges, 81, 88
  Biscuit Forest, USA, 150, 152–58, 160, 161, 162, 165
  deer management, NZ, 277, 279, 280, 284
  Indigenous management, 126–28, 131, 133, 138
  kangaroos, 92
  learning, 15
  Mitta Mitta River flows, 62
  modelling, 173, 174, 175, 176, 182, 185, 186
  National Parks and Wildlife Service, 307, 309, 310, 315, 318, 319
  Ospika mountain goats, 49–50, 53
  Peak District National Park, England, 196
  Tasmanian wilderness, 227–56
  tussock grasses, NZ, 95, 96, 99–106, 110
Murray–Darling Basin, 59–68, 262, 263, 267

N
National Parks and Wildlife Service, NSW, 305–19
Native Vegetation Act, NSW, 75
New South Wales, Australia
  sustainable wildlife enterprise trial, 73–94
New Zealand
  deer affected forests, 275–92
  grassland management, 95–112
Northern Caribou Adaptive Management Project, 46–48

O
objectives, 19–20, 25, 44, 79, 119, 195, 212–13, 239, 310
old-growth forests, 43, 147, 148, 149, 159
operational principles, 63
orders of outcomes model, 109
organisational capacity, 7, 307, 318, 346
organisational culture, 53, 161, 189, 197, 250, 333, 336
organisational learning, 240, 241, 277, 333–35, 336
Ospika Mountain Goat Project, 39, 40, 48–51, 52, 53

P
Pacific Northwest
  Biscuit fire, Oregon 143–66
participation, 21, 84, 91, 111, 115, 284, 291
  stakeholders, 22, 177, 186, 191, 195, 197, 198, 269, 285, 287
participatory planning, Peak District National Park, England, 189–200
participatory processes, 106, 174, 177, 190
partnerships, 22, 40, 53, 71, 186, 198, 204
passive adaptive management, 33, 44
pastoralism
  kangaroo management, 73–94
  New Zealand, 98
Peak District National Park, UK, 189–200
people, adaptability, 323–36
performance management, 244–49, 252
personal epistemological beliefs, 330–33, 336
pine–lichen woodlands, 46–48, 52
political aspects, 79, 90, 93, 115, 163, 164, 230, 306, 315, 346
precautionary principle, 56–57, 313
predicting, 11, 17, 19, 26
prescriptions, 147, 247
problem framing, 100, 158, 159, 163, 194, 292, 343, 344, 346
problem solving, 12, 111, 146, 180, 193, 210, 223, 262, 287, 328
problems, 12–13, 16, 26, 77, 144
  defining scope, 175, 177
  definition, 191
  description, 190
protocols, 44, 47, 50, 102, 103, 156, 164, 186

Q
quality assurance, 215, 216, 217, 219, 220, 224

R
rabbits, 74, 76, 98, 325
rangelands, 39, 44–45, 52, 73–94
reductionist inquiry, 3, 4, 6
replicated experiments, 40, 152, 279
reporting template, 241–42, 243, 247
resilience, 35, 82, 87, 334
risk, 17, 21, 23, 24, 56, 67, 70, 143, 162, 163, 222
risk assessment, 184
risk aversion, 162, 165, 189, 344, 346
risk management, 20, 210
risk-taking, 54, 345, 346
river management, 59–68

Index   351

S
scale, 3, 12, 14, 25, 39, 79, 162, 184, 197
science skill, 11, 21
scientific inquiry, 3, 4, 65, 95
self-criticism, 71
Semi-Arid Lands research programme, 98
sensitivity assessment, 174, 177, 178, 180–81
sheep grazing, 74, 83, 96, 97, 102, 194
sheep replacement, 76–77
Signposts for Australian Agriculture, 203–07
silviculture, 40, 46
skills, 326
social learning, 115, 173, 176, 177, 178, 179, 180, 189, 190, 195
Species at Risk Act, Canada, 46, 48
spotted owl, 143, 147, 148
stakeholders
    Australian Indigenous management, 124
    deer management, NZ, 277, 278, 281, 285, 287
    environmental flows, 268, 269, 270
    kangaroo management, 79, 85
    modelling, 173, 177, 178, 182, 186
    Peak District National Park, UK, 189, 190, 191–96, 197, 199, 200
    Tasmanian wilderness, 228, 233, 236, 246, 247, 248, 249, 252, 254
    tussock grassland management, NZ, 95, 100, 102, 103, 108, 111
standards, 176, 210–11, 244–46, 288, 315–16
stasis, 3
statistical measures, 47, 145
sustainable development, 60, 76, 82, 83, 92, 110, 120, 203, 204
sustainable wildlife enterprise trial, 73–94
Swedish wetlands, 299–301
systems
    adaptive management, 231
    environmental management, 209–25
    information management, 236, 250
    integrated, 100, 101, 105, 110
    quality assurance, 215, 216, 217, 219, 220, 224

T
Tasmanian Wilderness World Heritage Area, 227–56

threshold, 35
top down, 193
tourism, 79, 97, 111, 117, 118, 120, 134, 190, 192, 306
trade-offs, 81, 103, 176, 177, 182, 194, 198, 271, 306
traditional knowledge, 120–40
transparency, 63, 64, 173, 186, 228, 234, 241, 246, 249, 271
trials
    Ospika mountain goats, 39, 40, 48–51, 52, 53
    sustainable wildlife enterprise, 73–94
    water management, 63, 64, 65
trust, 66–67, 70–71, 87, 101, 291, 299, 342
tussock grasslands, NZ, 95–111

U
United Kingdom
    Peak District National Park, 189–200
United States
    Biscuit fire, Oregon, 143–66

V
values, 115, 143, 144, 239, 251, 265, 268, 302, 309
vegetation management, 176, 178, 183–85

W
water management Mitta Mitta River, SE Australia, 59–68
web-based tools, 204, 207, 236
websites, 43, 45, 105, 111, 287
weed management, 98, 99, 104, 105, 107, 125, 129, 139, 313
wetlands, 147, 264, 267, 299–301, 328, 329, 333
'wicked' issues, 60, 287
wildfires, Pacific Northwest forests, 143–66
wildlife
    management by Indigenous Australians, 117–40
World Heritage Site, Tasmania, 229, 230

Printed in the United States
153941LV00010B/28/P